CLARENDON LAW SERIES

Edited by
TONY HONORÉ AND JOSEPH RAZ

CLARENDON LAW SERIES

Edited by

TONY HONORÉ AND JOSEPH RAZ

CLARENDON LAW SERIES

ACT AND CRIME

*The Philosophy of Action and its
Implications for Criminal Law*

MICHAEL S. MOORE

CLARENDON PRESS · OXFORD
1993

Oxford University Press, Walton Street, Oxford OX2 6DP
Oxford New York Toronto
Delhi Bombay Calcutta Madras Karachi
Kuala Lumpur Singapore Hong Kong Tokyo
Nairobi Dar es Salaam Cape Town
Melbourne Auckland Madrid
and associated companies in
Berlin Ibadan

Oxford is a trade mark of Oxford University Press

Published in the United States
by Oxford University Press Inc., New York

British Library Cataloguing in Publication Data
Data available

Library of Congress Cataloging in Publication Data
Moore, Michael S., 1943–
Act and crime: the philosophy of action and its implications for
criminal law / Michael S. Moore.
(Clarendon law series)
Includes bibliographical references and index.
1. Criminal act. 2. Criminal liability. 3. Criminal law—
Philosophy. I. Title. II. Series.
K5055.M66 1993 345'.04—dc20 [342.54] 92-41296
ISBN 0-19-825791-0

Set by Hope Services (Abingdon) Ltd.
Printed in Great Britain
on acid-free paper by
Bookcraft Ltd., Midsomer Norton, Bath

For Heidi

PREFACE

The present title of this book indicates the sort of question that it asks: what implications are there for the criminal law from the philosophy of action? The former title of the book, 'Bodies in Motion', was more descriptive of the answer: that all actions that the criminal law cares about are no more than the bodily motions of persons. The original title, unfortunately, duplicates a widely watched aerobics programme on American television and video, and, thinking that a casual cataloguer might mistake this for a book on physical education, I opted for the present title.

That this book was written at all is due to the comment of an anonymous referee for the Oxford University Press on another of my books that Oxford is publishing *Placing Blame: A Theory of the Criminal Law*. The referee, while generally favourable, sniffed that *Placing Blame* had no real theory of action. Having cut my philosophical teeth on the philosophy of action (with the 'little red books' of the R. F. Holland Series in Philosophical Psychology of the 1950s and 1960s), I was sufficiently miffed by the criticism that I was moved to write a large chapter on action theory for *Placing Blame*. The more I got into writing such a chapter, however, the more I was impressed by the volume and the quality of what had transpired in the philosophy of action since 1970. It soon became apparent that no chapter would accommodate the issues needing to be addressed, and the present book was conceived as its successor.

So my first thanks should go to the anonymous referee of *Placing Blame*: whoever you are, thanks. Fortunately there is no similar anonymity to the referee Oxford chose for the present volume, Antony Duff of the University of Stirling Philosophy Department. Antony generously gave the kind of detailed comments, point by point for each chapter, that one hopes for only from one's friendliest and most interested colleagues. He also sent me a copy of his report, non-anonymously, to enable a dialogue to develop. Equally detailed and helpful were the written comments of Leo Katz, then at Michigan and now a colleague at the University of Pennsylvania Law School. Leo's unabashed love of a good puzzle led him to pose numerous such puzzles for me as I rewrote the book with his and Antony's comments side by side. Also very detailed and thoughtful were the comments I received from Michael Corrado. These were received after I had rewritten the manuscript but none the less proved invaluable as I gave the manuscript its final edit.

I have also had the good fortune to receive many helpful suggestions from commentators and participants at numerous gatherings of legal academics and philosophers. Chapters 1–3 were first presented at a legal-theory workshop at the University of Pennsylvania Law School and, later, to the Faculty Workshops of the University of Iowa College of Law and the Vanderbilt University School of Law. Bill Ewald was the commentator at the Penn Workshop, giving generously of his time and considerable philosophical expertise in making the chapters better. Chapters 4 and 14 were defended at Russell Hardin's Ethics Discussion Group at the University of Chicago, Chapter 5, at the Philosophy Department at the University of Illinois–Chicago, and Chapter 6, at the Philosophy Department of the University of Iowa. Chapters 7 and 8 were the subject of a Philosophy Club discussion at Northwestern University's Law School, and Chapters 10 and 11 were presented to the Faculty Workshop of that same school. Chapters 12–14 were presented to a luncheon discussion at the Chicago–Kent Law School of the Illinois Institute of Technology.

Each of these presentations and discussions has given me numerous suggestions and criticisms. I know that I have not responded to all, but I have appreciated the prod to thought that each caused.

Others who have read and commented on all or part of the manuscript have been Larry Alexander, Jay Hillsman, Heidi Hurd, Scott McDonald, Herbert Morris, and Stephen Morse. Their suggestions and encouragement I have deeply appreciated.

I also inflicted two seminars of law students to various chapters of the book in various early stages of completion. The written criticisms, and oral follow-ups, of Ben Alexander, Glen Bernstein, George Brencher, David Cohen, Daniel Epstein, Joel Friedlander, Marc Jerome, Ira Kaufman, Stephen Kershnar, Andrea Koppelman, Michael Lichtenstein, Jeff Rackow, Greg

Roer, Alex Seldin, Keven Smith, and J. P. Suarez of the University of Pennsylvania Law School, and of John Brewer, Elan Carr, Tim Ewald, Marcy Friedman, Tom Gilson, Daniel Hurtado, Ed Kristof, Bob Marischen, Sarah Marmor, Dan Ojeda, David Stagman, and Steve Yarosh of the Northwestern University Law School, were very helpful to me as I rewrote each chapter of the book.

Daniel Epstein was my only research assistant on this book. His thorough and organized research into the legal materials on the location of actions (Chapter 11) and on the type/token distinction (Chapter 12) gave those discussions whatever legal expertise that they possess. The remaining research was greatly aided by the research staff at the University of Pennsylvania Law Library. That staff patiently collected from a wide variety of libraries besides their own the materials needed. They have made my favourite research tool—the telephone—a very efficient instrument in the production of this book.

The bulk of this book was written over the course of two summers, 1990 and 1991. The very generous research grants of the University of Pennsylvania Law School for these periods made possible the exclusive attention such concentrated writing demands.

Every word of this book was inputted into computer programs by my secretary at Penn, Betty Burks. She has graciously tolerated my curmudgeonly attachment to fountain pens and long yellow legal pads by transforming illegible manuscripts into readable texts. Her dedication to the task has been supererogatory and truly an act of friendship.

Lastly, the person to whom this book is dedicated, my wife and fellow legal philosopher, Heidi Hurd, has shared more hours than all others combined in talking through the arguments of this book and making them better even when she disagreed with their conclusions. I know that she is as relieved as I am to have this book launched so that our philosphical conversation can turn to other topics. Thanks are neither needed nor adequate as a response to the kind of partnership her contribution to this book has represented.

M.M

Iowa City
February 1992

CONTENTS

1

INTRODUCTION

The science of legislation has reached its highest form in criminal legislation. This should occasion little surprise inasmuch as criminal law has been the prime focus of philosopher-legislators since Bentham. From the beginnings Bentham himself made in this field, through the Livingston, Field, and Macaulay codes of the nineteenth century, to the Model Penal Code in America today, criminal law has been favoured by systematic thought about its structure and about how that structure can best be formalized in a code.[1]

Such a structure is not simply a matter of keeping the prohibitions of the criminal law from contradicting each other in what they require of citizens. That much is certainly required, but the Enlightenment ideal that motivated Bentham and his successors is much more ambitious than that. One can grasp this ideal by thinking about its alternative for a moment.

In today's highly regulated, industrialized society, a criminal code typically prohibits approximately 7,000 types of actions. A code could simply list such prohibitions and their penalties, grouped perhaps under headings corresponding to interests protected by such prohibitions ('Crimes against Bodily Integrity', 'Crimes against Property', etc.).

Such an unstructured code would satisfy Bentham's well-known penchant for giving notice to citizens of what the criminal law expects of them—or at least it would serve the notice-to-citizens function as well as its alternative. Such a code would not, however, serve judges and legislators as well.

---

[1] Sandy Kadish has written an illuminating series of articles detailing the flagship status accorded criminal-law codification. See Kadish, 'Codifiers of the Criminal Law: Wechsler's Predecessors', *Columbia Law Review*, 78 (1978), 1098–1144; id., 'Introduction', in id., *Indian Penal Code*, Legal Classics (Birmingham, Ala., 1987); id., 'The Model Penal Code's Historical Antecedents', *Rutgers Law Journal*, 19 (1988), 521–38; id., 'Act and Omission, *Mens Rea*, and Complicity: Approaches to Codification', *Criminal Law Forum*, 1 (1989), 65–89. For Jeremy Bentham's role more particularly, see Gerald Postema, *Bentham and the Common Law Tradition* (Oxford, 1986), 304–13.

Lacking from such an unstructured code would be any resources with which to prevent great incompleteness in its coverage even while having great redundancy in its provisions. Such incompleteness and redundancy would surface in six ways. First, there is the interpretive problem with respect to just what is included when one prohibits 'killing another human being', 'destroying the property of another', and 7,000 other things. Each prohibition in an unstructured code does not share features in common with all other prohibitions, so the same interpretive issues will require resolution for each prohibition.

There are actually two interpretive problems. The first concerns the lack of any common source to all actions prohibited by such an unstructured code. Whether one can 'kill' by omission or while asleep, for example, will have to be resolved even though it has earlier been resolved whether one can 'destroy' by omission or while asleep. Since killing and destroying share no common nature, being different types of acts contained in different criminal-law prohibitions, resolving such questions about one act-type would say nothing about resolving them about the other.

The second interpretive problem has to do with properties of actions beyond those concerning the source of actions in omissions, bodily movements while asleep, and the like. I refer to the properties an act must have to be a killing, a destroying, etc. If there is nothing common to such properties, then resolving interpretive questions of whether any act that caused death is a killing will not resolve questions of whether any act that causes property destruction is a destroying. Such relational questions will have to be addressed anew for each criminal prohibition. It is possible, for example, that I may *kill* by shoving a victim out where his killer may find him, yet I might not *destroy* by shoving someone else's property out on to the street where another may run over it with his truck. If the relation between my act and the victim's death is different from the relation between my act and the property's destruction, the seeming similarity of the interpretive questions is no guarantee of any actual similarity in the answers to them.

Thirdly, the criminal law for various reasons seeks to locate the actions that it prohibits. Jurisdiction to legislate, for example, is often said to depend on *where* the crime took place, as is the venue for the proper site of the trial. Likewise, simultaneity between act and mental state is required by criminal law, and this requires that we locate *when* the crime took place so that we can assess whether the culpable mental state was then present or not. If the actions prohibited by the criminal law shared no common structure, these location questions would have to be separately addressed for each prohibition. Just because the death of the victim might be held to be part of an act of killing, for example (so that the killing occurs at places and times that included wherever death occurred), would not mean that destruction of property was also part of an act of destroying; destroyings might take place where and when some initial act was done, the causing of the destruction being only a *property* of such acts but no proper *parts* of them.

Fourthly, criminal codes specify mental states that must accompany actions before the latter can be punished as crimes. There are four main sorts of mental state often used to grade an offender's culpability: motivational states of desire, wish, or purpose; cognitive states of belief; conative states of intending or willing; and those substitutes for any true mental states called states of negligence. If the 7,000 types of action prohibited by a typical criminal code have nothing in common, then each aspect of each type of action will have to be separately addressed by some mental-state requirement. Consider the crimes of killing a policeman and destroying the property of another. One aspect of the first is that the victim was a policeman; one aspect of the second is that the property destroyed belonged to another. If these aspects are not the same in some way that is relevant to the question of what mental-state requirement should attach to each, then a code must specify what mental state is required for each such aspect of each of the 7,000 types of acts that are prohibited. Whereas if these aspects of action are in some relevant way similar, a code-drafter could require across the board that one mental state was sufficient. One might say, for example, that belief (that the victim was a policeman, that the property belong to another) was sufficient for liability of the most serious kind, so that one need not have been motivated by these circumstances in order to be subject to such most serious liability.

Fifthly, there is what might be called the 'overlapping statutes' problem. Given 7,000 prohibitions, there will be many cases where any given act by a defendant will instantiate several prohibited types of act. The question then arises whether the range of punishment fixed for violation of each prohibition is to be cumulated on top of the others. A given act of shooting may cause a wounding of another, such wounding may cause a disfiguring of another by putting out an eye, and the act may also cause death; it is then an act of discharging a firearm, wounding, maiming, and killing, four act-types prohibited by most criminal codes. If there is no relation between these four types of actions—so that one cannot say that they are or are not the same for purposes of barring cumulative punishment—then the drafter of a criminal code must specify the answer for all possible combinations of prohibited act-types. For a code containing 7,000 types of acts, this is a staggering task, for one would have to compare each prohibition with each of its 7,000 companions to assess whether or not such offences are to be punished cumulatively when a single act instantiates both. That is a minimum of 49,000,000 pair-wise comparisons, even if we put aside the possibility that when more than two crimes are committed by the same act, the pair-wise comparisons might change.

Sixthly and lastly, unstructured codes require judges to develop 7,000 distinct answers to what is commonly called the 'unit of offence' problem. This problem arises because it is often unclear whether a given defendant has done a prohibited type of action once, twice, or several times over some interval of time. If there is no commonality to actions, then there can be no general answer to the question 'How many acts did the defendant do over any given interval of time?' In default of there being any such general answer to this question, courts must ask and answer it for each distinct act-type prohibited by the criminal law.

The criminal law thus needs some structure if its codification is to be possible and if adjudication under such codes is to be non-arbitrary. More specifically, it needs some general doctrines—doctrines applying to all types of action prohibited by a criminal code—in order to avoid an ungodly redundancy and a woeful incompleteness. There are three sorts of doctrine answering to this need in Anglo-American criminal law. The first is that there can be no criminal offence without the doing of a voluntary act. The second is that all crimes consist not only of a *mens rea* requirement but also a requirement of *actus reus*. The third is that no one should be prosecuted or punished more than once for the same crime. The first is commonly called the voluntary-act requirement, the second, the *actus reus* requirement, and the third, the double-jeopardy requirement.

Such general doctrines each purport to answer two of the above-described six quandaries. The voluntary-act doctrine purports to answer the first and third needs of a systematic code, for if all actions are but one type of thing—which is what the voluntary-act doctrine presupposes—then there can be no unique interpretive problems about an action's source or its spatio-temporal location. The *actus reus* doctrine purports to answer the second and fourth needs. If all complex descriptions of action share a few recurrent features—which is what the *actus reus* doctrine presupposes—then code-drafters can with linguistic economy specify both what properties are required for a complex action to instantiate some prohibited act-type, and what *mens rea* is required with respect to each such property. Similarly, the double-jeopardy doctrine purports to answer the fifth and sixth concerns. For this doctrine presupposes that both act-types (which are abstract universals) and act-tokens (which are event-particulars), can, in general, be individuated for purposes of preventing disproportionate punishment. Such doctrine thus relieves code-drafters from the unmanageable task of spelling out either: how the range of punishment affixed to each prohibited act-type varies depending on whether the act instantiates (an)other prohibited act-type(s); or how big or small the unit of offence is for each prohibited act-type.

Unfortunately the important unifying functions of these three general doctrines has become obscured by the pall of scepticism that has descended upon them. In criminal-law scholarship there is little agreement about what any of these requirements come to, or about how they are related. About *actus reus*, to begin with, Jerome Hall once complained about the 'considerable confusion' introduced into criminal-law theory by the differing concepts other writers have employed.[2] Similarly, Josh Dressler, apparently influenced by Hall, finds 'the phrase [*actus reus*] has no single accepted criminal law meaning'.[3] Meir Dan-Cohen also finds that 'no single . . . definition prevails nor is the term used uni-

---

[2] Jerome Hall, *General Principles of Criminal Law*, 2nd edn. (Indianapolis, 1960), 222–31

[3] Joshua Dressler, *Understanding Criminal Law* (New York, 1987), 63.

formly among lawyers'.[4] Because 'the term comprises a set of elements with no obvious common denominator other than . . . that
they do not include the offender's state of mind', Dan-Cohen
concludes that 'it is impossible to give a comprehensive affirmative definition of the *actus reus*';[5] he therefore settles for 'a residual definition: the *actus reus* designates all the elements of the
criminal offence except the *mens rea*'.[6] As Glanville Williams
shows in detail, this negative definition simply lumps a number of
disparate items together.[7]

The voluntary-act requirement is thought to be in even worse
shape. There are three levels of scepticism here: doctrinal, moral,
and metaphysical. To begin with, a number of criminal-law theorists deny that Anglo-American criminal law has an act requirement. Dan-Cohen urges that this is a 'requirement' that is
'honored mainly in the breach',[8] as does Douglas Husak.[9] They
refer to crimes of status, omission, or possession as obvious
counter-examples to any supposed act requirement. A somewhat
different denial of there being any act requirement comes from
Mark Kelman.[10] On Kelman's view, one can always find, or not
find, an act on which to predicate criminal liability, depending on
how narrowly or broadly one frames the time period during
which one looks; any act 'requirement' thus becomes illusory for
Kelman.

A second version of this doctrinal scepticism charges that if
there is an act requirement in Anglo-American criminal law, it is
not *an* act requirement; rather, there are several, because there
are several concepts of act at work. As George Fletcher asks, 'An
act as opposed to what? Sometimes the concept of a human act
is contrasted with an "omission"; sometimes with a status or condition; sometimes with acting involuntarily as in cases of hypnotism and sleepwalking.'[11] And, as Fletcher does not add, but

[4] Meir Dan-Cohen, '*Actus Reus*', in S. Kadish (ed.), *Encyclopedia of Crime and Justice* (New York, 1983), 15.
[5] Ibid. 15–16.   [6] Ibid. 16.
[7] Glanville Williams, *Criminal Law: The General Part* 2nd edn. (London, 1961), 18–21.   [8] Dan-Cohen, '*Actus Reus*', 16.
[9] Douglas Husak, *Philosophy of the Criminal Law* (Totowa, NJ, 1989), 9–11, 83–97.
[10] Mark Kelman, 'Interpretive Construction in the Substantive Criminal Law', *Stanford Law Review*, 33 (1981), 591–673; id., *A Guide to Critical Legal Studies* (Cambridge, Mass., 1987), 15–185.
[11] George Fletcher, *Rethinking Criminal Law* (Boston, 1978), 420.

others do,[12] sometimes 'act' is fourthly opposed to mental states like wishes and intentions. The sceptical conclusion is that 'the' act requirement is in reality four separate requirements: the requirement that there be more than a pure omission, more than a mere status or condition, more than an involuntary bodily movement, and more than a mental state. These four requirements may be united by a common name but little else.

This fractionating of the act requirement may go even further. Some have thought that even when we consider each of these four act requirements, one at a time, there may be only an illusory unity. J. L. Austin famously talked about 'excluder' words, words like 'real' or 'free' (his examples) that seem to name something but actually are used to exclude the application of words like 'illusion' or 'coercion'.[13] Hart argued that 'voluntary' was an excluder word, so that if one focused on the supposedly univocal requirement that a bodily movement be *voluntary* to be an act, what really was being said was that a heterogeneous range of conditions was being excluded, conditions like unconsciousness, hypnotic states, sleep, etc.[14] In which case the supposed requirement of a voluntary act (versus involuntary movement) was in reality a large number of different requirements: the requirement that the actor not be asleep, not be unconscious, not be hypnotized, not be suffering a reflex movement, etc.[15]

J. L. Austin thought that even this might not be a finely enough grained approach to action.[16] As Austin noted, the general phrase 'doing an action' is an abstract expression only philosophers and lawyers use; ordinary people talk of killings, letter-postings, telephonings, etc. Austin queried whether we do not 'oversimplify metaphysics' by assuming that all actions share some essential feature(s).[17] We come easily to think that 'all

---

[12] Dan-Cohen, *'Actus Reus'*; Dressler, *Understanding Criminal Law*; Patrick Fitzgerald, 'Voluntary and Involuntary Acts', in A. Guest (ed.), *Essays in Jurisprudence* (Oxford, 1961).

[13] J. L. Austin, 'A Plea for Excuses', *Proceedings of the Aristotelian Society*, 57 (1956), 1–30.

[14] H. L. A. Hart, 'Ascription of Responsibility and Rights', *Proceedings of the Aristotelian Society*, 49 (1949), 171–94.

[15] Model Penal Code, §2.01(2) flirts with this negative way of defining act before it gives a positive, general characterization in terms of 'effort or determination of the actor, either conscious or habitual'. American Law Institute, *Model Penal Code*, (Philadelphia, 1962), §2.01(2)(*d*).

[16] Austin, 'A Plea for Excuses', 4–5.   [17] Ibid.

"actions" are, as actions (meaning what?), equal, composing a quarrel with striking a watch, winning a war with sneezing'.[18] To pursue Austin's query very far would be to dissolve any general act requirement running throughout the criminal law into the special act requirement of each statute that prohibits actions of mayhem, rape, arson, etc.

The moral criticisms of the act requirement focus on the moral justifiability of punishing people for their acts. Sometimes the criticism is that the act requirement is *over*-inclusive in what it makes eligible for punishment, the argument being that persons with disadvantaged social backgrounds, certain mental instabilities, and the like do not freely choose to do wrong even if formally they satisfy the act requirement. As often the criticism of the act requirement is that it is *under*-inclusive. With regard to situations where the actor possesses the ability easily to prevent some harm but intentionally omits to do so, the argument is that such an actor is just as culpable as is the actor who, by his acts, causes the condition of peril to start with. What is the moral difference, this argument asks, between the actor who omits to turn on the respirator with the intention that the patient die, and the actor who acts to turn off the respirator with the intention that the patient die? Do not both actors freely *choose* both that the patient die and that their own behaviour is such that that death is sure to occur?[19]

Even more radical scepticism about the moral justifiability of the act doctrine stems from the insight that morality is concerned with much besides our acts. We judge ourselves and others morally by who we are—what character and emotions we possess—as well as by what we *do*. Indeed, this insight leads some to conclude that bad acts are only moral proxies for what really matters to morality, bad character. A well-conceived criminal code, accordingly, would reduce the act requirement to a kind of evidentiary status: bad acts are usually good evidence of bad character, but their presence is not sufficient to infer bad character and their absence is not sufficient to infer good character. A criminal code that treats acts as sufficient in these two directions

[18] Austin, 'A Plea for Excuses', 5.
[19] James Rachels, 'Active and Passive Euthanasia', *New England Journal of Medicine*, 292 (1975), 78–80. See generally Jonathan Glover, *Causing Death and Saving Lives* (Harmondsworth, 1977), 92–112.

is to be criticized for sacrificing a closer fit with morality in the name of administrative convenience.

Apart from these scepticisms about there being any doctrinally coherent or morally justifiable act requirement, there is also some scepticism about the existence of acts themselves. On this view even if criminal law has an act requirement, there are no items in the world to answer to it. Herbert Hart, for example, once urged that the verbs of action did not refer to anything because the utterances in which such verbs were used were not descriptive utterances; rather, such sentences were used to ascribe responsibility so that action verbs expressed the conclusion that the 'actor' was responsible but did not describe a ground for responsibility.[20] As Patrick Fitzgerald once put Hart's point here, 'in ordinary speech the word "act", together with such allied expressions as "A did it", is used not so much to describe what has happened, as to ascribe responsibility'.[21] On this view, one might as well look in the world for 'ouches', 'dints', and 'sakes', as for acts, since all such words do not refer to things but have other functions in our language.

A more currently fashionable way of denying that acts even exist is part of what I have called the 'interpretive turn'.[22] On this view, whether we call something an act depends on the interpretive stance we, the observers, take to it. It does not depend on anything instrinsic to it. Critical legal-studies types like Kelman often believe this,[23] but even mainstream criminal-law theorists such as George Fletcher have imbibed some of this *verstehen* view. In understanding action, Fletcher tells us, we must eschew the 'causal understanding' appropriate to natural events like avalanches, and focus on the 'interdependence of human subjects and modes of understanding that arise from human interaction'.[24] In this 'perception of human acting as a form of intersubjective understanding' we interpret something to be an action only when 'we can perceive a purpose in what he or she is doing'.[25] It is *our* perception of a purpose that *is* the difference between action and non-action, for Fletcher.[26]

---

[20] Hart, 'Ascription of Responsibility and Rights'.

[21] Fitzgerald, 'Voluntary and Involuntary Acts', 383.

[22] M. Moore, 'The Interpretive Turn in Modern Theory: A Turn For the Worse?', *Stanford Law Review*, 41 (1989), 871–957.

[23] See Kelman, *A Guide to Critical Legal Studies*.

[24] Fletcher, *Rethinking Criminal Law*, 434.   [25] Ibid. 436.   [26] Ibid.

Not surprisingly, the supposed muddiness of the criminal law's act and *actus reus* requirements entails that the relationship between the two requirements is also unclear. Some assume that the act requirement and the *actus reus* requirement are one and the same thing.[27] Others urge that the act requirement is part of the *actus reus* requirement, but that the latter has other necessary features as well. Yet others have thought that the two requirements have almost nothing to do with each other.

The double-jeopardy requirement is also often thought to be in poor shape with respect to its doctrinal coherence, its moral point, and its metaphysical presuppositions. In the United States there are several different doctrinal tests: the 'same evidence' test, the 'same intent' test, the 'same act or transaction' test, etc. In addition, there are four different authoritative sources for these doctrines: the federal constitution, most state constitutions, a variety of state statutes, and the common-law merger doctrine. In addition, with respect to any of these sources and doctrines, there are three different contexts of application where the doctrines are differently applied; the same-evidence test, for example, means one thing when the issue is multiple *punishments* for multiple counts tried in a single trial but it has a different meaning when the issue is multiple *prosecutions* for multiple charges brought in successive trials. Some conclude from these characteristics of American double-jeopardy doctrine that there is no coherent double-jeopardy requirement in the American legal system.

Some sceptics go further and explain the doctrinal incoherence of American double-jeopardy law in terms of the lack of any unitary moral point to the double-jeopardy prohibitions. Because of the differing policies behind separable applications of the clause, Peter Westen, for example, urges:

in order to decide what the words of the fifth amendment mean, one must first ascertain what purposes they serve. When one does, one will discover . . . that the purposes are 'separate,' and, accordingly, that the meanings of the constituent terms change in accord with the several purposes for which they are invoked.[28]

---

[27] e.g. Fitzgerald, 'Voluntary and Involuntary Acts'; Husak, *Philosophy of the Criminal Law*, 9–11.

[28] Peter Westen, 'The Three Faces of Double Jeopardy: Reflections on Government Appeals of Criminal Sentence', *Michigan Law Review*, 78 (1980), 1001–65: 1004 n. 11.

Such policy-based scepticism extends to there being any univocal meaning to double jeopardy's crucial concept of the 'same offence':

> To try to formulate a single definition of 'same offence' for these three separate purposes would produce a statement of such abstract generality as to be of no usefulness in resolving actual cases.[29]

In addition to these doctrinal and moral doubts, many think that there can be no coherent double-jeopardy requirement because metaphysically there are no definite identity conditions to either actions or offences. To apply any double-jeopardy prohibition, in other words, requires that one be able to say when two somewhat differently worded offences are none the less 'the same' for double-jeopardy purposes, and it requires that one be able to say when two differently described acts done on some occasion by an accused are in reality 'the same'. Yet the first of these sameness enquiries requires us to be able to individuate *types* of actions (offences), which requires us to individuate universals— which (in the famous characterization of certain critics of the Scholastics in the Middle Ages) is like counting how many angels can dance on the head of a pin. Moreover, the second of these sameness enquiries requires us to be able to individuate particular acts. The sceptical thought here is that acts have no natural boundaries, that whether one sees one act or several depends on the eye of the beholder. J. L. Austin expressed such a scepticism when he queried: 'What is *an* or *one* or *the* action? For we can generally split up what might be called one action in several distinct ways, into different *stretches* or *phases* or *stages*.'[30] Many legal theorists have followed Austin's sort of scepticism here. Thus, John Salmond urged that 'an act has no *natural* boundaries any more than an event or place has. Its limits must be artificially defined for the purpose in hand for the time being. . . . To ask what act a man has done is like asking in what place he lives.'[31] Similarly, Larry Simon concludes (citing and relying upon Austin) that 'any sequence of conduct can be defined as an

---

[29] Ibid. n. 12.
[30] Austin, 'A Plea for Excuses', 27.
[31] John Salmond, *Jurisprudence*, 11th edn. (London, 1957), 401–2.

"act" . . . Whether any span of conduct is an act depends entirely upon the verb in the question we ask.'[32]

I thus assume that the act, *actus reus*, and double-jeopardy requirements can use some clarification and resuscitation. If we use the above summarized sceptical assaults on the three requirements as our guide, we can usefully separate the questions any theory of action suitable for criminal-law purposes should ask and answer. Some of these are questions about criminal-law doctrines and their coherence, some are about the moral point justifying such doctrines, and some are about the things such doctrines purport to require, namely, acts and actions. I begin with the act requirement:

1. (a) Does the criminal law have an act requirement as a prerequisite to liability to punishment?
   (b) If so, is it one requirement, several requirements, or as many requirements as there are different verbs of action used in the particular prohibitions of the special part of the criminal law?[33]
2. (a) Should the criminal law make acting a prerequisite to criminal liability?
   (b) To the extent that it should not do so—as, for example, about certain kinds of omissions—how can such departures from the act requirement be justified if the requirement itself is justified?
3. (a) If there is and should be a univocal act requirement as part of the definition of all crimes, do things called acts actually exist in the world such that the requirement could be satisfied?
   (b) If so, what is the nature of such acts? Are they, for example, events of a certain kind, and if so are they mental events, bodily movements, or even events in the physical world like the death of a person (when caused by a stabbing, say)?

[32] Note, 'Twice in Jeopardy', *Yale Law Journal*, 75 (1965), 262–321: 276.

[33] The phrase 'special part' is a term of art in criminal-law theory. The special part of the criminal law is 'special' in the sense that it consists of the various particular prohibitions contained in statutes prohibiting mayhem, murder, etc. The act requirement is thought to be part of criminal law's 'general part' in so far as it applies to all crimes, whatever their special nature. On the distinction, and different ways of drawing it, see M. Moore, *A Theory of Criminal Law Theories*, in D. Friedmann (ed.), *Tel Aviv Studies in Law* (Tel Aviv, 1990), 140–44.

There is a parallel set of questions to be asked of the *actus reus* requirement:

1. (*a*) Does the criminal law have an *actus reus* requirement as a prerequisite to liability to punishment?
   (*b*) If so, is it one requirement, several, or as many different requirements as there are different verbs of action used in the particular prohibitions of the special part of the criminal law?
2. Should the criminal law make the doing of some complex action that instantiates some complex action description contained in some statute in force at the time the act was done a prerequisite of criminal liability?
3. (*a*) If there is and should be a univocal *actus reus* requirement, do things actually exist in the world (let us provisionally call them 'complex actions') such that the requirement could be satisfied?
   (*b*) If so, what is the nature of such complex actions? Are they, for example, events of a certain kind, and if so, how do such events relate to other events like the movements of human bodies?

Once we have answered these two sets of questions, we should be in a position also to answer the parallel questions about the double-jeopardy requirement:

1. (*a*) Does the criminal law have a double-jeopardy requirement?
   (*b*) If so, is it one requirement, several, or as diverse as the actions or offences that might be said to be 'the same'?
2. (*a*) Should the criminal law prohibit being punished twice or even prosecuted twice for acts or offences that are in some sense 'the same'?
   (*b*) If so, do the values that justify such a prohibition have sufficient commonality that they justify one doctrinal prohibition, several, or many?
3. If there is and should be a univocal double-jeopardy requirement, are the identity conditions of both actions and the offences that prohibit them sufficiently precise to allow the requirement to be non-arbitrarily applied?

There are other questions we could ask in a philosophical theory of action, but the above are the most relevant to the criminal lawyer. Answering such questions may help us to rediscover the potential for elegance and intellectual integrity that a criminal code may possess, a potential that so charmed our forebears in legal philosophy. I shall begin with the questions about the criminal law's act requirement, the morality that justifies such a requirement, and the metaphysical nature of the acts which it requires.

PART I

BASIC ACTS AND THE ACT REQUIREMENT

2

THE DOCTRINAL UNITY OF THE ACT REQUIREMENT

1. DOES CRIMINAL LAW HAVE AN ACT REQUIREMENT?

1.1. Supposed Counter-examples to There Being Any Act Requirement

Crimes of possession, of omission, of status, and of mental acts, give rise to one of two responses by criminal-law theorists sceptical of the act requirement. One response is to deny that there is any act requirement, using these four kinds of crime as counter-examples to prove the point. Presupposed in this response is the idea that there are crimes of possession, of omission, of status, and of mental action. The other response is to deny that there is *an* act requirement that is univocal; rather, there are four act requirements whose content is to bar liability for omissions, statuses, mental acts, or involuntary bodily movements. Presupposed in this response is the idea that there are no crimes of omission, status, mental action, or involuntary movement (because the four act requirements bar liability for these). Although I shall deal with each response, treating the first in this section and the next in the succeeding section, it is important to see that a theorist cannot consistently hold both scepticisms about the act requirement. For there either are crimes of omission, status, and mental action, or there are not.

1.1.1. Crimes of Mental State.

The answer to the first sceptical response, in so far as that response is directed at crimes of mental state, is to deny its presupposition that such crimes exist. As a descriptive matter, it is not the case that Anglo-American criminal codes criminalize mental states or mental actions. The definition of treason as including the compassing of the death of the king was once an example of the former, but even that crime has long been construed to require an overt act that is a step towards the overthrow of the government.[1] Similarly, the inchoate crimes of solicitation, attempt, and conspiracy require acts, even if the acts required are sometimes acts of speaking. Uttering a thought is different from merely thinking it, and the former is a familiar kind of act we call 'speech-acts'. As Glanville Williams has noted, 'speaking or writing is an act . . . indeed, almost any crime can be committed by mere words, for it may be committed by the accused ordering an innocent agent (e.g. a child under eight) to do the act'.[2]

There is no inevitability to this contingent fact about Anglo-American criminal law. We could punish people for thinking bad thoughts, harbouring immoral desires or emotions, or for the mental actions of imagining how another could be made to suffer. Indeed, it was not unknown under Roman law to punish even those whose desire for the death of the Emperor was expressed only through the mental events of dreaming (of the Emperor's being killed).[3] As it happens, Anglo-American criminal-law doctrines do not contain prohibitions of thinking, desiring, dreaming, and other mental states and acts.

It might be objected that we in effect do punish people for thoughts alone when we define crimes in a way not requiring the doing of any wrongful act. In attempts, for example, the acts that amount to an attempt (versus 'mere preparation') may well be innocent on their face—such as walking towards a building with matches in one's pocket and the day's newspaper under one arm. Yet such acts become criminal—say, attempted arson—if done with a culpable intention, such as the intent to burn the building with the paper and matches. The objection is that in such instances we are in reality punishing for the intention alone.

Such an objection misses the formal, doctrinal nature of the question here examined. The question is not whether our criminal law either does or should punish only when there is some *morally wrongful* action, nor even whether legislatures are motivated to attach sanctions to certain actions because of the badness or dangerousness of those actions or because of the culpability of the

---

[1] On the history of the act requirement in treason, see Williams, *Criminal Law*, 3; Fletcher, *Rethinking Criminal Law*, 209–18.

[2] Williams, *Criminal Law*, 2.

[3] Related in Franz Scholz, *Schlaf und Traum* (Leipzig, 1887), trans. H. M. Jewett, *Sleep and Dreams* (New York, 1893), 62.

mental states with which such actions are done. The question is only whether our criminal law requires that some action be done before liability to punishment attaches. Even for crimes like the Federal Interstate Travel Act[4]—where the act prohibited (travelling interstate) is wholly innocent but is made criminal when accompanied by an intention to bribe officials or do other corrupt acts—the answer to the doctrinal question is that some act much be done before criminal liability attaches.[5]

1.1.2. Crimes of Status. As we shall see in Chapters 4 to 6, a state differs from an act in two ways: (1) like all events, acts tend to be of short duration, while states can be quite long-lasting or even permanent; (2) human acts essentially involve the choice (or willing) of the actor in a way that states of that actor do not. Statutes criminalizing the *use* of drugs or of alcohol punish acts; statutes criminalizing being addicted to drugs or alcohol, or being intoxicated by such substances, punish states and, because of that, are deemed to be crimes of status.[6]

Admittedly, crimes of status, such as being a vagrant, being a common drunkard, or being a common scold, were once more prevalent than they are today. As the late Herbert Packer noted, however, 'laws of this sort are in fact very much on the way out'.[7] Not only are such laws rarely enacted today, but in America a number of constitutional infirmities are regularly found to afflict such laws so that even where they do remain on the books they are not valid.[8]

[4] 18 U.S.C. §1952 makes it illegal to travel on any facility in interstate commerce.

[5] In one respect the answer is not as formal as might be thought, for even crimes like that defined by the Federal Interstate Travel Act do not treat the act that must be done in order to trigger liability as a mere formality. Such statutes do not say, for example, that it is a crime to intend to bribe an official while at the same time whistling Dixie, combing one's hair, or washing one's cat. *That* would be to treat the act requirement as an empty formality. What such statutes rather require by way of an act is some act that executes, however slightly, the culpable intention. Travelling interstate in order to bribe an official satisfies such an act requirement; washing one's cat while intending to bribe an official does not. In addition, normatively I think statutes like the Interstate Travel Act are not a good idea—for there is no *wrongful* act being done. See Moore, 'A Theory of Criminal Law Theories', 174–9.

[6] *Robinson* v. *California*, 370 U.S. 660 (1862); *Powell* v *Texas*, 392 U.S 514 (1968).

[7] Herbert Packer, *The Limits of the Criminal Sanction* (Stanford, Calif.), 78.

[8] *Robinson* v. *California*, 370 U.S. 660 (1962) (status of being an addict may

About the only status for which one can be deprived of liberty is being diseased. If one is mentally ill and a danger to others, one may be civilly committed; if one is afflicted with a communicable disease, one may be quarantined against one's will. But such loss of liberty is not punishment, for it is not imposed for punitive purposes;[9] such liability thus need not be accounted for in examining criminal-law doctrines for their coherence.[10]

1.1.3. Crimes of Possession. Crimes of possession may seem to be more troublesome, because there certainly are constitutionally valid statutes criminalizing possession of: burglar's tools, certain weapons, obscene materials, stolen goods, and controlled substances like marijuana. Many sceptics suppose that here, at least, there are plain counter-examples to any supposed act requirement running throughout the criminal law.

One response to this is to redefine what an act is in order to include possession within it. Hyman Gross, for example, defends the act requirement, asserting that 'crimes of possession, like all other crimes, can be committed only when there is an act'.[11] What Gross means by 'act', however, is gerrymandered so that states of possession can be included: 'An act of possession is . . . rather special . . . these acts . . . consist of a state of affairs, rather than an event . . . a person may be responsible for the way things stand no less than for what happens and . . . acts include both these grounds of responsibility.'[12] To my mind, this isn't to

not constitutionally be punished); *State* v. *Palendrano*, 120 NJ Super. 336, 293 A.2d 747 (1972) (status of being a common scold may not constitutionally be punished); *Papachristou* v. *City of Jacksonville*, 405 U.S. 156 (1972) (status of being a vagrant may not constitutionally be punished).

[9] I have elsewhere defended the view that criminal law is a functional kind, so that one decides whether or not some particular doctrine is or is not a doctrine of *criminal* law by the (punitive) purpose of the doctrine, not by any structural features. Moore, 'A Theory of Criminal Law Theories'. A tax is not a fine because of the lack of punitive purpose, however similar a tax may be to a fine structurally.

[10] I omit any discussion of the extent to which we take various states into account in deciding upon severity of sentence. Admittedly, when we increase sentences because of an offender being a recidivist, or because he lacks remorse, we are punishing for status. What we are not doing by increasing sentences for these reasons is making statuses criminal. Acts remain our trigger for liability, in other words, even though the extent of liability will often be affected by various states or properties of the defendant.

[11] Hyman Gross, *A Theory of Criminal Justice* (New York, 1979), 66.

[12] Ibid.

defend the act requirement; it is to obliterate it. Gross like Humpty-Dumpty may mean what he pleases by the word 'act', but he may not make the word 'act' mean what he pleases, namely, 'a state of affairs, rather than an event'. Once acts are equated with states, then in no meaningful or useful sense of the word is there an act requirement in the criminal law.

The more traditional response is to note that possession crimes are generally defined so that either an act (of acquiring possession) or an omission (to rid oneself of possession) are prerequisites to liability.[13] Thus, it is not the *state* of possessing that is being punished, but either the act of taking possession, or (in the cases where the defendant comes into possession without doing anything) the omission to rid oneself of possession. Possession crimes, so construed, present no counter-examples to the act requirement, or at least none greater than that presented by omissions generally.

The objection often voiced to this traditional accommodation of possession crimes to the act requirement is that we aren't really punishing acts of acquisition of, or omissions to dispose of, things like burglary tools, stolen goods, etc. Really, we are punishing the possession itself, however the crimes are formally construed.[14] Yet this objection is confused. If its idea of what is 'really' being punished in possession crimes is motivational, it is partly correct: those who pass statutes making possession of various items criminal are not motivated by any supposed badness either of the act of taking possession or of the omissions to rid oneself of possession. But then, such legislators are not typically motivated by any supposed badness of *possessing* these items either. The typical motivation behind possession statutes is a preventive one: it is to prevent future burglaries that there is a crime called possession of burglary tools; it is to prevent violence that there are various crimes for possessing certain weapons by certain classes of people; it is to prevent use or sale of drugs that possession of them is criminalized; etc. Sometimes the motivation behind possession statutes is an evidentiary one: the fence in possession of stolen property presumably purchased the property from a thief (and thereby did an act greatly encouraging theft), but we can't prove it; the movie producer in possession of

---

[13] Williams, *Criminal Law*, 8; Model Penal Code §2.01(4).
[14] See e.g. Husak, *Philosophy of the Criminal Law*, 22.

obscene films presumably participated in the making of them, but one can't prove it; etc. In either case, we do not really punish possession because that state is bad or harmful, but only as a proxy—either as a proxy for past acts (which we can't prove) or as a proxy for propensities for future acts (which we can't punish because they haven't yet happened).

Seeing all this about the typical motivations for possession crimes does not make them counter-examples to the act requirement. The compatability of possession crimes with that requirement is a formal matter, and, formally, possession is defined so as to include an act or an omission.[15] What the insights about typical legislative motivation should do is raise serious doubts about the wisdom of enacting possession statutes. For if there is no *wrongful* act in acquiring, maintaining, or failing to rid oneself of possession of items like guns, drugs, obscene materials, etc., then on one well-regarded theory of criminal legislation these acts should not be made criminal.[16] Faced openly, impatience (for future crimes) and inability to prove guilt (for past crimes) are not comfortable rationales for criminalizing conduct. Faced openly, most crimes of possession perhaps should not be crimes, not because there is no act, but because there is no *wrongful* act being punished.[17]

1.1.4. Crimes of Omission. Yet to be accounted for are crimes of omission. Here, too, it is sometimes argued that 'act' can be redefined so as to include omissions. This redefinitional strategy was first deployed by Bentham, who included what he called 'negative acts' in his category of acts.[18] Bentham was followed in this respect by John Austin[19] and by John Salmond.[20]

---

[15] Model Penal Code, §2.01(4). See generally Wayne LaFave and A. Scott, *Criminal Law*, 2nd edn. (St Paul, Minn., 1986), 200 n. 47.

[16] This is what is often called the 'legal moralist' theory of what ought to be criminalized, or what I have more recently called a 'non-exclusionary' theory. See M. Moore, 'Sandelian Anti-liberalism', *Californian Law Review*, 77 (1989), 539–51.

[17] See Fletcher, *Rethinking Criminal Law*, 197–205, for criticism of punishing possessions.

[18] Jeremy Bentham, *Introduction to the Principles of Morals and Legislation* (Buffalo, NY, 1988), 72 n. 18.

[19] John Austin, *Lectures on Jurisprudence*, 5th edn. (London, 1885), i. 366. (The 19th-century jurisprude John Austin is to be distinguished from the 20th-century ordinary-language philosopher J. L. Austin.) Because Austin later took

The redefinitional strategy with respect to including omissions as acts is not quite so bald an assertion that we can 'make words mean what we please' as it is with respect to possession, because ordinary speech often accommodates such a usage. Not telling anyone about a certain matter may quite idiomatically be described in the active mood as 'keeping her secret', not rescuing a drowning man, as 'ignoring the pleas of the drowning man', not filing an income tax return, as 'defrauding the government', and not feeding one's own baby, as 'starving him to death'. None the less, the redefinitional strategy that would include omissions as acts has to justify its inclusiveness by showing how omissions are like commissions, i.e. what is the nature of 'actness' that unites them? It won't do simply to say, as Salmond did, that it would be 'inconvenient' to restrict the term 'act' to acts of commission for that would leave us 'without a name for the genus' that includes acts both of commission and of omission as its species.[21] For the question is, what nature do omissions and commissions share that makes them members of the same genus, no matter what the label?

Bentham recognized that he had to justify his 'extensive and . . . inconsistent signification here given to the word *act*', and urged in his defence that many omissions are preceded by 'acts of will' just as are positive acts.[22] There are two problems with this defense by Bentham of his 'extensive signification' of the word 'act'. Most fundamentally, it makes *mental* acting (willing) sufficient for acting in the legal sense of the word. Such an identification makes impossible any exclusion of pure states of mind by the criminal law's act requirement (as John Austin came to recognize in his rejection of Bentham's 'internal acts').[23] Failing to exclude mental acts from legal liability would be inconsistent with our attributions of moral responsibility, where we blame people for their overt physical actions, not for what they think (no matter how actively they will their thoughts). Secondly,

---

back his idea that there were internal acts, or what we would call mental acts (ibid. 420), and because his stated reason for including certain omissions as 'negative acts' was that they, like internal acts, were determinations of the will, he should have later taken back the idea that there could be negative acts.

[20] Salmond, *Jurisprudence*, 400.

[21] Ibid.

[22] Bentham, *Introduction to the Principles of Morals and Legislation*, 72 n. 1.

[23] Austin, *Lectures on Jurisprudence*, i. 420.

even if one concedes the metaphysical, moral, and legal plausibility of the identification, such identification is not sufficient to justify Bentham's extensive sense of the word 'act'. As Bentham himself recognized, many omissions are not willed yet they are punishable as 'acts' under Bentham's extensive signification of the word.[24] Negligent omissions, for example, are not willed omissions because one's mind was not directed to the situation calling for the action omitted. So something besides a supposed willing that is common to both acts and omissions must justify Bentham's 'extensive signification' of the word 'act'.

'Act', then, cannot simply be redefined so as to make true the assertion 'the criminal law always requires an act'. We thus must look at the criminal law that we have and ask how often, if ever, does it truly punish a mere omission to act? In order to pursue this question meaningfully, we must know what to look for in criminal prohibitions when we seek to ascertain whether they punish acts or omissions; in other words, we need to have before us some intuitive distinction between acts and omissions. Since there is some considerable disagreement as to how to frame this distinction, we must pause long enough to satisfy ourselves we have a defensible distinction with which to ask our doctrinal question (of whether our law punishes omissions).

There are four leading conceptualizations of the act/omission distinction. Since the first two are pretty obvious non-starters, I shall put them aside quickly. The first of these is the suggestion that we can have some grammatical test for the distinction: if it is idiomatic English to describe the relation between a person and a harm with an actively voiced verb, then there is action.[25] For example, 'He took her money', 'She hit him with the stick', etc. The implausibility of the suggestion stems from the readiness of idiomatic English to so describe many intuitively clear omissions: 'He ignored the pleas of the drowning man', 'She kept her

---

[24] Bentham, *Introduction to the Principles of Morals and Legislation*, 72 n. 1.

[25] George Fletcher makes such a suggestion in his *Rethinking Criminal Law*. Some such ordinary-language intuition may have originally motivated Philippa Foot to have thought that not feeding a starving beggar in order to harvest his organs when he died was a *doing* rather than a mere *allowing*. See Foot, 'The Problem of Abortion and the Doctrine of Double Effect', *Oxford Review*, 5 (1967), 5–15. She apparently retracts the example in 'Morality, Action, and Outcome', in T. Honderich (ed.), *Morality and Objectivity* (London, 1985), 25, 37 n. 6.

secret', 'She broke her promise', etc. Such examples suggest that there is no reason to suppose that idiomatic English tracks any interesting distinction here.

The second suggestion about the act/omission distinction, equally wide of the mark, is that the distinction is a moral distinction, not a metaphysical one. The idea is that we will *say* something is an act or an omission depending on our conclusions about the responsibility of the actor/omitter; if she is responsible for the harm, then she acted, and if she is not responsible, then she is only an omitter.[26] The suggestion, of course, abandons the distinction between acts and omissions, appropriating the labels to refer misleadingly to some other distinction we think to be more relevant morally.

The grain of truth to the otherwise empty moral approach is that we do want an act/omission distinction that has the potential at least to carry some moral freight. The moral freight relevant here is the moral difference between our negative obligations not to make the world worse and our positive obligations to make it better. Many sense, as do I, the very real difference in the force of these two kinds of obligation. Indeed, some would urge that the only obligations we have are negative ones; though we would be more virtuous to make the world better, such supererogation is in no way obligatory. Others who reject this extreme libertarian view of the matter none the less show how different is the moral force of these two kinds of obligations: negative obligations are much less susceptible to being overridden by good consequences than are positive obligations. For example, we may easily override our positive obligation to save one drowning baby by the necessity of directing our rescue efforts to the saving of two other babies elsewhere; but we may not so easily override our negative obligation not to kill the one baby even if the only way to save two others is by yanking the rope away from the one.[27]

---

[26] See P. J. Fitzgerald, 'Voluntary and Involuntary Acts', in A. G. Guest (ed.), *Oxford Essays in Jurisprudence* (ser. 1; Oxford, 1961), 11.

[27] The distinction between positive and negative obligations is suggested by Philippa Foot. See Foot, 'The Problem of Abortion and the Doctrine of Double Effect'; id., 'Killing and Letting Die', in Jo Garfield (ed.), *Abortion: Moral and Legal Perspectives* (Amherst, Mass., 1984); id., 'Morality, Action, and Outcome'. See also Warren Quinn, 'Actions, Intentions, and Consequences: The Doctrine of Doing and Allowing', *Philosophical Review*, 98 (1989), 287–312. On the effect of

The last two conceptualizations of the act/omission distinction answer to this need to draw the distinction in a way that marks the moral difference between positive and negative obligations. The third answers this need very directly and is what I shall call the 'baseline' notion of the distinction. On the baseline view, whatever makes the world morally worse (from some baseline state of affairs) is an action; whatever does no more than return things to the baseline state of affairs is an omission.

Nowhere does one see this conceptualization of the distinction more at work than in the discussions of 'active' versus 'passive' euthanasia.[28] Killing an otherwise terminally ill patient with her consent is held to be a passive failure to save if what is done does no more than return the patient to some baseline condition. The Catholic version of this is to take the baseline to be the condition the patient would have been in without 'extraordinary' or 'heroic' medical treatment. The more recent baseline adopted by some courts and commentators has been the 'natural' condition the patient would have been in without any medical treatment, where 'medical treatment' includes intravenous (as opposed to oral) nutrition and hydration.[29] In either case, the disconnection of various devices is held not to be an action because such disconnections do no more than return the patient to the condition she would have been in had the treatments not been undertaken.

The intuitive appeal of the baseline notion stems from its direct fit with the moral distinction between when we make the world worse and when we only fail to make it better. If we have no obligation to undertake rescue measures, but do so and then discontinue the effort, how have we made anyone worse off?

Despite this intuitive appeal, however, the baseline notion fails us in the crucial selection of what to count as the status quo ante (to which we may return without 'acting'). The artificiality of the 'extraordinary medical treatment' and the 'medical (as opposed to non-medical) feeding' lines reveals what was problematic all along: there is no state of affairs that is obviously natural or fitting to be picked as the appropriate baseline. What is to be

the act/omission distinction on the burden of justifying apparent breaches of our obligations, see also M. Moore, 'Torture and the Balance of Evils', *Israel Law Review*, 23 (1989), 280–344.

[28] As in Rachels, 'Active and Passive Euthanasia'.

[29] *Barber* v. *Superior Court*, 147 Cal. App. 2d 1006, 195 Cal. Rptr. 484 (1983).

said, for example, of the defendant who administers poison to a ship's captain, the captain not being able to sail his ship and dying one week later of the poison—is this a mere omission should it turn out that the ship on which the captain sailed was lost with all hands during the intervening week? Why isn't the baseline here the death by drowning the ship's captain would have suffered when the ship sank, had he been on it, as easily as it is the ship's captain's being alive at the end of the week, had he been both on shore and not poisoned? Analogously, in the euthanasia setting, if the baseline for judging the act/omission line is the condition the patient would have been in without extraordinary medical treatment, then why is not the intruder who unplugs the patient from the respirator, with the intention and the effect that the patient dies, only returning the patient to his baseline and, thus, only omitting to save him?[30]

What these examples seem to suggest is that the baseline conception of the distinction must be relativized to individual actors: if *we* intervene to improve someone's condition, then *we* may return them to the status quo ante without 'acting'; but the baseline for anyone else is triggered by the condition of the world when they first intervened—in the case of the ship's captain, when he was healthy and on shore, and in the case of the patient, when he was breathing comfortably with the help of the respirator.

Yet this individualization of baselines won't do either. Suppose I throw a rope to an otherwise clearly drowning man. Is the relevant baseline ever after that he was dying when I first intervened? May I thus throw him back in the water, or shoot him a year later, and be said only to have omitted to save him?

I conclude that the baseline notion is hopeless as an explication of the act/omission distinction. What the 'active' versus 'passive' euthanasia discussion has revealed is how such a manipulable distinction can be used to smuggle notions of justification in under the guise of being 'omissions'. We think that medical personnel may disconnect respirators, whereas intruders may not, because we think the first but not the second is justified in (actively) causing death in this way.

---

[30] See Sanford Kadish and Stephen Schulhofer (eds.), *Criminal Law and Its Processes*, 5th edn. (Boston, 1989), 214 n. 1.

That brings us to the fourth and last conceptualization of the distinction: omissions are simply absent actions. An omission to save life is not some kind of ghostly act of saving life, and certainly not some ghostly kind of killing. It is literally nothing at all.[31] An omission to save X at some time t is just the absence of any instantiation of the type of action 'saving X'. On the metaphysical view of actions developed in Chapter 4–6, actions are event-particulars of a certain kind, namely, willed bodily movements. Anticipating the results of that analysis, we can then say that omissions are the absence of any willed bodily movements. An omission by A to save X from drowning at t is just the absence of any willed bodily movements by A at t, which bodily movements would have had the property of causing X to survive the peril he faced from drowning.

This conceptualization of the act/omission distinction is capable of carrying the moral freight earlier described (marking the difference between supererogatory virtues and positive duties, on the one hand, and negative duties, on the other). If, as I shall argue in Chapter 10, omissions do not *cause* anything, then when I omit to prevent some harm I do not make the world worse; I

---

[31] I take this to be the generic notion of an omission employed in the extensive philosophical literature on the topic. I say the generic notion, because many philosophers use it as the beginnings of more discrete notions of omissions. One might add that one was expected, or duty-bound, not to omit; or that one had the capacity not to omit; or that one intended to omit; or that one adopted some other action as a means to leaving something undone; etc. These and other distinctions then give rise to conceptions of 'refraining', 'allowing', 'letting happen', 'resisting', 'normative not-doing', 'forbearance', 'displacement refrainings'. See Arthur Danto, 'Freedom and Forbearance', in K. Lehrer (ed.), *Freedom and Determinism* (New York, 1966); G. H. Von Wright, *Norm and Action* (New York, 1963), ch. 3, §8; Myles Brand, 'The Language of Not Doing', *American Philosophical Quarterly*, 8 (1971), 45–53; Douglas Walton, 'Omitting, Refraining, and Letting Happen', *American Philosophical Quarterly*, 17 (1980), 319–26; Frederick Siegler, 'Omissions', *Analysis*, 28 (1968), 99–106; P. J. Fitzgerald, 'Acting and Refraining', *Analysis*, 27 (1967), 133–9; Gilbert Ryle, 'Negative "Actions"', *Hermathena*, 81 (1973), 81–93; Judith Thomson, *Acts and Other Events* (Ithaca, NY, 1977), 212–18; Bruce Vermazen, 'Negative Acts', in B. Vermazen and M. Hintikka (eds.), *Essays on Davidson, 'Actions and Events'* (Oxford, 1985). I adopt the more generic notion because it is the most interesting notion metaphysically and because it marks *one* important moral distinction. I shall deal briefly with displacement refrainings and resistings in Ch. 5, and with intended omissions in Chs. 5 and 10.

only fail to make it better. Only when I cause that harm to occur—through my actions—do I worsen the world.[32]

The movement/non-movement conceptualization of the act/omission distinction often garners two objections: one is a moral objection, to the effect that there is no moral difference between supererogation and obligation, or between positive and negative obligations, and that therefore the movement/non-movement conceptualization of the act/omission distinction is without moral significance. The second is a conceptual objection, to the effect that the movement/non-movement conceptualization of the distinction does not match our intuitive (and moral) employments of the distinction. I shall deal with the moral objection in Chapter 3. Here I wish to defuse the conceptual objection.

The conceptual objection is this: suppose A does not throw B a rope when B is in the water (due, let us suppose, to C's pushing B in) and when B would have been saved had he been thrown the rope. If B drowns, we intuitively think that A omitted to save him by not throwing the rope. Yet suppose A was not motionless during this time; he was, we may suppose, shouting with glee and dancing a jig on the dock to celebrate the fact that his old enemy, B, was about to die. By the movement/non-movement criterion, has A therefore not acted? The response of course is to admit that A's shouting and dancing are actions but to deny that they are causally relevant to B's death; therefore, A did not *kill* B, he only omitted to save B from death. Now the real objection: the shouting and dancing motions are causally relevant to B's death, for if, contrary to fact, A had not shouted and danced, he would have thrown the rope and B would not have drowned. A's shouting and dancing, the argument goes, are as causally relevant to B's death as are the bodily movements of C in pushing

[32] The Model Penal Code conceptualizes the distinction in terms of bodily movement/non-bodily movement, as do some philosophers. See §1.13(4), defining an omission as a failure to act, and §1.13(2), defining act as a bodily movement; see also Vermazen, 'Negative Acts'. When conjoined with the metaphysical insight that omissions do not cause harms, the movement/non-movement conceptualization can be seen as one sort of baseline conceptualization. It is the limiting case where each particular intervention starts a new baseline, so that acts (that cause the prohibited harm) always make the world worse whereas omissions (that cause nothing) only fail to make the world better.

B into the water in the first place; all such bodily movements are equally *necessary* for the death of B to occur.[33]

It is surprising how many ingenious and complicated proposals have been spawned by this objection. Jonathan Bennett, for example, proposes to distinguish between explanations for events like B's death in terms of 'strong facts' about behaviour versus 'weak facts' about behaviour. C's pushing B into the water is a strong fact explaining B's death because its contribution to B's death could be accomplished in only a limited number of ways; A's dancing and shouting is only a weak fact explaining B's death because their contribution to B's death could have been accomplished by any number of other actions.[34] Analogously, the late Warren Quinn distinguished between direct and indirect contributions to harms like B's death: C's pushing directly contributes, but A's shouting and dancing do not because they contribute only by causing A to omit to throw the rope, which then causes B to drown.[35] Analogously, Leo Katz distinguishes C's action from A's inaction by asking a counterfactual question about the existence of either actor: would B still have drowned even if C or A had not existed? Since the answer is 'no' for C, 'yes' for A, Katz characterized C's movement as actions and A's movements as omissions.[36]

The back of the objection can be broken more easily than by these complicated proposals. For the objection presupposes an illegitimate notion of causation. It presupposes: (1) the necessary condition (or counterfactual) analysis of causation; and (2) that there is no answer to the well-known indeterminacy of counterfactuals.[37] A's dancing and shouting *cause* B's death only in the sense that they are necessary conditions of B's death, i.e. but for their occurrence, B would not have died. Further, A's dancing and shouting are necessary conditions of B's death only if the

[33] For the objection, see Jonathan Bennett. 'Morality and Consequences', *The Tanner Lectures on Human Values*, 2 (1981), 45–116; Leo Katz, *Bad Acts and Guilty Minds* (Chicago, 1987), 141–2.

[34] Bennett, 'Morality and Consequences'.

[35] Quinn, 'Actions, Intentions, and Consequences'.

[36] Katz, *Bad Acts and Guilty Minds*, 143.

[37] For a brief introduction to the indeterminacy of counterfactuals, see Moore, 'Thomson's Preliminaries about Causation and Rights', *Chicago–Kent Law Review*, 67 (1987), 497–521. The classical statement of the problem is in Nelson Goodman, *Fact, Fiction, and Forecast*, 4th edn. (Cambridge, Mass., 1983), *passim*.

possible world in which we test the counterfactual is a world in which A's dancing and shouting are replaced, not with A doing nothing, but with A throwing the rope that saves B.

We should reject both of these presuppositions. As I argue in Chapter 10, we should be wary of analysing causation in terms of necessary conditions (or counterfactuals more generally). And, even if we do adopt that problematic analysis of causation, we have no reason to test counterfactuals like that involving A's dancing and shouting in a possible world in which those actions are only replaced by acts of A saving B. If we were to adopt, for example, David Lewis's analysis of counterfactuals, we would test them in the possible world that is 'closest' to the actual world in which we live.[38] Whatever 'closest' is taken to mean, it surely does not mean 'the world where A's hatred of B is gone, his desire to be a good Samaritan is enhanced, and where he therefore does just the act that saves B'. The possible worlds in which A does nothing, or smiles to himself, or claps his hands in glee, are 'closer' to the actual world where A dances and shouts than is the possible world in which A has a remade personality and thus saves B.

I conclude that there is no conceptual objection to be made to the movement/non-movement conceptualization of the act/omission distinction. I shall accordingly adopt that conceptualization of the distinction. That allows us, finally, to ask whether Anglo-American criminal-law doctrine criminalizes omissions as well as actions. This is not itself as straightforward a question as it may seem, because of what I shall call the 'embedded omission' problem. Consider the example of a statute making criminal the failure to obtain a professional licence, say, to practise medicine. It may seem that such a statute punishes an omission, not an action. Yet a more accurate description of what such statutes criminalize, Graham Hughes notes, is: 'practicing medicine without a license', which describes an *act* that is done in the particular circumstance that no licence has been obtained.[39] Hughes would generalize this example to all (or at least the 'great bulk') of crimes of omission. As put by Hyman Gross, who follows Hughes in this strategy: there is only an 'illusion that liability for

[38] David Lewis, 'Causation', *Journal of Philosophy*, 70 (1973), 556–7.
[38] Graham Hughes, 'Criminal Omissions', *Yale Law Journal*, 67 (1958), 590–637: 598.

an omission . . . is simply liability for failing to do what is required'.[40] In fact, 'crimes of omission are committed only when specified acts are done . . . Even though liability is imposed *because* something was not done, liability nevertheless is *for doing* certain things without doing certain other things.'[41]

Hughes and Gross give many plausible examples where apparent criminalization of omissions has a hidden act requirement. It is not criminal simply to fail to file an income tax return; what is criminal is to earn income and fail to file the return. It is not criminal simply to omit to wear seat-belts; what is criminal is to operate or ride as a passenger in an automobile without wearing seat-belts. It is not criminal simply to omit to obtain the woman's consent to intercourse; what is criminal is to have intercourse in the circumstance that consent has not been obtained. Etc. Yet not all omissions punishable under Anglo-American law have such hidden act requirements. The examples just given are plausible examples of there being hidden act requirements because one complex action that takes place at one discrete moment in time is prohibited by such statutes. The omission then becomes a circumstance existing at the time the act is done (e.g. driving) that makes the complex action (e.g. driving without a licence) the type of action that is prohibited.

The plausibility of so viewing liability for omission evaporates when we cannot view the omission as a circumstance that is simultaneously present at the doing of some positive action. Thus, we hold parents criminally liable for not feeding their children, husbands liable for not protecting their wives from danger.[42] The only positive act done in such cases by the parent or husband is the one Hughes singles out: there may be a 'voluntary assumption of the burden by the individual' by 'his entrance into the relationship or his possibly intentional, fathering of children'.[43] Yet the omissions to feed and protect occur long after these acts by husbands and fathers, so that one cannot view the non-performance of the omitted activity as a circumstance in which another activity is done. The most one could claim would be that such defendants married or fathered children in the cir-

---

[40] Gross, *A Theory of Criminal Justice*, 63.   [41] Ibid. 65.

[42] On the 'status' or 'relationship' exception to a general lack of omission liability, see generally Dressler, *Understanding Criminal Law*, 82.

[43] Hughes, 'Criminal Omissions', 599.

cumstance that they did not intend to protect their wives or feed their children, but that is not at all the crime for which neglectful spouses and parents are punished under the well-established status exception for omission liability.

A closer question is afforded by doctrines imposing liability for omissions that occur after any positive acts by the accused but where none the less a plausible case can be made that in reality we are punishing for the earlier act, not the subsequent omission. Consider two such doctrines: the undertaking and the having-caused-the-condition-of-peril doctrines.[44] With regard to the undertaking exception, current doctrine criminalizes omitting to rescue once one has undertaken a rescue, even where absent the undertaking one had no legal duty to rescue. The Hughes–Gross analysis of this doctrine should go like this: where someone by positive act undertakes to rescue another (say by throwing a rope to one drowning), when he abandons the enterprise (by not tying the rope down, say), we hold him liable for his earlier act, not his subsequent omission. For if he had not thrown the rope, someone else would have thrown the rope or otherwise undertaken to rescue; or the swimmer himself would have made greater efforts; thus, by the defendant's earlier *act* of undertaking to rescue, he has caused the death.

Sometimes this analysis will go through, but sometimes it will not. Suppose that the rescue undertaken does not make the rescuee any worse off then before—there were no other potential rescuers, and there were no efforts the swimmer himself could have made that he didn't make because of the thrown rope. Now, although the undertaking is still an act, that act doesn't cause the harm. If we hold the abandoning rescuer liable, it thus can't be for the earlier act. Moreover, even in cases where the earlier act is causally relevant, there is a difference between holding the rescuer liable for the earlier act and holding him liable for the later omission to follow through. For one thing, the later omission will more easily satisfy any *proximate*-cause requirement. More importantly, the *mens rea* present at the two different times will often be different, as where the rescuer undertakes to rescue with no thought of not saving the person in peril but who, on seeing that that person is his old enemy, deliberately and

---

[44] See generally Dressler, *Understanding Criminal Law*, 83.

intentionally omits to follow through on the rescue. In the last case supposed, if we were truly punishing for the earlier act, there should be no liability because there was no *mens rea* at the earlier time; yet the undertaking exception generates liability, which must therefore be for the subsequent, culpable omission.

A like analysis applies to the causing-of-the-condition-of-peril doctrine. The doctrine is that those who innocently or culpably cause the condition of peril—as by intentionally, negligently, or innocently bumping someone into the water—are liable if they omit to rescue when they can do so at minimal risk to themselves. It is sometimes thought that in such cases liability is actually imposed for the earlier act of bumping, not the later omission to rescue.[45] Yet again, often there will be a difference in the *mens rea* of the bumper versus the *mens rea* of the non-rescuer: the bump could be non-culpable, yet the omission a knowing failure to help an old enemy. Given that there is liability in such cases, it can only be for the subsequent omission, not for the earlier, innocent act.

I conclude that the Hughes–Gross kind of reconciliation project cannot save the act requirement from the existence of some counter-examples of true omission liability without action. The most one can say of the Anglo-American criminal law's act requirement is that either an act or an omission is required for liability, with the proviso that true omission liability is exceptional rather than customary.

This raises the possibility that perhaps Anglo-American criminal law is mistaken in those rare instances where it genuinely imposes omission liability. Perhaps there ought not to be such crimes, recognizing that presently they do exist. Because such a query is part of the larger normative question of whether there should be an act requirement at all, I shall defer consideration of it until we have further clarified just what the act requirement is.

---

[45] Suggested in Kadish and Schulhofer (eds.), *Criminal Law and Its Processes*, 210. The position is argued for explicitly in Epstein, 'A Theory of Strict Liability', *Journal of Legal Studies*, 2 (1973), 151–204: 192; E. Mack, 'Bad Samaritanism and the Causation of Harm', *Philosophy and Public Affairs*, 9 (1980), 230–59: 240–1, 242–3.

1.2. Some Critical Legal Silliness about the Act Requirement

The other scepticism about the existence of any act requirement that I shall consider is that of Mark Kelman.[46] Kelman urges that the act requirement is vacuous in the sense that the requirement can be manipulated to yield whichever result one wants. To see the argument, consider one of Kelman's examples, *People* v. *Decina*.[47] Decina was an epileptic who had killed someone while driving his automobile during an epileptic seizure. Although the seizure and the movements of Decina's body that it caused were not voluntary acts, none the less the court held Decina liable for manslaughter because of his earlier voluntary act of getting into the car and driving while he was not in the grip of a seizure. Kelman's point: whether there is a voluntary act depends on how broadly a court is willing to look for one; a narrow time frame (e.g. at the time of the accident) results in no voluntary act; a broad time frame (e.g. during the entire drive) results in there being a voluntary act; and there is no principled way for a court to choose a broad or a narrow time frame. This is why Kelman condemns criminal law's act requirement with his 'winking dismissal': 'You know what? There's not much to it.'[48]

The truth is that there is not much to Kelman's point, although we can use its obvious falseness to make (an equally obvious) clarification of the criminal law's act requirement. If there were a 'time-framing' choice to be made in criminal cases, Kelman is right in his observation that there would be no principled way to make it. But where did Kelman get his assumption that there is such a choice to be made? Every competent teacher of elementary criminal law that I know teaches the act requirement in the following way: if, from the big bang that apparently began this show to the heat death of the universe that will end it, the court can find a voluntary act by the defendant, accompanied

[46] See Kelman, 'Interpretive Construction in the Substantive Criminal Law', 600–5, 618–20, 637–40; id., *A Guide to Critical Legal Studies*, 92–3. Kelman's form of scepticism about the act requirement continues to have unwarranted influence even beyond the cosy confines of critical legal studies, and so is worth dispatching. See e.g. J. M. Balkin, 'The Rhetoric of Responsibility', *Virginia Law Review*, 76 (1990), 197–263: 228–30; Larry Alexander, 'Reconsidering the Relationship among Voluntary Acts, Strict Liability, and Negligence in Criminal Law', *Social Philosophy and Policy*, 7 (1990), 84–104.

[47] 2 N.Y. 2d 133, 147 N.Y. S. 2d 558, 138 N.E. 2d 799 (1956).

[48] Kelman, 'Interpretive Construction in the Substantive Criminal Law', 637.

at that time by whatever culpable *mens rea* that is required, which act in fact and proximately causes some legally prohibited state of affairs, then the defendant is prima facie liable for that legal harm.[49] There is no 'time-framing' choice here. If there is *any* point in time where the act and *mens rea* requirements are simultaneously satisfied, and from which the requisite causal relations exist to some legally prohibited state of affairs, then the defendant is prima facie liable. The presupposition of Kelman's entire analysis is simply (and obviously) false.

Consider *Decina* again. The New York court rightly decided that Decina's bodily movements at the time of the accident were not acts, and that Decina's movements beginning to drive were acts. The court did not, however, arbitrarily focus on the earlier time because it had arbitrarily chosen a broad time-frame in which to look for a voluntary act. Rather, the court looked at all possible times and found one where Decina not only acted (in beginning to drive), but did so recklessly (in light of prior seizures he was aware of the risk to others posed by his driving), which reckless act caused the victim's death.

Contrast *Decina* to another case that Kelman discusses, a case where liability was not found, *Martin* v. *State*.[50] Martin was accused of the crime of being drunk in public. Martin got drunk in his own private residence, which is not 'in public' within the meaning of the statute. He was bodily carried out into the street by the police, however, and they arrested him for being drunk in public. Kelman thinks that the Alabama court could justify its decision (of no voluntary act by Martin) only by a 'narrow time-framing'; for a broad time-framing would reveal earlier acts by Martin that were voluntary, namely, the taking of drinks. What Kelman overlooks is that those earlier acts by Martin were not the proximate cause of his being drunk in public. The police officers' intentional placing of Martin in a public place constitutes an intervening cause on anyone's reading of that notion,[51]

[49] See e.g. Dressler, *Understanding Criminal Law*, 171; P. Robinson, 'Causing the Conditions of One's Own Defense: A Study in the Limits of Theory in Criminal Law Doctrine', *Virginia Law Review*, 71 (1985), 1–63. The reaction to Robinson's paper by the leading Anglo-American and German criminal-law scholars when it was first presented (at the Criminal Theory Conference, Max Planck Institute, Freiburg), was 'Of course—that is what we all teach.'
[50] 31 Ala. App. 334, 17 So. 2d 427 (1944).
[51] See H. L. A. Hart and A. Honoré, *Causation in the Law*, 2nd edn. (Oxford, 1985), for the leading exposition of the idea of an intervening cause.

making Martin not a proximate cause of the legally prohibited state of affairs. In addition, had the Alabama statute required any *mens rea* with respect to the element of public place, as it should have, Martin's earlier acts of drinking in his home were unaccompanied by such *mens rea* and were thus ineligible to be the basis for his conviction for that reason too.

The only intelligible point that Kelman could be making here is the familiar worry about the proximate-cause requirement: it is vague, ambiguous, or in some other way hopelessly mushy, so that such requirement does not meaningfully restrict the time during which a voluntary act by a defendant may be found.[52] To *that* familiar worry there are two answers. One is that, so construed, Kelman's point cannot be that courts can willy-nilly choose whether to find an act or not. Even if the proximate-cause requirement is totally empty of meaning, that only would result in a rather more extensive liability than that the courts currently impose, namely, defendants would be prima facie liable whenever their voluntary act in fact caused a legally prohibited result when that act was accompanied by a culpable *mens rea*. That might make Martin liable under a strict-liability statute, but not otherwise. Secondly, the vagueness/mushiness of the proximate-cause tests is no reason at all to think that any connection can or cannot be said to be one of proximate causation. To think this would be to repeat the medieval fallacy about heaps: since 'heapness' is a vague notion, one can never say that the subtraction (or addition) of one stone unmakes (or makes) a heap; therefore, everything (or nothing) is a heap.

Kelman's form of scepticism is worth rebutting only because it allows us to see clearly how the act requirement works in Anglo-American criminal law. The requirement is only that there be some act of the defendant's, done at any time whatsoever, so long as it is connected in the requisite ways (causation and intentionality) to the state of affairs the law prohibits. There is no requirement that one first isolate some point in time at which one asks, 'Did the defendant do an act *then*?'

---

[52] Larry Alexander so takes Kelman's point, in Alexander, 'Reconsidering the Relationship among Voluntary Acts, Strict Liability, and Negligence in Criminal Law', *Social Philosophy and Policy*, 7 (1990) 92.

2. DOES CRIMINAL LAW HAVE ONE, SEVERAL, OR MANY ACT REQUIREMENTS?

As I noted in the Introduction, sometimes 'the' act requirement is thought to be, in reality, four separate requirements, matching the four things excluded by it: status, mental state, omission, and involuntary bodily movement. If one thinks that these four things have little in common with one another, one might conclude that there is not one (positive) requirement but four (negative) requirements: no one shall be punished for being in a certain status, for mental states alone, or for involuntary bodily movements, and only rarely may one be punished for an omission unaccompanied by any act.

It is the beginning of wisdom here to see that the unconnectedness of status, omission, mental state, and involuntary bodily movements is not much of a reason for thinking that the opposite of each of these four things is not in reality one thing with a unified nature. For the properties by virtue of which an act is opposed to each of these four things could be different, and yet each property be an essential property of 'actness'. Imagine a zoologist told you that there was something: that was not a plant or a mineral; that was not a bird or a reptile; that was not small; that was not a meat-eater; and that could not reproduce with anything but elephants. Suppose she called this thing 'elephant'. Would you have reason to believe that there was no one thing called 'elephant' just because you correctly observed that plants, minerals, birds, reptiles, bigness, meat-eating, and non-elephant reproductive capacity have little to do with one another?

Just as we could tease out a definition of the unitary nature of elephant by thinking of the opposite of the properties elephants don't possess, perhaps we can figure out what concept of action the law employs by thinking of the opposite of the four things acts are not. What is the opposite of status? 'Opposite' here cannot mean 'contradictory', because many things are not statuses. 'Opposite' here means 'contrary', a thing that is not a status. Since many things are contraries of status, we have some choice here. If we take 'status' to be the legal terminology for 'state', a familiar contrary of 'state' in metaphysics is 'event'. A state is a more or less enduring property of an object, whereas an event is a change in the nature of some object(s), their qualities or relations, over some relatively discrete interval of time. If this is the right opposition, then the act requirement of criminal law requires the existence of some kind of event. What kind of event? Not a mental event, an event that occurs in a person's mind like a sensation, an act of imagination, or a thought. A customary opposite of 'mental event' of a person is 'physical event' of a person. The physical event of persons are the events that take place within or with their bodies—involuntary movements of the peristaltic gut, nerve impulses in the central nervous system, or movements of the fingers, for example. If this is the right opposition, then the act requirement of criminal law requires the existence of some kind of physical event to occur within or otherwise involving a person's body. What kind of human bodily event? Not an involuntary bodily movement. The usual opposite of 'involuntary' is 'voluntary', by which we mean 'willed' or 'of one's own volition'.[53] If this is the right opposition, then the act requirement is a requirement that there be a willed bodily movement.

One may notice that we haven't yet used omissions as we grope our way around the elephant. This is because to grasp what an omission is requires that we already know the nature of the thing omitted, namely, an act. For 'omission' as we earlier analysed it is the contradictory of 'act', not a mere contrary; moreover, of the contradictory pair 'act/omission', it is 'act' that has the primary meaning.[54] That is, an omission is just a not-action; to omit is not to do. Omissions have no nature except that of being the absence of the actions omitted, just as absent elephants have no nature save that of elephants that aren't present.

We end up, then, with the notion of an act as a willed bodily movement that is indeed the opposite (contrary or contradictory)

[53] 'Involuntary' has another accepted sense in both ordinary speech and the law: we idiomatically say that the person who acts under the threats of others 'acted involuntarily'. Still, such a person clearly *acts*, so this cannot be the sense of 'involuntary' relevant to the act requirement. For the two senses of 'involuntary' in ordinary speech, see Gilbert Ryle, *The Concept of Mind* (London, 1949), 69–74; in law, see Fitzgerald, 'Voluntary and Involuntary Acts', 12.

[54] J. L. Austin was perhaps the first to see that often one term in a contradictory pair of terms has primary meaning, the other member of the pair taking its meaning wholly by negating the first member of the pair. Thus, some have argued that 'healthy' only means 'not ill'. See M. Moore, *Law and Psychiatry: Rethinking the Relationship* (Cambridge, 1984), 116.

of status, mental event, omission, and involuntary bodily movement. If the two parts of this definition—bodily movement and
willings—themselves have a unitary nature, then so does the
criminal law's requirement that there be a willed bodily movement (that is, act). So far I have not attempted such a metaphysical showing, nor have I given any normative justification for why
the criminal law should require willed bodily movements for liability. (Although I shall shortly do both.) I haven't even shown
that willed bodily movement is the conception of act required by
the criminal law's act requirement. All the above discussion is
designed to show is that a commonly accepted reason for reaching the conclusion that there cannot be a unitary act requirement
in the criminal law—because there are four different opposites of
act that are excluded by the requirement—is not a good reason.

There are two similar arguments that merit brief mention here,
both because they are similar to, and because they are false in
much the same way as, the argument just considered. I refer to
the ordinary-language arguments of Herbert Hart and J. L.
Austin mentioned in the first chapter. Early in his career Hart
advanced the view that the *mens rea* requirement of the criminal
law was not a unitary requirement:

What is meant by the mental element in criminal liability (mens rea) is
only to be understood by considering certain defenses or exceptions,
such as Mistake of Fact, Accident, Coercion . . . The fact that these are
admitted as defenses or exceptions constitutes the cash value of [the
mens rea requirement]. But in pursuit of the will-o-the-wisp of a general
formula, legal theorists have sought to impose a spurious unity . . . upon
these heterogeneous defenses and exceptions, suggesting that they are
admitted as merely evidence of the absence of . . . two elements ('fore-
sight' and 'voluntariness') . . . it is easy to succumb to the illusion that
an accurate and satisfying 'definition' can be formulated with the aid of
notions like 'voluntariness' because the logical character of words like
'voluntary' is anomalous and ill-understood. They are treated in such
definitions as words having positive force, yet . . . the word 'voluntary'
in fact serves to exclude a heterogeneous range of cases such as physical
compulsion, coercion by threats, accidents, mistakes, etc., and not to
designate a mental state . . .[55]

[55] Hart, 'Ascription of Responsibility and Rights', 179-80.

Hart's analysis of *mens rea* (or 'voluntariness' in an extended sense) has been extended to 'voluntary' in the sense here relevant, namely, as a synonym for 'willed' or 'volitional'. Thus, Patrick Fitzgerald conducts his analysis of 'voluntary' by considering: physical forces of nature; physical force used by another; epileptic convulsions; unconsciousness; sleep; hypoglycemic episodes; hypnotism; and the like.[56] Similarly, the American Law Institute defines 'voluntary' mostly by instancing situations that are not voluntary: reflexes, convulsions, and bodily movements during sleep, while unconscious, under hypnosis, or under post-hypnotic suggestion.[57] If one were to pursue Hart's analysis here, it would be to conclude that these excluded conditions were all there was to 'willed', or 'voluntary' in its relevant sense.

Such a conclusion would be as unwarranted here as it was with regard to the like claim made about omissions, statuses, mental acts, and involuntary bodily movements. Just because there are opposites of 'voluntary' that do not themselves have much to do with each other is no reason to believe that there is no unitary thing to which each of these conditions is opposite. That remains true even if our legal or ordinary usage of a word like 'voluntary' is as Hart, Fitzgerald, and the American Law Institute describe: we guide our usage by the excluding conditions, not by any nature to the thing excluded by them. 'Voluntary' may still refer to a unitary property, act, or state, even so.[58]

There is another argument from ordinary usage that is different from the argument just surveyed, which emphasizes the supposedly 'excluder' role of words like 'voluntary' or 'will'. As quoted earlier, in Chapter 1, J. L. Austin correctly observed that most of our verbs of action are verbs of quite diverse character: 'killing', 'castling a king', 'telephoning a friend', 'winning a war', etc.[59] On their face they do not seem to share much with each other, nor, as Austin noted, are they easily assimilated 'one and all to the supposedly most obvious and easy cases, such as . . . moving fingers'[60] Herbert Hart supplemented Austin's observations about ordinary lay usage with his own observation about customary legal usage of action verbs:

Not only do they [judges] not refer to muscular contractions or 'volitions' or desires for them, but they do not speak as if they were faced with any general doctrine that . . . voluntary movements or omissions are . . . necessary for responsibility. Instead they discuss the meaning of the words in the statutes which they are considering, e.g., words like 'driving' used in [the statute] making driving dangerously an offence.[61]

The inference we are supposed to draw from these facts of ordinary and legal usage is that there is no act requirement running throughout the criminal law; only as many act requirements as there are distinct verbs of action used in the prohibitions of the special part of some jurisdiction's criminal code.

We are not yet in a position to answer this argument of Austin's and Hart's fully, for to do so requires that we work through the metaphysics of complex actions (e.g. killings) and show how such actions are related to basic acts (e.g. moving one's finger). What we can do here is see that the ordinary-language observations that prompted Hart and Austin are an insufficient basis on which to conclude that there can be no univocal act requirement underlying the diverse action prohibitions of the special part of the criminal law.

Ordinary usage of the verbs of action developed for ordinary uses, including the ascription of moral responsibility. Among those ordinary uses was not an attempted systematization of the conditions of either moral responsibility or legal liability. It is thus quite open to the moral or legal theorist to propose such a hidden systematization, even if the concepts and principles employed in doing so are quite alien to ordinary ways of thinking and speaking. No observations about the bountiful diversity of ordinary usage of action verbs can preclude the claim that underneath such diversity there is none the less a hidden unity. It is surely *not* a 'fatal defect in any account of action' that it is 'quite at variance with the ordinary man's experience and the way in which his own actions appear to him', or that it ignores 'the simple but important truth . . . that when we deliberate and think about actions, we do so not in terms of muscular move-

[61] H. L. A. Hart, *Punishment and Responsibility* (Oxford, 1968), 108.

ments but in the [quite different and diverse] ordinary terminology of actions'.[62] To think that this is a fatal defect would be like thinking that the quite diverse things ordinarily said about the planet Venus prior to the discoveries of Babylonian astronomy—things like 'It is the star that appears in the morning', and 'It is the star that appears in the evening'—could preclude someone discovering a hidden unity, namely, that these 'stars' were in reality one and the same thing, namely, the second planet from the sun.

In the context of assessing the divergence between the ordinary, idiomatic usage of complex action verbs and the unitary account of acts in terms of willed bodily movements, Herbert Morris once observed that 'it is obviously one thing to criticize a metaphysical position and quite another to criticize a definition introduced for some limited purpose'.[63] Yet the ordinary-language scepticism about the possibility of there being a unitary act requirement does neither. In particular, it does not, as Morris supposed, criticize the metaphysics presupposed by some unitary definition of act. There are serious metaphysical critiques that we shall examine—about whether the *act* of moving my fingers can be identical to the *bodily movement* of my finger moving, whether my mental state can *cause* an action, whether there are 'volitions', or states of 'willing', that can cause actions, whether basic acts like moving my fingers can be identical to more complex actions like killing. But nothing in ordinary usage of the verbs of action will answer such questions one way or the other. J. L. Austin perhaps understood this when he conceded that ordinary usage 'embodies . . . something better than the metaphysics of the Stone Age' but that 'certainly . . . ordinary language is *not* the last word' about such metaphysics.[64]

---

[62] Ibid. 101–2.

[63] Herbert Morris (ed.), *Freedom and Responsibility* (Palo Alto, Calif., 1961), 107.

[64] Austin, 'A Plea for Excuses', 11.

3

THE ORTHODOX VIEW OF THE ACT REQUIREMENT AND ITS NORMATIVE DEFENCE

It is time we left descriptive arguments based on the existing doctrines of the criminal law for normative arguments justifying why we should retain, modify, or eliminate the criminal law's act requirement. In order to make the normative discussion more concrete, it is time I describe in more detail *what* act requirement it is that I think worth considering on normative grounds. I shall do so in terms of what is usually regarded as the orthodox criminal-law theory of the act requirement.[1]

1. PRELIMINARY OVERVIEW OF THE ACT REQUIREMENT

As I would unpack it, there are four theses to this theory, as it was propounded by Bentham, (John) Austin, Holmes, and Walter Cook, and carried on into our own time by the American Law Institute's Model Penal Code.[2] The four theses are as follows. First, what I shall call the identity thesis holds that the acts required for criminal liability are partially identical to events of a certain kind, namely, bodily movements like moving one's finger or tongue. Such bodily movements on which the act requirement focuses are the simplest things one knows how to do as a means to achieving some end. Raising one's arm, for example, is usually such a simple act because one doesn't do (or know how to do)

---

[1] So characterized by Hart, *Punishment and Responsibility*, 97, and by Gross, *A Theory of Criminal Justice*, 49.

[2] Bentham, *Introduction to the Principles of Morals and Legislation*; Austin, *Lectures on Jurisprudence*; Oliver Wendell Holmes, *The Common Law* (Boston, 1881), 54, 91; W. W. Cook, 'Act, Intention, and Motive', *Yale Law Journal*, 26 (1917), 645–63; Model Penal Code, §2.01. Herbert Hart notes other proponents of the doctrine in English criminal-law theory. See Hart, *Punishment and Responsibility*, 99 n. 24.

some even more simple act in order to do it. If, however, one raises one's arm by moving one's foot on a pulley arrangement attached to the arm, then raising the arm on that occasion is not the act focused on by the act requirement of criminal law. *A fortiori*, if one's raising one's arm on some occasion is to signal the start of a race, so that in moving one's arm one was starting a race, the latter act is also not the act focused on by the criminal law's act requirement; for starting a race is far removed from the simplest act one knows how to do.

Secondly, the only acts that exist are the simple acts on which the act requirement focuses. Although there are complex acts of killing, hitting, scaring, telephoning, and the like, and although criminal codes invariably use these complex action descriptions in their substantive prohibitions, these acts in reality are never anything but some acts of bodily movement. I shall call this the exclusivity thesis.

Thirdly, not just any bodily movements (that are the simplest things one knows how to do) are acts satisfying the act requirement. A reflex movement of the leg, for example, is not an act no matter how identical it is behaviourally to simple leg movements that are acts. Thus, the third thesis—what I shall call the mental cause thesis—requires that such bodily movements must be caused by a certain mental event or state. Such event or state is variously styled as an act of willing, a volition, a desire, a simple intention, or a choice. Because all of these ordinary expressions have connotations inappropriate to this third thesis when it is most favourably construed, I shall use the least ordinary of the terms, 'volition'.

The three theses thus far, taken together, assert that the criminal law's act requirement requires that there be a simple bodily movement that is caused by a volition before criminal liability attaches, and that such a movement is all the action a person ever performs. This is the positive core of the orthodox view of the criminal law's act requirement. The fourth thesis seeks to accommodate this univocal act requirement of the general part of the criminal law to the multiple, complex action descriptions used in the various prohibitions of the special part. Statutes almost never use simple action descriptions like 'moving one's finger'; they prohibit actions described as 'starting a race without a permit', 'killing', 'disfiguring', 'removing another's property', etc.

For the act requirement just described to fit the criminal law as we know it, something must be said about how such diverse, complex action prohibitions are related to the univocal, simple act requirement. This is the burden of the fourth thesis.

This thesis asserts that any complex action description used in the special part of the criminal law is equivalent to (and thus can be replaced by) a description of some simple act (as defined above) of the accused causing a prohibited state of affairs. Murder statutes prohibit the complex action of killing another, for example. The fourth thesis—what I shall call the equivalence thesis—asserts that such a prohibition is equivalent to a prohibition that forbids 'any simple act that causes the death of another'.

There are many promissory notes issued in this highly abbreviated account of the act requirement. It needs to be clarified considerably in light of possible variations of each of its four theses. It needs a metaphysical defence in the sense that one must show that the entities, qualities, and relations posited to exist by the requirement do in fact exist. And it needs a doctrinal defence, namely, that this is the act requirement of Anglo-American criminal law. Before I do any of these things, however, my aim in this chapter is to show why we should care about such a requirement. What justifies the idea that before one can be punished for any crime whatsoever, one must have performed some simple bodily movement caused by one's volition?

2. A NORMATIVE DEFENCE OF THE ACT REQUIREMENT

What justification one gives for any particular prerequisite to criminal liability depends on one's theory of punishment. A utilitarian, who thinks that we punish in order to prevent crime via incapacitation of the dangerous and deterrence of the deterrable, ought to impose liability on, but only on, the dangerous or the deterrable. A retributivist, who thinks that we punish in order to give those who culpably do wrong their just deserts, ought to impose liability on, but only on, culpable wrongdoers. A mixed-punishment theorist, who thinks we are justified in punishing only when both crime prevention and just deserts will be achieved, ought to impose liability on, but only on, those who are either deterrable or dangerous, and who are culpable wrong-doers.[3]

Utilitarians and mixed theorists have made some gestures towards justifying the criminal law's act requirement. Those who only wish and fantasize criminal acts, but don't actually do them, aren't dangerous; those whose (involuntary) clumsiness cause others harm aren't deterrable; etc. Yet it is not obvious that these generalizations hold. Mightn't 'accident-prone' individuals be dangerous, and thus subject to preventive detention on utilitarian grounds? Mightn't such classes of individuals be somewhat deterrable, at least to the extent that they could take some pre-cautions against their dangerous tendencies? And even if they themselves are not deterrable, mightn't the criminal law gain an increment of general deterrence by making such persons liable anyway, because then those voluntarily causing harm will know that there is no possibility of *pretending* to have involuntarily caused it?[4]

Since very few people are purely utilitarians about punishment, I shall not pursue these familiar queries here. For both retribu-tivists and mixed theorists, the act requirement will be sufficiently justified if there is no culpable wrongdoing without such an act. It is to that moral question that I thus turn.

The standard approach to that question is to fractionate it into four questions: Why don't we punish statuses? Why don't we punish mental acts and mental states? Why don't we punish involuntary movements? And why don't we (by and large) punish omissions? As we have seen, these excluding conditions do not uniquely determine the content of the act requirement, however consistent with (and suggestive of) it they may be. To justify such exclusions from punishment, accordingly, may miss the justifica-tion of the act requirement itself.

We may justify excluding punishment for mental acts, for example, on grounds of privacy, of liberty, and of administrative

---

[3] I explore this taxonomy of punishment theories in M. Moore, 'Closest Ret-ributivism', *USC Cites* (Spring–Summer 1982), 9–16. I defend the retributive the-ory of punishment there and in M. Moore, 'The Moral Worth of Retribution', in F. Schoeman (ed.), *Character, Responsibility, and the Emotions* (Cambridge, 1987).

[4] These arguments are nicely deployed by Herbert Hart in *Punishment and Responsibility*, 40–3.

ease: mental acts being private, proof of them requires intrusive investigative methods by the state; no one wants to live in a state where they are not free at least to think whatever they please; good evidence of mental states is rare, and the required lines (between mental acts and passive states, between intentions and wishes) are fuzzy, so that the effort to administer a system that punished mental acts would be large, as would the possibility of error.

Likewise, we may justify excluding punishment for omissions on grounds of liberty, mental tranquillity, and a too extensive liability: affirmative regulation by the state is a greater infringement of liberty than negative prohibition, so the state should not punish even those omissions that are morally culpable; no one knows what they have *not* done as well as they know what they have done, so for citizen repose we ought not to make omissions punishable; only one or a few people typically cause harm, whereas many people omit to prevent harm, so to keep most of the populace non-criminal we shouldn't punish omissions.

Likewise, we may justify excluding statuses from punishment because they seem an illegitimate short cut around either proof problems or jurisdictional problems: punishing addiction, for example, may be a way to punish past use for drugs when we are unable to prove past use, or at least when we are unable to prove past use within the jurisdiction.

Likewise, we may justify excluding involuntary movements from punishment because it would be unfair to punish someone for 'doing' something that he could not help: reflex movements, somnambulistic movements, movements due to post-hypnotic suggestions, etc. cannot fairly be punished because those whose movements they are couldn't avoid 'doing' them.

Some of these arguments are quite persuasive for the exclusions from punishment that they justify. Yet even considered collectively, they do not justify the act requirement because they do not tell us why those, and only those, who act are deserving of punishment. Rather, they give us many reasons why we might well decline to punish such persons even if they are morally responsible and thus deserve punishment.

To be sure, the last of these reasons, the fairness reason, is sometimes put forward as a reason justifying the act requirement in general and not just the exclusion of the punishment of involuntary movements.[5] The argument is that only if we require acts (simple bodily movements caused by volitions) for criminal liability do we give citizens the kind of opportunity to avoid criminal sanctions that fairness demands. Yet fairness cannot justify the act requirement. In the first place, such a rationale does not justify any other exclusion from punishment besides the involuntary-movement exclusion. Fairness does not preclude the punishment of omissions, when the omitter has the capacity not to omit to do the required act. Fairness does not preclude the punishment of those mental events that are actively called forth ('mental acts').[6] Fairness does not preclude the punishment of those statuses of which one of two things is true: either the person could have helped getting into the status, such as being a lung cancer patient when the cancer is caused by smoking; or the person could easily get out of the status, such as the Christian Scientist being terminally ill, but only because she refuses normally successful, standard medical treatment.

In the second place, fairness by its nature cannot justify why we should punish people who act. It can only justify why we shouldn't punish people who do not act. While the latter may be enough to justify the act requirement to the mixed theorist about punishment, it isn't nearly enough for the retributivist. To justify the act requirement for the latter is to show why acting makes one responsible as much as it is to show why not acting makes one not responsible.

To make the requisite positive showing takes us to some fundamental questions about the general shape of morality, on the one hand, and about our natures as moral agents, on the other. On what may fairly be called the classical view of morality, morality is primarily concerned with the virtue of individual persons. 'What is virtue?' and 'How may I achieve it?' are the central questions that a virtue-based ethics will ask about such a virtue-based morality. On the classical view, actions are the

---

[5] See e.g. Fitzgerald, 'Voluntary and Involuntary Acts'.

[6] A 'mental act' is my active doing in the sense that I can will myself, for example, to try to remember a name. Other mental states are not willable however: 'I will myself to want that', or 'I will myself to believe that', or 'I will that I love you' can only be uttered by those who have yet to understand the basically passive nature of these mental states. See Brian O'Shaughnessy, *The Will*, 2 vols. (Cambridge, 1980), i. 1–37. On mental acts generally, see Peter Geach, *Mental Acts* (London, 1957).

subjects of moral norms (prohibitions and permissions) only as subsidiary means of promoting persons' virtue. Thus, Aristotle discusses justice not primarily as an obligation each person owes to others, but as a virtue in oneself that each should cultivate; one should do just actions, not because such actions are required by morality's 'categorical imperatives', but because doing such actions helps to make one a just person.[7]

Such a virtue-based ethics goes hand in hand with a character-based view of personal identity. If morality deals primarily with virtue, not with obligations, then the aspects of persons that are morally relevant will be those generally enduring traits of themselves we call their characters, not those temporally isolated events we call their acts.[8]

On the virtue-based conception of ethics, it is surely a puzzle why one must *act* in order to be morally responsible. As George Sher observes: 'Why, if persons are constituted by their preferences and abilities but not by their actions, do they more often deserve things for their actions? Indeed, if actions are too short-lived to be constituents of selves, then why assign them *any* independent standing as desert-bases?'[9] A standard answer is that we don't; rather bad acts express bad character, so that nominal responsibility for actions is in reality responsibility for character. Acts on this view are proxies for character because they typically express the actor's character; when we punish bad acts, accordingly, we are really punishing bad character.[10] Such an 'expressive function' of acts leaves unanswered two questions. First, what about bad acts that are 'out of character', that is, that are not expressive of bad character—do we not rightly punish such acts even though there is (by hypothesis) no bad character'. Secondly, what about bad character that, as luck would have it, is never afforded the opportunity to express itself in bad action—do we not rightly refrain from punishing an individual for his awful but unexpressed character?[11]

[7] Aristotle, *Nicomachean Ethics*, bk. v.

[8] These two conceptions of ethics, with their differing focuses on character versus action, are discussed by me in M. Moore, 'Choice, Character, and Excuse', *Social Philosophy and Policy*, 7 (1990), 29–58.

[9] George Sher, *Desert* (Princeton, NJ, 1987), 171.

[10] Peter Arenella, 'Character, Choice and Moral Agency: The Relevance of Character to Our Moral Culpability Judgments', *Social Philosophy and Policy*, 7 (1990), 59–83.

[11] See Moore, 'Choice, Character, and Excuse'.

An affirmative answer to the first of these questions reveals an alternative view of morality that owes more to both Kant and the utilitarians than to Aristotle. On this view, morality is fundamentally a matter of norms. Such norms do not have as their subjects virtuous character. Rather, such norms prohibit or permit either complex actions (like killings) or states of affairs (like deaths) being caused by persons. Whatever differences there are between 'deontologists' and 'consequentialists' about the general shape of morality's norms, on this they agree: one's obligations are *not* derivative from some more fundamental imperative that one ought to nourish one's own virtuous character. Quite the contrary: one ought to nourish a virtuous character, not as an end in itself, but as a means of making it easier for one to *do* what one ought to do.

The aspect of persons that this latter view of morality emphasizes is consciousness. It is the Lockean view that we each are one whole person because of the experience of unified consciousness that each of us has. Such continuity of conscious experience includes choices made at different points in our lives, no matter how in or out of character such choices may be. It is the choice to do evil on a particular occasion that makes a person morally responsible for any wrong that flows from such choice, irrespective of whether such choice expresses bad character or not.

The sufficiency of action for responsibility falls out easily from this non-classical view of morality. On the deontological version, morality prohibits the doing of various complex actions like killing, lying, etc.; to perform such complex actions is to do moral wrong, on such a view. In succeeding chapters I will argue that each such complex action is nothing other than a willed bodily movement. Anticipating the result of that argument, it is thus easy to see that the willing of certain movements (that is, the ones that violate moral norms) is to do moral wrong and thus deserve punishment. On the consequentialist version, morality is primarily concerned with bad states of affairs but derivatively concerned with persons not causing such bad states of affairs to occur. Once one makes the metaphysical showing that persons cause states of affairs through their *acts* (bodily movements caused by volitions), then one has shown that those bodily movements that violate derivative moral norms are sufficient bases to attribute moral wrongdoing and deserved punishment.

Since morality does contain both deontological norms and the consequentialist principle,[12] the doing of an action made wrong by such norms or principle is a sufficient basis for blame. Less clear is whether the doing of an action is *necessary* for the kind of blameworthiness properly the subject of criminal punishment. Two possibilities question such necessity. One is the possibility adverted to before: perhaps each of us is blameworthy for our character, including our emotional make-up, and should be punished for bad character as well as for bad actions. In addition to the consequentialist principle and to the agent-relative norms of deontology, morality also does certainly contain those ideals of character we call the virtues. Why should a retributivist punishment scheme not punish for lack of virtue as well as for lack of conformity to morality's (obviously action-oriented) obligations?

Perhaps the first answer that leaps to mind is that fairness bars punishment of people for being the sort of people that they are. For one might plausibly think that the emotions people feel, the dispositions to evil that tempt them, are not matters within their control. I cannot will away the feeling of bad emotions as I can will away the doing of bad actions. Yet this I take not to be much of an objection. Moral blame inevitably requires the existence of moral luck. Inevitably, that is, we must and should blame people for facts about them that they did not will into existence. Although we can choose our actions, we usually do not choose the desires that cause such choices; and even if we choose our desires, we do not choose the factors that cause us to make such choices. Ultimately, in short, we are blamed for choices or characters that are caused by factors that are themselves unchosen.[13]

If this seems harsh, think of the moral judgements we constantly make in everyday life. We judge people morally for not feeling grief when they should, for not loving their children or their parents enough, for entertaining lustful thoughts and impulses, for being unduly distrustful or suspicious of others, etc. The morality of the virtues plays an important part in our moral life, and it is easy to be so enamoured of our traditional criminal-

[12] I defend a version of a deontological theory in Moore, 'Torture and the Balance of Evils', while recognizing that such deontological norms operate in the shadow of the consequentialist principle.

[13] See M. Moore, 'Causation and the Excuses', *California Law Review*, 73 (1985), 201–59; also id., 'Choice, Character, and Excuse', where I compare the stopping-points of a choice- versus a character-oriented morality.

law practice of punishing only for actions that we forget the familiar daily-life practice of blaming for character.

So the question recurs: why does not a retributive punishment scheme punish for that lack of virtue that morality plainly blames? Utilitarians about punishment have an easy answer here, stemming from people's lack of control over the emotions they feel, the impulses that tempt them, etc. What citizens cannot control the criminal law cannot control through the use of penal sanctions. Forward-looking punishment schemes thus have reason enough not to punish for bad character.

Retributivism, however, is a backward-looking punishment scheme. For a retributivist, it is a sufficient reason to punish that the person deserve it, even if such punishment does not change future behaviour at all. The retributivist answer thus has to lie in the nature of desert. Desert must be such that we deserve punishment for violations of morality's norms of obligation, but we do not deserve punishment for violation of morality's ideals of virtuous character. Why should this be so?

Consider Robespierre's execution in one of Molière's plays. Robespierre asks for what he is being punished, since, as a good revolutionary, he has done nothing wrong. The answer given by Molière: 'because you lack grace'. You are not, in other words, a well-formed person, and for that you shall be executed.

Such an answer is jarring because no one deserves to be punished for lacking grace—or for lacking any other virtue. Those who lack virtue, but do not exhibit it through bad actions, have done no wrong. They have wronged themselves, by being less worthy persons than they should be; but they have not wronged anyone else. Friends, lovers, and others who care deeply about another may be disappointed in his failures of virtue; but not even they have been wronged.

Without wrongdoing, what is there to exact retribution for? Robespierre might punish himself for lacking grace, much as Gide's Lafcadio stabs himself slightly for each defect of virtue he discovers in himself.[14] But Robespierre doesn't *deserve* punishment even from himself; and certainly no one else has the right to give it to him even if he did deserve it, in the absence of his wronging anyone else.

[14] André Gide, *Lafcadio's Adventures* (New York, 1928).

Our criminal law thus rightly shies away from punishing bad character, even though we (equally rightly) judge people morally by their characters in non-legal contexts. Conceding this, there remains one other query on the necessity of action to a just punishment. This query concerns bad *choice*, not bad character. I refer to intentional omissions. When we choose not to rescue our fellows even though we could do so at no risk to ourselves, we are not blamed for bad character. Rather, we are blamed for making a bad choice on a given occasion. Yet the act requirement as it was defined in Chapter 2 by and large excludes criminal liability for these blameworthy choices. The query is how the exclusion of this class of blameworthy individuals from punishment can be justified, even if we concede that we should not punish for bad character.

As we hinted at in Chapter 2, the answer lies in the differential force of our negative obligations not to make the world worse, on the one hand, and either our positive obligations, or our supererogatory ideals, to make it better, on the other. Let us draw out these distinctions within our morality, beginning with supererogatory/obligatory distinction.[15]

Much of the harms that we could (but do not) prevent do not represent failures by us to do our duties. For we have no duties to prevent many of the world's evils. Our own individual interests in leading lives of beauty, interest, creativity, and enjoyment justify us in not sacrificing those attributes of our lives in the saintly quest to make the world better. Consider the choice faced by Camus's interlocutor in *The Fall*: a woman has fallen (or thrown herself) into the Seine, and he could risk his own life by jumping into the dark waters and trying to save her. Camus's character understandably feels guilty that he did not choose to rescue, but such guilt should not be taken to indicate a failure of duty. Even if the chance of his own drowning was relatively small, and the chance of successful rescue correspondingly high, he had no obligation to put himself at such risk. His felt guilt betokens a negative judgement about himself, namely, that he did not live up to the lofty ideals that he professed to admire (as a tireless seeker of social justice). Such negative judgement is a truthful one: he did not behave as virtuously as he might have.

---

[15] On this last distinction, see generally J. O. Urmson, 'Saints and Heroes', in A. I. Melden (ed.), *Essays in Moral Philosophy* (Ithaca, NY, 1967).

But such judgement is about a failure of supererogation, not a failure of obligation.

Consider also the well-known example given by Macaulay to justify why the penal code he proposed for India did not punish omissions:

> It will hardly be maintained that a surgeon ought to be treated as a murderer for refusing to go from Calcutta to Meerut to perform an operation, although it should be absolutely certain that this surgeon was the only person in India who could perform it, and that if it were not performed the person who required it would die.[16]

It is not obligatory for the surgeon to go even though the journey is not risky, just inconvenient to his other interests. Were he to sacrifice such interests and make such journeys, it might well be saintly of him to do so. He would deserve our praise and admiration for his Mother Teresa-like virtue. Yet his not going is not an occasion for blame, for he breached no obligation.

Notice that the risk and inconvenience justifying *not* doing certain life-saving actions would hardly justify (active) killings. If a gunman threatens us with loss of our own life if we do not kill a woman on a bridge over the Seine, yet we could easily escape that threat by jumping into the dark waters and have a good chance of surviving, we cannot justify killing by the risk of drowning were we to jump. Likewise, if a gunman in Calcutta threatens a surgeon with death unless he then and there carves up an innocent person, yet the surgeon could easily escape the situation by hopping on a non-stop, express train to Meerut, the surgeon cannot justify killing by the inconvenience of travelling to Meerut. We are obligated not to kill others, and while we may justify exceptions to that obligation, these kinds of risks do not make out such a justification. We are not, in general, obligated to prevent harm to others, and while it might be virtuous to do so on many occasions, morality itself permits us to be non-virtuous.

The criminal law ought not to punish this class of intentional omissions for the same reason as was given to justify non-punishment for those who have non-virtuous emotions and

---

[16] T. Macaulay, 'Notes on the Indian Penal Code, 1837', in *Works*, vii (New York, 1897), 494.

character, namely, that no retributive justice is achieved by punishing lack of virtue. Indeed here it is even clearer that it is lack of retribution that justifies non-punishment of these moral defects. For fairness and utility *would* be served by such punishment. Choices not to help are just that: choices that the actor had the opportunity to make other than the way he made them. If we were to punish such choices, an actor would have a fair opportunity to avoid criminal sanctions. Because of such opportunity to choose otherwise, such choices are also deterrable.

Unlike defects in our emotions or in other aspects of our character, non-virtuous omissions do have an impact on others. Some harm to them that could have been prevented is not prevented. Therefore, it is more tempting to think that the intentional omitter wrongs his 'victims' and thus deserves punishment on retributive grounds. Yet once we concede that for this class of omissions there was no obligation to prevent harm, then there can be no wrong done in omitting to prevent harm. The absence of obligation, after all, just is the presence of a permission; and what morality permits us to do cannot be morally wrong. (It may well be morally virtuous for us not to accept the permission, but it cannot be wrong to accept it.) And without wrongdoing, there is no retributive justification for punishing.[17]

It is not plausible to think that all intentional omissions are not morally wrong. We do have some positive duties, and breach of these duties is not just lacking in virtue but is also wrong. Should we not punish at least this subclass of intentional omissions? Prima facie, the answer is surely 'yes' for a retributivist, for retributive justice is achieved by punishing such culpably chosen, wrongful breaches of duty. It takes other values besides retributive justice to justify *not* punishing intentional breaches of positive duties.

To see what sorts of values ought to constrain the achievement of retributive justice here, we need some plausible examples of situations where we have positive duties not to omit to act in certain ways. Parents have positive duties (not to omit) to care for their children in various ways. Strangers also have positive duties to prevent serious harm befalling children when such strangers

---

[17] Although we sometimes punish for culpable action even in the absence of wrongdoing (e.g., attempts), with this class of omission there is no culpability, only lack of virtue.

can prevent such harm by no more risk or effort than the throwing of a rope. Retributive justice would be achieved by punishing both classes of omitters, yet why might a criminal code nonetheless distinguish between them, as indeed Anglo-American criminal codes typically do? The answer lies in the value we accord to persons' *liberty* to make the wrong choice. To require parents to save their children diminishes their liberty to a certain extent; but to require strangers to save all children in peril diminishes their liberty to a greater extent. Furthermore, on the hard-to-justify but commonly felt assumption that we owe more to those near and dear to us than we owe to strangers, our positive duties towards our own children are stronger than they are to children as such; put another way, we do more wrong in failing to save our own child than we do in failing to save someone else's child.

Putting these factors together, it may be that the greater infringement of liberty that punishing strangers would entail justifies the allowance of the less wrongful violation of the (less powerful) positive duty going unpunished; whereas with parents, the balance tips the other way: the duty is stronger, and the wrong for its violation accordingly greater, so that such wrong going unpunished more greatly flouts the achievement of retributive justice, while at the same time the costs to liberty of achieving such justice through punishment of parents are less.

I say all of this about this example rather provisionally, because, while I am sure the balance is different in these two cases, I am unsure that the balance is different enough that a rightly conceived penal code would not punish both sorts of intentional omission. Since Anglo-American criminal codes do draw precisely this distinction, however, I use it to illustrate how some such distinction is to be justified on retributive grounds.

One might object that if the diminution of liberty can outweigh the achievement of retributive justice for many wrongful omissions, the same should be true for an equally large class of wrongful actions. Thus, parents can be prohibited from killing their children, because the achieving of retributive justice outweighs the diminution of liberty such punishment entails; but, with strangers, the balance is the other way: the diminution of liberty is greater by prohibiting them from killing any children, and the wrong done by such killings is less, so that non-punishment of such killings sacrifices less retributive justice.

Since the conclusion of this *reductio* is absurd, we had better be able to avoid the analogy which generates it. This we can easily do, however, by adverting to the very real difference there is in the moral force of our negative versus our positive duties.[18] We do much more wrong when we kill than when we fail to save, even when such failure violates a positive duty to prevent death. Because negative duties are so much stronger than their positive counterparts, the liberty diminished is almost never of sufficient value to outweigh the retributive justice achieved by punishing such wrongs. The only sort of cases where the interests of liberty are high enough to justify violation of negative duties are cases like abortion, when the woman's liberty to be wrong outweighs the injustice of such wrongful action going unpunished.[19]

Many people of course dispute that there is this large difference between the moral force of our negative versus our positive duties. James Rachels gives a well-known comparison between the uncle who lets his young nephew drown in a bath-tub in order that the uncle will inherit the nephew's fortune, and the uncle who holds the nephew's head under water for the same reason.[20] We are supposed to conclude that the two uncles make equally wrongful choices, even though the first only violates a positive duty while the second violates a negative duty.

There are a surprisingly large number of delicate dances around this sort of comparison, done by those conceding the equivalence of wrong here but showing why that cannot be generalized to an equivalence in wrong between actions and omissions.[21] Such delicacy is surprising because the straightforward response is much more intuitive: the first uncle is much less deserving of punishment than is the second because the first did much less wrong. Wrongful as it is to let the child drown, it is much more wrongful to drown the child. Drowning it makes the

---

[18] The analogy may also be broken if negative prohibitions diminish liberty more than do positive requirements.

[19] Judy Thomson's famed violinist may be another example of where we may think the interest of liberty outweighs the retributive value of a wrongful action going unpunished. Thomson, 'A Defense of Abortion', *Philosophy and Public Affairs*, 1 (1971), 47–66.

[20] Rachels, 'Active and Passive Euthanasia'.

[21] See Quinn, 'Actions, Intentions, and Consequences'; F. Kamm, 'Harming, Not Aiding, and Positive Rights', *Philosophy and Public Affairs*, 15 (1986), 5–11.

world a worse place, whereas not preventing its drowning only fails to improve the world.

Such blunt disagreement about intuitive cases does not of course do much to convince those not already convinced of the truth of one's position. But in this context we may rightly place the burden of proof on those like Rachels who would have us change a deeply felt distinction, one that shows up in real-life homicide verdicts as well as in philosophy articles.[22] Rachels and his ilk present such hypotheticals as starting-points for more extensive liability, yet if such hypotheticals do not strike one at all like they strike Rachels, that is reason enough to wait for a better argument.

The upshot of this discussion is to leave us normatively in the same position as we were in descriptively at the end of the preceding chapter. There is and should be an act requirement, although it is and should be subject to an exception in the case of certain omissions. Normatively, the retributive value of punishment justifies why we should punish actions and that value (sometimes in conjunction with the values of fairness and of liberty) also justifies why we should punish *only* actions and not character, emotions, or omissions. The only exception to this is for those omissions that violate our duties sufficiently that the injustice of not punishing such wrongs outweighs the diminution of liberty such punishment entails.

---

[22] Jury verdicts in homicide cases where liability is predicated on an omission show a marked tendency to find only manslaughter despite the fact that the omission is done with knowledge that the victim will die (which is a *mens rea* sufficient for murder, not manslaughter). Those few statutes that do impose omission liability upon strangers also punish such wrongs much less severely than actions that cause death. See e.g. Vermont Statutes Annotated, title 12, §519, ($100 fine for failure to rescue strangers even when such rescue can be done 'without danger or peril').

4

THE METAPHYSICS OF BASIC ACTS I: THE EXISTENCE OF ACTIONS

Before defending directly the metaphysics presupposed by the identity, exclusivity, and mental-cause theses of the orthodox view of action, it is necessary to deal with the broader metaphysical objection that, put crudely, actions do not exist. For such a blunderbuss objection can be used to undermine the effort to defend the metaphysics of the orthodox view of action before that defence can even get under way. Antony Duff, for example, has imbibed a large dose of such scepticism about actions:

To ask which is 'the action itself' is as absurd as to ask what 'the event itself' is when the roof of a house is damaged in a storm—is there just one event (the roof being damaged); or are there 'really' many events (each individual tile being damaged)? Actions and events are identified and individuated only by our descriptions of them: what someone does can be described in various ways, drawing different distinctions between 'the action' and its circumstances or consequences; and which of these possible descriptions we offer depends not on some objective truth about what 'the action' really is (since there is no such truth), but on our own interests (and on the vocabulary available to us) . . . it is absurd to ask which description picks out 'the action itself', since there is no such thing.[1]

[1] Antony Duff, *Intention, Agency, and Criminal Liability* (Oxford, 1990), 41. Two things can be learned by attending to this statement of metaphysical scepticism by Duff. One is how pervasively influential such a scepticism would be for the theory of action developed in this book: not only would such scepticism reject the idea that acts are willed bodily movements (Chs. 5, 6, 10, 11), Duff's nominal target here; it would also undercut any fixity to the distinction between acts, on the one hand, and consequences and circumstances, on the other (Chs. 7 and 8), and it would render unanswerable the questions of act-token individuation raised in Chs. 11 and 14. The second thing to be learned is how difficult it is for a sceptic like Duff to state his scepticism without presupposing its refutation: for to what is Duff referring when he speaks of 'what someone does' or when he allows that that (thing?) 'can be described in various ways'? As Donald Davidson has noted, 'All this talk of descriptions and redescriptions makes sense . . . only on

We need to distinguish three routes by which one might arrive at this blunderbuss scepticism about actions. First, if the language used to 'describe' human action in reality has exclusively non-referential usages, then we can dispense with any assumption that human actions exist; secondly, if the existence of events generally is in doubt, then we should equally doubt the existence of that kind of event we call a human action; and thirdly, if we think that our concept of human actions requires such things to have some characteristic, and if we think that the kind of events that we call human actions lack just that characteristic, then we would conclude that there were no human actions but only certain events that are badly misdescribed. In this chapter I will discuss each of these scepticisms seriatim.

1. DO PURPORTED DESCRIPTIONS OF ACTION NOT REFER?

A great deal of sophisticated work has been done detailing how actions are described. Some of that work we will utilize in Chapter 8. Here the question is more preliminary: do seeming descriptions of action actually describe or refer to anything? Or can we, simply from getting clear the 'logical form' of action descriptions, see that we need not commit ourselves ontologically to the existence of actions?

Consider the sentence (1) 'X moved his arm oddly'. If we nominalize the verb, we can say (2) 'The movement of X's arm yesterday was odd' (a perfect, derived nominal); or, alternatively, (3) 'X's moving his arm yesterday was odd' (an imperfect, gerundial nominal).[2] If we focus on the former nominalization for now, it seems indisputable that in uttering (2) we have referred to something—a human action—with the grammatical subject of our sentence. If we can always so paraphrase our usage of verbs of action in sentences like (1), it seems clear that we are committed to there being actions by our use of such verbs.

the assumption that there are *bona fide* entities to be described and redescribed.'
Davidson, *Essays on Actions and Events* (Oxford, 1980), 165.

[2] On nominals, see Robert Lees, *The Grammar of English Nominalizations* (Bloomington, Ind., 1963). On the application of the semantics of English nominals to actions and other events, see Zeno Vendler's 'Facts and Events', in Z. Vendler, *Linguistics in Philosophy* (Ithaca, NY, 1967), and Jonathan Bennett, *Events and Their Names* (Indianapolis, 1988), 4–12.

Rather remarkably, for a time a group of Oxford ordinary-language philosophers denied that action entities were referred to by the usage of action language as in sentence (1). This view proceeded from J. L. Austin's observation that words could have 'jobs' besides describing things.[3] Wittgenstein's example: saying 'ouch'. 'Ouch' did not describe pain, nor did it describe anything at all; rather, typical use of the word was to *express* one's pain.[4] Similarly, H. L. A. Hart argued, action sentences like (1) did not *describe* anything' they *ascribed* responsibility to persons for states of affairs.[5] If the movement of an arm were a bad or illegal thing to do, sentence (1) on this view would ascribe responsibility to X for that result. Hart's 'ascriptivist' account of action sentences passed into criminal-law theory, even though Hart himself later came to abandon it. Patrick Fitzgerald, for example, once subscribed to Hart's view, observing that 'in ordinary speech the word "act," together with such allied expressions as, "A did it," is used not so much to describe what has happened, as to ascribe responsibility'.[6]

If ascriptivism were right, there could be no such thing as a theory of human action, for we would have no reason to think that there was a class of action things about which we could have a theory. That action sentences seemed to refer to entities would give us such a reason no more than usage of 'sake' (in 'for the sake of'), 'dint' (in 'by dint of'), and 'ouch' give us reason to think that dints, sakes, and ouches exist and demand a theory about their nature.[7]

But ascriptivism is surely false, as Peter Geach[8] (and later, Herbert Hart himself)[9] pointed out. Just because action sentences may typically be used to do one job does not mean they do not

---

[3] 'Jobs' was the early Oxford terminology for the speech-acts one could do in using words. See generally J. L. Austin, *How to Do Things with Words*, 2nd edn. (Cambridge, Mass., 1975).

[4] On Wittgenstein's expressive theory of first-person mental-state avowals, see Stuart Hampshire and H. L. A. Hart, 'Decision, Intention, and Certainty', *Mind*, 67 (1958), 1–12.

[5] Hart, 'Ascription of Responsibility and Rights'.

[6] Fitzgerald, 'Voluntary and Involuntary Acts', 383.

[7] 'Dints' and 'sakes' were Quine's examples of words that, although apparently nouns, did not occur in referential position, a linguistic fact one could ascertain simply because of the fused nature of their typical use. Willard Quine, *Word and Object* (Cambridge, Mass., 1960).

[8] Peter Geach, 'Ascriptivism', *Philosophical Review*, 69 (1960), 221–5.

[9] Hart, *Punishment and Responsibility*, p. v.

always perform another job. 'Murderer', for example, may always be used to ascribe responsibility, as in 'He is a murderer', yet that is *no* argument that 'murderer' does not predicate a property of the persons to whom it truly applies, namely, one of intentional killing. To think that words can only do one job was the mistake of early forms of emotivism in meta-ethics, a mistake garnering its own label as the 'speech-act fallacy'.[10] Moreover, there is every reason to think that action sentences like (1) make perfectly good sense even when neither moral nor legal responsibility is at issue. If that is so, then even if the ascriptivist can commit the speech-act fallacy and get away with it, he is still left with no account of the meaning or use of such sentences.

The alternative form of language-based scepticism about action prevalent in criminal-law theory is what I called in the first chapter 'interpretivism'. The interpretivist about action thinks that we, the observers (interpreters) create actions by our interpretations. Actions do not antecedently exist, waiting to be referred to by action descriptions. As George Fletcher puts it, we interpret something to be an action only when 'we can perceive a purpose in what he or she is doing'.[11] Similarly, Antony Duff believes, 'particular acts (raising my arm) have their character and meaning as actions only in virtue of their role within a wider structure of action and context (signalling to turn left, or waving to a friend)'.[12]

A more precise rendering of this form of scepticism about action might be as follows. First, action sentences are held to create Intentional contexts. Just as one does not believe, desire, or intend *simpliciter*, so one does not just seek, predict, agree, solicit, or promise—one seeks (to find one's friend), one predicts (that it will rain tomorrow), one agrees (to rob the bank), one solicits (aid in robbing the bank), one promises (to come back tomorrow). Such action constructions take Intentional objects no less than do believe, desire, intend, and many mental-state constructions.[13] Secondly, this argument continues, the requirements

---

[10] See William Lycan, *Judgement and Justification* (Cambridge, 1988), 204.

[11] Fletcher, *Rethinking Criminal Law*, 436.

[12] Duff, *Intention, Agency, and Criminal Liability*, 134.

[13] For an introductory explication of Intentionality, and its application to the verbs of action as well as to mental-state verbs and nouns, see Daniel Dennett, *Content and Consciousness* (London, 1969), 23.

that must be satisfied in order to ascribe content to such Intentional constructions are incompatible with states with such contents existing in the world. Rather, whether we see some swatch of the world as a seeking, or an intending, etc. depends on what interpretive stance we, the observers, take towards the phenomenon. If we take what Dan Dennett calls 'the Intentional stance' towards it, we will see it as a seeking or an intending, but if we take the design stance, or the physical stance, we will not so see it—and, Dennett adds, there is no fact of the matter about it or about what stance we take, only pragmatic justifications for preferring one stance to another.[14]

My own realism about mental states like intending resists the second of these steps,[15] as does the metaphysics of many philosophers of mind. But we need not canvass what is surely *the* problem in the philosophy of mind, the ascription of content to Intentional states. For present purposes it is enough to reject the first step in the interpretivist's sceptical argument. Only a subclass of complex action verbs take Intentional objects, but many more do not. 'Seek' does, but 'kill', 'rape', 'transport', 'burn', 'hit', do not. Only some of what I shall in Chapters 7 and 8 call 'intentionally complex' action verbs create Intentional contexts, and these are a small fraction of the verbs we use to talk about actions. Moreover, I eventually shall defend the thesis that every complex action is identical to some basic action of bodily movement. No basic action verbs create Intentional contexts, so that Intentionality is a feature only of some *descriptions* of some actions; it is not even a feature of any action itself.

I have no doubt attributed greater precision to the claims of interpretivists about action than many of them would recognize. Yet, to the extent the above two-step argument is not their argument, such interpretivists rely either on some hazy, dualist's metaphysics that make human actions a special, non-physical data for the *Geisteswissenschaften*, or on some linguistic dualism that thoroughly severs any connection between action concepts and the concepts of natural science.[16] Although either of these

[14] Daniel Dennett, *Brainstorms* (Montgomery, Vt., 1978); id., *The Intentional Stance* (Cambridge, Mass., 1987).
[15] M. Moore, 'Mind, Brain, and the Unconscious', in P. Clark and C. Wright (eds.), *Mind, Psychoanalysis, and Science* (Oxford, 1988).
[16] I take both Fletcher's and Duff's interpretivism (*supra*, nn. 11 and 12) to be species of the linguistic dualism characteristic of ordinary-language philosophy. I

moves would make actions look pretty ghostly to the rest of us, there are decisive reasons not to accept either of these claims about how the universe—or even our talk about the universe—is fundamentally sundered.

2. THE SUPPOSEDLY SHADOWY EXISTENCE OF EVENTS

If we leave linguistic bases for scepticism about the existence of human action, and do our metaphysics more straightforwardly, we might start with the feeling that human actions do not exist in the robust sense that physical objects, for example, do exist. Since this scepticism about human action is but the application of a more general scepticism about events, I shall examine this scepticism as it is aimed at its broader target.

It is a deliverance of common sense that existence is most robust for physical objects like tables, chairs, the sun, or the ocean liner, *Queen Elizabeth II*. Such physical objects plainly exist as particulars, even if everything else's existence has to be argued for. One of Bertrand Russell's arguments for the existence of universals like redness, heat, weight, or intelligence was a kind of crossing the Rubicon argument. To talk of separate particulars at all, Russell observed, we had to admit some universals, namely, relational properties; so why not go on to sack Rome itself, i.e. admit non-relational properties as well?[17]

In addition to particular physical objects and their properties, many have argued that there must be a third broad category of existents: propositions. Propositions are about both the particular things referred to by singular terms and the properties referred to by predicates; they are what sentences (which are constructed out of singular terms and predicates) express.

In this three-part inventory of the ultimate furniture of the universe there is thus far no mention of events. Events most intuitively seem to fit as particulars of some kind. Events do not seem to be at all the same thing as propositions, which may be

---

explore these and other interpretivisms in Moore, 'The Interpretive Turn in Modern Theory'.

[17] Bertrand Russell, *The Problems of Philosophy* (Oxford, 1912).

about events but do not themselves seem to *be* events.[18] *Facts* are propositional, but events are not. The fact that X moved his arm yesterday is different from the event referred to as 'the movement of X's arm yesterday'—even though both can be expressed in nominalizations of the verb 'to move'. The fact of X's moving his arm yesterday is propositional; such fact locutions create Intentional contexts so that 'X's moving his arm' is the content of the 'fact that' operator. Such propositional content may refer to the event (of the moving by X of his arm) or to some aspect of that event, but that does not make facts the same as events.[19]

Are events properties of some kind? The fact that a certain event occurred with respect to some object is a property of that object. Because X moved his arm yesterday, X today has the property of being a yesterday arm-mover. That, however, does not make the movement event itself a property of X's. Even those who most closely identify events with properties, such as Jaegwon Kim[20] and Alvin Goldman,[21] do not treat events *as* properties. Rather, events on the Kim–Goldman view, are instances (or exemplifications) of properties. Events are thus particulars, not universals, being dated occurrences of timelessly existing universals. The event of the moving by X of his arm yesterday is an exemplification of the property, being an arm movement. If one were to identify the particular event that was X's arm movement yesterday with the property of being an arm movement, that has the counter-intuitive consequence that

[18] Roderick Chisholm has argued that events are propositions in his *Person and Object* (La Salle, Ill., 1976), and in 'Events and Propositions', *Nous*, 4 (1970), 15–24, and 'States of Affairs Again', *Nous*, 5 (1971), 179–89. See also Neil Wilson, 'Facts, Events, and Their Identity Conditions', *Philosophical Studies*, 25 (1974), 303–21, and Wilfrid Sellars, 'Actions and Events', *Nous*, 7 (1973), 179–202: 181, 197–98, for other articulations of this view. One can hold this view only if (like Sellars explicitly) one identifies events with facts about events. See Sellars, 'Actions and Events', 195–7.

[19] For an excellent discussion of the distinction (and of the importance of making the distinction) between facts and events, see Bennett, *Events and Their Names*. See also Anthony Quinton, 'Objects and Events', *Mind*, 88 (1979), 197–314: 206.

[20] Jaegwon Kim, 'On the Psycho-Physical Identity Theory', *American Philosophical Quarterly*, 3 (1966), 227–35; id., 'Causation, Nomic Subsumption, and the Concept of Event', *Journal of Philosophy*, 70 (1973), 217–36; id., 'Events as Property Exemplifications', in M. Brand and D. Walton (eds.), *Action Theory* (Dordrecht, 1976).

[21] Alvin Goldman, *A Theory of Human Action* (Englewood Cliffs, NJ, 1970).

X's movement of his arm today is the very same event as his arm movement yesterday—not just the same *kind* of event, but literally the same individual event.

By elimination we come then to the common-sense view that events are particulars. They are not physical objects, so what sort of particulars are they thought to be? Fundamentally, events seem to be *changes* in objects. By 'change', we mean pretty much what Aristotle meant by the word in the *Physics*:

There are three classes of things in connection with which we speak of change: the 'that which', the 'that in respect of which', and the 'that during which'. I mean that there must be something that changes, e.g. a man or gold, and it must change in respect of something, e.g. its place or some property, and during something, for all change takes place during a time.[22]

This is also a standard modern view of events. Lawrence Lombard, for example, summarizes his view of events in a manner very much reminiscent of Aristotle:

Events are non-relational changes in objects; and for an object to change it must, roughly, have a static property at one time and lack it at another. To be an event, then, is just to be a having and then a lacking by an object of a static property at a time . . .[23]

There are five aspects of this standard account of events that may cut into one's confidence that such entities exist in any very robust sense of the word. One stems from the dependence of events on there being properties of things. If one has some uncertainty about the ontological status of properties themselves, events may look suspect too. Nominalists about universals may have a hard time countenancing events even though events are particulars, not universals; for event-particulars are changes in (static) properties, and such changes are themselves instances of (dynamic) properties.[24] All this reliance on universals might be unpalatable to a staunch nominalist. Yet such nominalists might

---

[22] Aristotle, *Physics*, bk. v, ch. 4 (227–621).

[23] Lawrence Lombard, *Events: A Metaphysical Study* (London, 1986), 220.

[24] For such an account, see ibid. A static property as Lombard defines it is simply one that an object can have without changing; a dynamic property, by contrast, is one the possession of which implies change (ibid. 104).

find some way to think of events that does not depend on properties. Quine, for example, thinks of events as being the material content of a space-time zone.[25] As Bennett recognizes, 'a Quinean event is . . . uniquely determined by the zone, with no need to mention properties at all'.[26] Nominalism about properties is thus not much of a basis for being sceptical about events. Not only are most people not nominalists, but nominalists (witness Quine) will find a way to accommodate events even as they studiously avoid ontological commitment to universals.

A second kind of scepticism about events comes from quarters opposed to those of the nominalist, even though this scepticism too proceeds from the relation between events and properties. Recall the standard account of events: events are changes, and changes are the having and then the lacking of a static property at a time; therefore, an event is itself the instance of a kind of property possessed by an object, namely, what Lombard calls a dynamic property.[27] The event consisting of X's moving of his arm yesterday involves the static properties of the arm's spatial location: it was here, it was not there. Change in those static properties could be: it was not here; it was there. As it was moving, X's arm then possessed the dynamic property of being a movement from here to there. All well and good, but the worry is this: physical objects and the (static or dynamic) properties they possess unproblematically exist; but events require there to be a third kind of thing—not the physical object (a particular), not the property that that object possesses (a universal), but an instance of that (dynamic) property that is a particular instance of a universal. The objection is not that this is some kind of category mistake (thinking that universals could have instances); rather, it is the Occam's razor objection that the only instances of properties we need introduce are the particular objects that possess such properties on particular occasions. The objection is that events (as property instances not identical to the physical objects whose properties they are) are ontologically excessive.[28]

[25] Willard Quine, 'Events and Reification', in E. LePore and B. McLaughlin (eds.), *Actions and Events: Perspectives on the Philosophy of Donald Davidson* (Oxford, 1983).
[26] Bennett, *Events and Their Names*, 104.
[27] Lombard, *Events: A Metaphysical Study*, 105.
[28] For doubts along these lines, see Bennett, *Events and Their Names*, 90; Moore, 'Thomson's Preliminaries about Causation and Rights', 521.

There are two ways to deal with this worry. One is to adopt the set-theoretic notion of an ordered triple, saying that an event is a triple, a class whose members are an object, a property, and a time. Events then become themselves abstract entities—sets or classes—that some find less ontologically problematic than property instances. Of course, one would have to say more about the semantic intuitions that guide the ordering of the triples. One would not want to say that the event of X's moving of his arm yesterday occurs if the class ⟨X's arm, moving, yesterday⟩ occurs; rather, as Kim notes in pursuing this line, X's arm must *have* the dynamic property, moving, *at* the time, yesterday.[29]

The second way to attack the worry of ontological excess is less dramatic but more salvaging of the idea that events are instances of properties. One could concede that only physical objects exemplify properties, that events themselves do not. Still, events could be the *exemplifyings* of dynamic properties. As Lawrence Lombard puts this distinction:

What exemplifies a property is a thing that has it. The rock's falling, however, does not have the property of falling; the rock has, and hence exemplifies, that property. The event, however, has, and hence exemplifies, the property of being a falling. Events are not exemplifications of properties that objects have at times; they are, rather, the exemplifyings of properties that events have at times.[30]

This distinction may help to alleviate one's sense of ontological excess, or double counting, for it separates the instance of a universal (an exemplifying) from the object that possesses (exemplifies) that universal. Since universals surely do have instances, if such instances are not the objects possessing them on particular occasions, they must be something else—property instances, or what we call events.

A third scepticism about events also begins with the supposition that events, if they existed, would be related to objects, properties, and times as the standard view sets forth. Events, thus conceived, are supervenient on these three more fundamental items in the sense that: (1) any variance in the object, the

[29] Kim, 'Events as Property Exemplifications', 161. See Bennett, *Events and Their Names*, 90–2; Terence Hogan, 'The Case against Events', *Philosophical Review*, 87 (1978), 28–47: 30.
[30] Lombard, *Events*, 52.

property, or the time must result in a variance in the event, and vice versa; and (2) such co-variance between events and objects, properties, and times is itself explained by the asymmetrical dependence of events on objects, properties, or times.[31] (That is, with regard to 'asymmetrical dependence', there would be no events if there were no objects, properties, or times, but there could be objects, properties, and times, but no events—a very boring world, to be sure, but a possible world none the less.)

This 'supervenience' of events on more basic items in our ontology leads some to scepticism about events. 'Events are not basic items in the universe' and 'they should not be included in any fundamental ontology', Jonathan Bennett concludes,[32] because of their supervenience on more basic items. As Bennett also recognizes, this might be put more dramatically as '*Basically* there are no events', or '*strictly* speaking there are no events'.[33]

Although some have denied that events are supervenient on objects, properties, times, I am inclined to grant the supervenience thesis. This in no way makes me sceptical about the existence of events. I also think that moral qualities supervene on natural properties, and that mental states supervene on physical states. And in none of these instances should supervenience convince one that events, moral qualities, or mental states do not exist.

On what is perhaps now the going, standard account of how we figure out what exists, we ask: what must we assume to exist in order to make the best explanation (of something else that we provisionally hold to exist) true? On this account, we have good reason to believe that events, moral qualities, and mental states exist because they figure in the best explanations we have about various other items. Indeed, of these three classes of things, events figure most prominently in our causal explanations of just

---

[31] Supervenience has been a much discussed topic in relating mental states to brain states in the philosophy of mind and in relating moral properties to natural properties in ethics. See Jaegwon Kim, 'Supervenience and Nomological Incommmensurables', *American Philosophical Quarterly*, 15 (1978), 149–56; id., 'Causality, Identity, and Supervenience in the Mind–Body Problem', *Midwest Studies in Philosophy*, 4 (1979), 31–49; id., 'Supervenience and Supervenient Causation', *Southern Journal of Philosophy*, Suppl., 22 (1984), 45–56. On supervenient relations between events and objects/properties/times, see Lombard, *Events*, 217–33.

[32] Bennett, *Events and Their Names*, 12.

[33] Ibid. 17.

about everything.[34] We thus have every reason to think that they exist.

The ontological suspicion engendered by supervenience stems from conflation of supervenience with reduction. A term-by-term reductionist thinks that he can in principle eliminate all talk of events (moral qualities, mental states) in favour of talk of the base entities supervened upon. The late Fred Skinner's logical behaviourism about mental states was of this form, holding mental states to be nothing more than complex dispositions to behaviour so that talk of them could be eliminated in favour of behavioural talk entirely. Yet a supervenience theorist is not a reductionist. Rather, in most cases of supervenience there is no possibility of reducing one level of discourse to another, because there are no (type–type) identities between one class of things and the other. On the contrary, the best explanations of various things will be in terms of events (moral qualities, mental states), with no hope of replacing these with better (and more ontologically parsimonious) explanations. Whether this is so, of course, is a matter of contingent fact. The only point here is that, if supervenience is true of events (as it seems to be true of moral qualities and mental states), that provides no reason to think that events do not exist, or are in any sense second-class citizens in our ontology.

The fourth and fifth worries about events stem from invidious comparisons of events to physical objects in both the temporal and the spatial dimensions. Locke stated the temporal worry about events: 'Substances alone of all our ideas have particular or proper names, whereby only one particular thing is signified . . . actions that perish in their birth, are not capable of a lasting duration, as [are] substances that are the actors.'[35] The worry is that events are too transient to exist in the robust sense in which physical objects exist.

The spatial worry is somewhat similar. Physical objects exclusively occupy the spatial regions in which they exist, in the sense that no two distinct physical objects can occupy exactly the same

---

[34] On the explanatory role of events, see Kim, 'Events as Property Exemplifications'; Davidson, *Essays on Actions and Events*, 165. Compare Hogan, 'The Case against Events', 31–5, 38–41.

[35] John Locke, *An Essay Concerning Human Understanding* (first pub. 1689; New York, 1965), iii. vi. 42.

region at the same time. Not so for events. Not only can many events occupy the same region at the same time, but events can 'co-occupy' the same region with a physical object. It is this inability to be excluded spatially by each other or by other things that makes an event seem less a thing itself. As Myles Brand notes, 'the seeming intangibility of events is due to their lack of fully occupying a region'.[36]

One way to answer both the temporal and the spatial worries is by attacking the assumed paradigmatic status of physical objects. Quine, for example, thinks of events as being the content of spatio-temporal zones, and then uses these to explain the *less* basic notion of a physical object.[37] Similarly, Donald Williams urges that at the deepest metaphysical level there are property instances (the common idea of events), and that both physical objects and properties are derivative collections of these.[38] Any metaphysical views like these completely demolish any idea that events suffer in comparison with physical objects.

Even if one doesn't attack the common-sense assumption that physical objects are our paradigmatic examples of existents, the temporal and spatial worries should not make one a sceptic about events. With regard to events' lack of 'durability' through time, events do have temporal location, like physical objects. My moving my arm occurs at a time. Moreover, events typically have a duration: they take time to occur. The movement of my arm, the sinking of the *Titanic*, etc., occur over an interval of time. To that extent one might think that events are 'durable'—more so than some short-lived physical objects like soap bubbles and unstable houses of cards, at any rate. Yet the whole comparison is misleading in ways that give rise to Locke's kind of sceptical worry. The interval of time over which an event occurs is not an interval during which the event persists; events are not durable like objects.[39] Events have temporal parts in a way that objects do not. The moving of my arm occurred over some temporal interval t_1–t_3, but the moving did not occur at some time t_2 in between; only part of the moving occurred then. Objects do not

[36] Myles Brand, *Intending and Acting* (Cambridge, Mass., 1984), 58.

[37] Quine, 'Events and Reification'.

[38] Donald Williams, 'The Elements of Being', *Review of Metaphysics*, 6 (1953), 3–18, 171–93. On Williams's view, see Quinton, 'Objects and Events', 211–14.

[39] P. M. S. Hacker, 'Events and Objects in Space and Time', *Mind*, 91 (1982), 1–19.

have that kind of relation to time. Quine's famous rabbit (as opposed to rabbit-slices in time) exists during the interval t_1–t_3, and it exists at any time t_2 in between. The whole idea of comparing length of temporal stays of events and physical objects is an odd idea.

The same is true for the idea of comparing kinds of spatial occupations by events and physical objects. Events don't 'occupy' spatial zones at all, so that it is a mistake to compare invidiously their ghostly (i.e. non-exclusive) occupation with the robust (i.e. exclusive) occupation of physical objects.[40] Events do have spatial location, but that is only because the physical objects (of whose properties such events are a change) have spatial location. The only occupying is that of the physical object itself.

Even if the spatio-temporal location of events is stripped of any misleading suggestions of instantaneous existence or of spatial occupation, there may be some lingering worry about the locatability of events. Does Sally in California telephoning Harry in New York occur during an interval including his talking or not? Is her act of telephoning located in California, New York, in the wire in between, or all three? Unless one can provide some answers to the locatability of events—answers that are more than arbitrary stipulations—one might still have some lingering scepticism that such things exist at all. We shall thus have occasion to examine this question later, in Chapter 10, when we have a bit more of the metaphysics of action before us. For now, we should conclude that there are no doubts about human actions that stem from some general doubt that events exist. Human actions thus potentially form the subject-matter for a theory, and we shall shortly turn to defending the orthodox theory of action earlier outlined.

3. HUMAN ACTIONS AS A NATURAL KIND OF EVENTS

Doubt about whether human actions form a natural kind (of events) is most persuasively answered by developing a theory about the nature of such a kind and seeing whether that theory holds up under critical fire. This is the task to which I shall

[40] Ibid.; see also Fred Detske, 'Can Events Move?'. *Mind*, 76 (1967), 479–92: 484–7.

shortly turn, in Chapters 5 and 6. There is one generic doubt about human actions that needs to be dispatched at the outset, however. This is the 'incompatibilist' doubt whether in a fully determined universe of events there could be a subclass of those events that cause other events to occur while themselves not being caused to occur. The rhetorical question 'How can there be uncaused causers?' is the incompatibilist's basis for denying the existence of human actions.

There are two premisses leading to the incompatibilist's conclusion that humans actions do not exist. One is what is standardly called the determinist premiss: all events are caused. The other is what is standardly called the incompatibilist premiss: to be an action, an event must be free, where 'free' means 'uncaused'. If all events are caused, and if an event may be a human action only if it is uncaused, necessarily there can be no such events as human actions.

Since the logic of this argument is impeccable, seemingly one has to reject either the determinist premiss or the incompatibilist[41] premiss in order to deny the conclusion. A route uncommonly popular amongst action theorists is to deny the determinist premiss.[42] This, however, forces one to defend the libertarian metaphysics that makes us all gods in an otherwise unbroken chain of causal relations. Such a metaphysics is so hard to defend that if this were the only way to defend the existence of actions, I too would be a sceptic. Fortunately there is a better way, which is to reject the incompatibilist premiss.

The incompatibilist believes that there is a necessary incompatibility between an event being caused and that event being an action. Actions must be uncaused to be actions; if the universe is a network of unbroken causal relations, then 'no room can be found for agents, for beings capable of initiating *new* causal chains'. On this view of agency and action, 'every attempt to account for actions within this deterministic picture . . . will destroy what is specific about agency'.[43]

---

[41] I have elsewhere charted other conceptual wiggles to avoid the conclusion that in a deterministic universe there are no actions (and no moral responsibility). See Moore, 'Causation and the Excuses'.

[42] See e.g. A. I. Melden, *Free Action* (London, 1961); Norman Malcolm, 'The Inconceivability of Mechanism', *Philosophical Review*, 77 (1968), 45–72; Richard Taylor, *Action and Purpose* (Englewood Cliffs, NJ, 1976); Carlos Moya, *The Philosophy of Action* (Oxford, 1990).   [43] Moya, *The Philosophy of Action*, 9.

What one wants to ask the incompatibilist about agency and action is how he justifies his view of agency. What justifies him in requiring that agents be free of causal determination on pain of not being agents at all? The answer of every incompatibilist with which I am familiar is a conceptual answer: our *concept* of action makes it analytic that an event must be free (uncaused) in order to be an action event. Yet when pressed to give the justification for believing this supposed conceptual truth about action, the answer is usually couched in terms of linguistic facts, notably ordinary-usage patterns of various related concepts. Such defence commonly includes, for example: the claim that to talk of one's *ability* to do other than one did, or to talk of what one *can* do, requires contra-causal freedom; the claim that our talk of *agency* or *authorship* of some harm only takes place where we do not causally explain the exercise of that agency by non-actional events; the claim that to talk of Smith's performing the *action* of killing Jones presupposes that some other event did not cause Smith's body to engage in those motions that in turn caused the death of Jones; the claim that we talk of *choice* only when we do not think that there are causes sufficient to bring about one decision rather than another. The claim in each case is that our ordinary usage of the concepts of ability, agency, choice, and action shows that such concepts are appropriately used only when we are confident that there are no causes for such acts or choice, or for such exercises of ability and agency.

There are two responses to this grounding of the conceptual claim in these supposed patterns of ordinary usage. The first is that the data of ordinary usage does not in fact support the claim.[44] We commonly speak of someone having the *ability* to run a mile in under four minutes despite knowing that both that ability, and its exercise on a particular occasion, are fully caused (by genetics, training, etc.). We also commonly speak of people being the *agents* of some untoward result, of their *choosing* and *acting* so as to bring it about, even when we have identified causes of their choices and their acts. For example, one of my former colleagues had served in the Truman Administration and

---

[44] See Moore, 'Causation and the Excuses'; see also Joel Feinberg, *Doing and Deserving* (Princeton, NJ, 1970); Bruce Aune, 'Can', in P. Edwards (ed.), *Encyclopedia of Philosophy* (New York, 1967); Daniel Dennett, *Elbow Room* (Cambridge, Mass., 1984).

would inevitably tell the same story if the right trigger for associations were produced. Saying the word 'Truman' caused him to tell the same old story, but that there is this causal nexus in no way relieves him from having *chosen* to tell the story, nor from that story having been his *action*, a exercise of his *agency*.[45]

In any event, even if the data of ordinary usage were all in favour of the incompatibilist's conceptual claim, such data are an insufficient basis on which to defend the claim. For think what an extraordinary claim it is. Imagine an analogous claim about Lake Michigan: our usage of the word 'lake' shows that for anything to be a lake, it must be uncaused; therefore, if Lake Michigan was caused by glaciation, it is not a lake. Explain its existence, in other words, and the lake (*qua* lake) disappears!

What is extraordinary is the priority given to supposed criteria (concepts) derived from ordinary usage. Suppose that there is a clash between our concept of an action (as uncaused) and our discovery that there are causes for what we took to be actions. Such a clash presents us with a choice in conceptualization: either we say there are no actions, sticking with our concept at the cost of ignoring the nature of the thing to which we thought we were referring; or we say that there are actions, but that they have a quite different nature from that we had supposed them to have. If actions are a natural kind of events, as lakes are a natural kind of object, we have every reason to prefer the second conceptualization. For this route preserves continuity between our ordinary concepts and the insights of an advancing science; this route does not require us to seal off as irrelevant (because not matching our concepts) the insights of scientists about surprising features of natural kinds.[46]

Do human actions form a natural kind? What is and is not a natural kind is itself a matter of scientific discovery, not of conceptual necessity. We could think that 'multiple sclerosis' names a natural kind, or that 'bachelor' does not, and be quite wrong in

---

[45] Nor from the responsibility for telling a boringly repetitive story. These kinds of counter-examples usually call forth distinctions by incompatibilists between sufficient causation and 'inclining' or 'partial' causation, alleging that only the latter are present in such counter examples. For continuation of the story, see Moore, 'Causation and the Excuses'.

[46] Hilary Putnam, 'The Meaning of Meaning', in H. Putnam, *Mind, Language, and Reality* (Cambridge, 1975); see also Moore, 'A Natural Law Theory of Interpretation'.

either case. To determine whether a kind has the sort of unified nature distinctive of natural kinds is to develop a theory of that nature. If such a theory is defensible, that will be the ultimate answer to the conceptual claim of the incompatibilist. It is to developing such a theory that I now turn.

5

THE METAPHYSICS OF BASIC ACTS II: THE IDENTITY OF ACTIONS WITH BODILY MOVEMENTS

1. THE IDENTITY THESIS: AT LEAST SOME ACTS (THE BASIC ONES) ARE IDENTICAL TO BODILY MOVEMENTS

1.1. The Meaning of the Identity Thesis

Oliver Wendell Holmes stated the two theses examined in this and the succeeding section of this chapter very succinctly: an act, Holmes said, 'is a willed muscular contraction, nothing more'.[1] Notice that there are two theses here. The first states an identity claim, identifying acts with a kind of muscle movement. The second states what I shall call the exclusivity claim, that acts are never anything more than such kinds of muscle movements.

Holmes followed John Austin in his view of what acts are, for Austin earlier had explicitly stated both theses. 'A voluntary movement of my body, or a movement which follows a volition, is an act.'[2] And: 'bodily movements are the only objects to which the term "acts" can be applied with perfect precision and propriety'.[3]

Although ultimately to defend one of these theses will be to defend the other, for ease of exposition it is helpful to examine them one at a time. I consider here the identity claim: at least some acts are bodily movements. This thesis has come in for much heavy weather in both philosophy and legal theory. Although in 1960 Hart could perhaps with accuracy describe this identity claim as part of the 'orthodox theory' of action in law,[4]

---

[1] Holmes, *The Common Law*, 73–4.
[2] Austin, *Lectures on Jurisprudence*, i. 415.   [3] Ibid.
[4] Hart, *Punishment and Responsibility*, 97–9. Douglas Husak has recently noted that 'despite criticism sometimes amounting almost to ridicule, the view that acts consist of bodily movements continues to attract widespread support'. Husak, *Philosophy of Criminal Law*, 112 n. 20.

even by 1960 such a claim was by and large rejected in philosophy. Before coming to the criticisms of the identity claim, however, we first need to clarify it. What do we mean by 'at least some acts', what do we mean by 'bodily movements', and what do we mean by 'identity' here?

The qualification '*at least some* acts' is intended to hold in abeyance the following kind of argument: 'The killing of Jones by Smith yesterday was certainly an act; yet that act of killing involved more than Smith moving his finger on the trigger of the gun with which he killed Smith; it involved the gun firing, the bullet striking Smith, and Smith dying, none of which can plausibly be identified as movements of Jones's body.' This is the kind of argument with which the exclusivity thesis must deal, where a proponent of the Austin–Holmes view must either deny that killings are acts or affirm that killings are bodily movements despite the foregoing argument.

The 'at least some' phrase is intended to restrict our gaze to what (provisionally) we shall think of as a subset of acts.[5] These are acts like moving your finger, raising your arm, winking your eyelids, moving your tongue and lips. These kinds of acts are often called 'basic', 'simple', or 'primitive' acts.[6] These labels are appropriate because the idea marking such acts off from others is that these are the simplest, most basic, or most primitive acts we do in order to do anything else. John Austin appreciated this

[5] The 'provisionally', because in Ch. 11 I shall argue that all acts are willed bodily movements, that is, basic acts. The basic/complex distinction then becomes a distinction between two sort of *description* of acts, not between two sorts of *action*. 'Moving my finger' and 'killing Jones' may refer to one and the same act, but the former is a basic description while the latter is a complex description. On this, see Jennifer Hornsby, *Actions* (London, 1980), 68. For now, I adopt the more idiomatic terminology of basic versus complex actions.

[6] The basic insight contained in the idea of a simple, basic, or primitive act may be found in H. A. Prichard, *Moral Obligation* (Oxford, 1949), 32; Melden, *Free Action*. It is developed in Arthur Danto, 'What We Can Do', *Journal of Philosophy*, 60 (1963), 435–45; id., 'Basic Actions', *Philosophical Quarterly*, 2 (1965), 141–8'; Feinberg, *Doing and Deserving*. The idea finds various modification in F. Stoutland, 'Basic Actions and Causality', *Journal of Philosophy*, 65 (1968), 467–75; Myles Brand, 'Danto on Basic Actions', *Nous*, 2 (1968), 187–90; Goldman, *A Theory of Human Action*, 63–72; Davidson, *Essays on Actions and Events* (Cambridge, 1973); David Pears, 'The Appropriate Causation of Intentional Basic Actions', *Critica*, 7 (1975), 39–69; Julia Annas, 'How Basic Are Basic Actions?', *Proceedings of the Aristotelian Society*, 78 (1978), 195–213. For a summary, see Hornsby, *Actions*, 33–88.

distinctive feature of this class of acts as well as any modern philosopher of action:

if I wish that my arm should rise, the desired movement of my arm immediately follows my wish. There is nothing to which I resort, nothing which I wish, as a means or instrument wherewith to attain my purpose. But if I wish to lift the book which is now lying before me, I wish certain movements of my bodily organs, and I employ these as a means, an instrument, for the accomplishment of my ultimate ends.[7]

A basic act as Austin implicitly defines it is an act that may be a means to some other end but that is itself not brought about by some other act that is its means. We may strike another by moving our arms, but we (typically) do not move our arms by doing something else; we just move them.[8]

A further clarification is in order here. The sorts of acts about which the identity claim is made are particular acts, standardly called 'act-tokens'. As C. S. Peirce (who reintroduced the type/token distinction to modern discussions) described such tokens:

A single event which happens once and whose identity is limited to that one happening or a single object or thing which is in some single place at any one instant of time, such event or thing being significant only as occurring just when and where it does . . . I will venture to call a token.[9]

'I moved my finger yesterday' refers to a particular act of finger-moving that I did at a particular time. In addition to such act-tokens, there are of course act-types. 'Moving one's fingers' refers to a type of act that people on various occasions do. Such act-types are universals like all properties, classes, sets, and kinds.

---

[7] Austin, *Lectures on Jurisprudence*, i. 413.

[8] We shall need to clarify the 'primitive means' idea of basicness later. It will turn out that this idea is epistemic in the sense that what means we (persons) adopt to realize our ends is a function of what we know we can do. Basic acts are thus not the causally most basic events that go on when we act; for there are causally *more* basic events (like nerve signals, muscle contractions, and brain events) that cause the movements of our limbs. Basic acts are, however, the most basic *means* we adopt in order to do more complex actions. On the distinction between 'epistemic' or 'teleological' basicness and causal basicness, see Hornsby, *Actions*, ch. 8, and Annas, 'How Basic Are Basic Actions?', 198–9.

[9] C. S. Peirce, *Collected Works* (New York, 1935), ii. 537. See generally Goldman, *A Theory of Human Action*, 10–15.

The identity thesis is not about act-types, in part because there are no act-types that are basic. Raising my arm is usually a basic act I do because I don't do anything else in order to do that. But not always: I can raise my arm by lifting it with the other arm, by pushing it hard against a doorway for several minutes and having it rise afterwards, etc. There is no *type* of act that is basic; only particular acts. The identity claim is thus that any particular basic act is identical to some bodily movement.

About the other side of the identity claim, bodily movements, some clarifications are also in order. To begin with, notice the difference between Holmes and Austin on that with which basic acts are identified: Austin usually (but not always) speaks of 'bodily movements' while Holmes talks about 'muscular contractions'. These are not the same things. That my arm muscles move is one type of event; that my entire arm moves is another. Certain arm muscle movements typically *cause* certain arm movements, which means that these are two distinct events (for nothing is the cause of itself).

These considerations militate towards Austin's idea of movement and away from Holmes's. For if we were to identify basic acts like raising one's arm on a particular occasion with movements of certain muscles on that occasion, then we would face the unpalatable conclusion that raising our arm caused the arm to move. For if the raising of one's arm *is* one's arm muscles moving, and if one's arm muscles moving caused one's arm to rise, then the raising of one's arm caused the arm to rise.[10] Although one might say this of some mental act like trying or willing to raise one's arm, I find this an implausible thing to say about the basic act of raising one's arm. To make the identity thesis more plausible, therefore, we should construe that with which basic acts are identified as the arm's rising, not the muscle movements that cause that rising. This allows the holder of such a thesis to say that my act of raising my arm on some occasion is

[10] If this is not intuitively obvious, consider the following: can it matter to the truth of the expression 'Venus obeys Kepler's laws of planetary motion', how I describe Venus? Surely not. Surely if the quoted statement is true, then so is this one: 'The second planet from the sun obeys Kepler's laws of planetary motion'. As Leibniz said, identicals (like Venus and the second planet from the sun) at any given time have all the same properties. Such 'indiscernibility' is close to being the very meaning of numerical identity. See Baruch Brody, *Identity and Essence* (Princeton, NJ, 1980).

identical to my arm rising on that occasion, rather than saying that the first *causes* the second.

Similar considerations debar various events that go on in the brain and central nervous system from being the bodily movements with which basic acts are to be identified. These are, as we shall see shortly, causes of bodily movements, not those movements themselves.

Another clarification here has to do with omissions and mental acts. Some philosophers sympathetic to the identity claim include both omissions and mental events in their definition of 'bodily movement'. For example, Donald Davidson: 'If we interpret the idea of a bodily movement generously, a case can be made for saying that all primitive actions are bodily movements. The generosity must be openhanded enough to encompass such "movements" as [the omission of] standing fast, and mental acts like deciding and computing.'[11] I see no reason to be this generous. If our account of actions develops some theoretical reason to accommodate such items—because, for example, willing is essential to acts and certain stillnesses and mental states can be willed—then we may be forced to be generous. But until then, let us be stingy: bodily movement means motion (not stillness), and the motion is of the body (not the mind).

A final clarification of 'bodily movement' is the most important of all. H. A. Prichard was perhaps the first to note the systematic ambiguity that exists for nominalizations of the verbs of action like 'move'.[12] The relevant ambiguity is not the active/passive ambiguity of sentences like 'I saw the shooting of the children', where 'the shooting of the children' may refer to the children doing the act of shooting, or it may refer to their getting shot.[13] The active/passive ambiguity only makes ambiguous who did the action versus who suffered the effects of an action done by another. Rather, the relevant ambiguity is between whether there is an action done by anyone or not. On this axis of ambiguity 'the moving of the hand' can refer to a movement by someone of their hand, or it can refer simply to the physical motion

---

[11] Davidson, *Essays on Actions and Events*, 49. For one explication of this passage, see Vermazen, 'Negative Acts'.
[12] Prichard, *Moral Obligation*, 190–1.
[13] The example is of what Quine calls the 'process/product' ambiguity. *Word and Object*, 130.

of the hand. Jennifer Hornsby labels these two senses of the word 'movement$_T$' and 'movement$_I$'.[14] Movement$_T$ is named for the grammatical fact that where 'moves' is used as a transitive verb (that is, it takes grammatical objects) it names an act, as in 'He moves his hand'. Movement$_I$ is named for the grammatical fact that where 'moves' is used intransitively, it names a bodily movement that is not necessarily an act, as in 'His hand moves'.

Thus the ambiguity: does the identity thesis seek to identify basic acts with movements$_T$ or with movements$_I$? Hornsby believes it to be the former, because 'if there is any hope of truth in an identification of actions with bodily movements, then they must be movements$_T$, not movements$_I$, that are actions'.[15] One might believe this only because one takes 'movement$_I$' to preclude action, so of course one would not identify acts with what precludes acts. Contrary to this assumption, however, 'moves' in its intransitive sense is non-committal about whether the event in question is or is not an act. 'His arm moves$_I$' does not preclude that he moves it; it just doesn't address itself to that issue. 'His arm moves$_I$' is an appropriate description that can be used to refer either to a part of his action of moving$_T$ his arm, or to an event that is not part of a human action, as when his arm is blown by the wind. Thus, it is not false (on the face of the language alone) to assert that each basic act is partially identical to some bodily movement$_I$.

My motive for so construing the identity claim is the triviality the alternative construction engenders. If basic acts are identified with bodily movements$_T$, then the identity thesis is much less informative than its proponents would have hoped. For such an identity claim becomes a non-reductive account of action. It doesn't reduce acts to some other kinds of events. It only asserts that all acts of one kind (basic ones) are also acts of another kind (movements$_T$). As a historical matter, this is not the (reductive) identity claim that interested Austin and Holmes.[16] Moreover, the reductive identity claim that did interest Austin and Holmes *is* the interesting one, for if true it shows how acts are part of the ordinary, physical world.

Finally, we need to say a word about the identity 'relation' itself if we are to understand the identity thesis. To begin with,

---

[14] Hornsby, *Actions*, 2–3.   [15] Ibid. 3.
[16] On this, see Brand, *Intending and Acting*, 7.

the identity spoken of is of course numerical and not qualitative identity. The claim is not 'each act has all the same properties as some corresponding bodily movement'. Rather, the claim is that each basic act is one and the same thing as some bodily movement. Secondly, as adverted to before, the identity claimed is between act-tokens and movement-tokens, not between act-types and movement-types. The claim is that any particular raising of my arm just is a particular movement of my arm upward, not that the type of act that is an arm-raising is identical to the type of bodily motion that is an arm's rising. Seeing this last point should clip the wings of a potential objection to the identity thesis before that objection can even get started. The objection is that many bodily movements are not acts. Falling off a table while unconscious, or the reflex movement of the leg in response to a tap on the knee, for example. Yet notice the identity thesis does not entail the contrary. The identity thesis does not assert that the *type* of events known as basic acts is identical to the *type* of events known as bodily movements$_I$. This broad type-identity claim is false, and is shown to be false by the counter-examples of movements$_I$ that are not basic acts. The only claim made by the token-identity thesis is that for each event that is a basic act, it is also a bodily movement$_I$.

The final clarification about the identity asserted to exist between basic actions and bodily movements is this: the identity is only a partial identity. Partial identities exist between particulars whenever two putatively distinct particulars either share a common part or one is part of the other.[17] An example of the first kind of partial identity is where two Siamese twins share a common organ; an example of the second kind of partial identity is where one county is part of one state.

The identity thesis that is worth defending here is the thesis that asserts there to be a partial identity of the second kind between basic acts and bodily movements$_I$. The thesis is that bodily movements$_I$, like the moving$_I$ of one's hand are parts of basic acts like the moving$_T$ of one's hand. Full identity exists, on this view, only between basic acts and the causal sequence volitions–cause–bodily movement$_I$; no full identity is asserted to exist between basic acts and bodily movements$_I$ by themselves.

---

[17] On the idea of partial identity, see David Armstrong, *A Theory of Universals* (Cambridge, 1978). 36–9.

Seeing the partial identity asserted to exist by the identity thesis alone reveals the rather artificial division between this thesis and the mental-cause thesis. For it is of course both these together that assert a complete identity to exist between actions and volitions causing bodily movements$_I$. Despite the artificiality, I find it helpful to isolate the identity thesis from the mental-cause thesis for ease of exposition. The danger is that in discussing the identity of basic acts to bodily movements$_I$ one may forget that it is only volitional causings of bodily movements$_I$ that are acts. Although for ease of separate exposition I shall talk of basic acts being identical to bodily movements$_I$, it is important to remember that the addition of the mental-cause thesis makes the identity a partial one only. One benefit of keeping this clarification in mind is that it defuses a potential worry for the identity thesis, namely, that basic acts like clapping one's hands occupy a different spatio-temporal region from bodily movements$_I$ like one's hands moving together. This is not an objection if one recalls that the identity asserted is between the *causal sequence* of a volition causing the hands moving$_I$ together, and the act of clapping; for these two events more plausibly occupy the same spatio-temporal region.

It is probably true that the proponents of the identity thesis in legal theory have not always asserted only a partial identity to exist between basic acts and bodily movements$_I$. The earlier quotations from Holmes and Austin, for example, are susceptible of two quite different interpretations: (1) actions are identical to bodily movements$_I$ whenever such movements are caused by volitions; or (2) actions are partially identical to bodily movements$_I$, partially identical to volitions, and only fully identical to the causal sequence volition–causes–bodily movement$_I$. A major difference between (1) and (2) is that volitions are *parts* of actions in (2); they are *causes* of actions according to (1). However unclear Holmes, Austin, and others may have been about the difference between (1) and (2), John Stuart Mill rather clearly saw that (2) was the right thesis. To his own question 'What is an action?' Mill answered:

Not one thing, but a series of two things; the state of mind called a volition, followed by an effect. The volition or intention to produce the effect, is one thing; the effect produced in consequence of the intention,

is another thing; the two together constitute the action. I form the purpose of instantly moving my arm; that is the state of my mind: my arm
(not being tied or paralytic) moves in obedience to my purpose; that is a
physical fact, consequent on a state of mind. The intention followed by
the fact, or (if we prefer the expression) the fact when preceded and
caused by the intention, is called the action of moving my arm.[18]

1.2 Post-Hart Legal Theory and the Identity Thesis

Having clarified the identity thesis, it remains to defend it. Post-
1960 legal theory has been quite condemnatory of this thesis. In
one way or another, many criminal-law theorists have each
rejected the thesis.[19] They have done so, however, mostly by mis-
understanding what it is the thesis asserts. There are five arguments that the critics collectively make.

The first is the argument from omissions. The argument is that
some acts are omissions to move one's body; so it cannot be true
of all acts, that they are each a bodily movement.[20] There are
two ways to treat this objection: as a metaphysical objection, to
the effect that there are some members of the kind, acts, that are
not bodily movements; or as a legal objection, to the effect that
it is 'inconvenient' to stipulate any meaning to 'act' (the word as
used in the act requirement of criminal law) that excludes omission, since people are sometimes liable for their omissions as well
as their commissions.

Taken metaphysically, this objection is not (yet) an argument.
Needed is some competing metaphysical description of what acts

[18] John Stuart Mill, *A System of Logic* 8th edn. (London, 1961), I. iii. 5. For
modern proponents of the partial identity between basic acts and bodily movements₁, see Brand, *Intending and Acting*, 15–16; Irving Thalberg, *Perception, Emotion and Action* (New Haven, Conn., 1977), 53–62; Thomson, *Acts and Other
Events*. (The last two theorists do not identify *complex* actions with volitionally
caused bodily movements, because they think that such complex actions have
parts other than volitions and bodily movements₁; this broader, 'componential'
view is rejected in Ch. 11, *infra*.).
[19] See Hart, *Punishment and Responsibility*, 101–4; Fletcher, *Rethinking Criminal Law*, 421–2; Hall, *General Principles of Criminal Law*, 178; Gross, *A Theory of
Criminal Justice*, 49–55; Husak, *Philosophy of Criminal Law*, 91; Eric D'Arcy,
Human Acts (Oxford, 1963), 4–10; Herbert Morris, 'Dean Pound's Jurisprudence',
Stanford Law Review, 13 (1960), 185–210; Alan White, *Grounds of Liability*
(Oxford, 1985), 31; D. O'Conner, 'The Voluntary Act', *Medical Science and the
Law*, 15 (1975), 31–6.
[20] Hart, *Punishment and Responsibility*, 100; Hall, *General Principles of Crimi-
nal Law*, 176; Gross, *A Theory of Criminal Justice*, 53; Dan-Cohen, '*Actus Reus*',
16.

are, which description must both be more plausible than the act-is-movement thesis and be such as to include omissions as acts. In addition, some conception of what omissions are must be developed that is different from that defended in Chapter 2, where I argued that omissions are simply absent actions; for *no* theory of action can be developed that includes as acts things which are not acts by that very theory. Consider in this regard the statement 'often to refrain from an act is no less an act than to commit one, because inhibition is coequally with excitation a nervous activity'.[21] Although this contains a rudimentary gesture towards a theory of action—as some kind of nervous activity— omissions cannot simply be absent actions (as is presupposed by describing them as being refrainings *to act*) and then be declared to be actions.

Admittedly, some philosophers have used conceptualizations of omissions different from mine to argue that such reconceived omissions are actions. Bruce Vermazen helpfully distinguishes what he calls 'resistings' and 'displacement refrainings' from other 'negative acts' (his generic label for what I would call omissions). A resisting occurs when an agent's body is about to be made to move by outside forces, but he keeps his body from moving by activating the appropriate muscles.[22] The example given by Vermazen of a displacement refraining is as follows:

Suppose that our agent, Andy, is confronted with a table laden with attractive hors d'œuvres, but has a pro-attitude toward not eating them and a belief that if he keeps his hands otherwise occupied say by twisting the buttons on his vest, the movements of his body will amount to not eating the hors d'œuvres; and so he twists his buttons.[23]

About such a case, we might say that Andy omits to eat by twisting his buttons.

About both such cases, we might well say that these sorts of omissions metaphysically are actions. Yet such concession in no way undermines the identity thesis. After all, in both cases there are bodily movements with which such negative acts are to be

---

[21] C. S. Sherrington, *Integrative Action of the Nervous System* (New York, 1906).
[22] Vermazen, 'Negative Acts', 95.
[23] Ibid.

identified.[24] *These* limited classes of 'omissions' can then be conceded to be actions without raising any objection to the identity thesis.[25]

Other philosophers wish to press the metaphysical branch of the objection further, however. Consider this example by Myles Brand: 'The policeman who keeps his arm at his side and does not shoot the fleeing youth refrains from shooting him. . . . Idiomatically, we would simply say that he refrains from shooting the fleeing youth *by* keeping his hand by his side. Refraining, then, is one type of action.'[26] Brand's conclusion is one to be resisted, for such refrainings are not actions. The policeman does not do anything in order not to shoot the fleeing youth; it is not as if he grabbed his gun hand with his other hand in order to prevent the shooting. The intention of the policeman not to raise his arm, motivated by the desire not to shoot the fleeing youth, does not make this refraining into something the policeman does. The policeman simply refrains from shooting and from raising his arm, an omission Brand has not shown us to be an action (and, thus, not something needed to be accounted for by the identity thesis).

These suggested counter-examples by Vermazen and Brand go further than is usually done by those who present omissions as counter-examples to the identity between actions and movements. More typically, it is simply assumed that any sort of omission is obviously an action, on some unspecified theory of action. Yet without both a plausible theory of the nature of actions as something other than movements, and without an equally plausible conception of omissions as something other than absent actions, the metaphysical version of the objection soundly begs the question of what acts are. To such an objection Austin and Holmes can simply reply, 'But omissions aren't acts, so the failure of our theory to include them as acts is not a defect but a virtue.'

[24] In the case of resistings, we need to reconstrue 'bodily movements' to include muscle-flexings; yet in these cases such muscle-flexings are just what the actor wills to do in order to keep his body from moving.

[25] A point on which Vermazen also is very insistent. See 'Negative Acts', 104.

[26] Brand, 'The Language of Not Doing', 45–6, 49. See also Annas, 'How Basic Are Basic Actions?', 204: 'what about omissions and cases where acting involves precisely *not* moving the body? Suppose the boy stands on the burning deck, whence all but he had fled? It is his *standing* which is his action, and this involves no bodily movement, unlike his coming to stand there.'

Taken legally, the omissions objection is at this stage of my argument irrelevant. I have already conceded that sometimes, if rarely, our criminal law does punish true omissions, and that the act requirement, accordingly, is not a requirement for crimes of omission. We are now engaged in trying to figure out what an act is, assuming the act requirement is generally, but not universally, applicable. We thus need objections showing us that acts are not what Austin and Holmes said they were, not objections urging that one stipulate a new meaning of the word 'act' for criminal-law purposes. Whether we should stipulate a new and broader meaning to the word 'act' so that our act requirement is universal but not univocal, or whether we should retain the narrower usage of 'act' so that our act requirement is univocal but not universal, is not a very interesting question. In any case, it is not the question asked here, which is the metaphysical question of what sorts of things acts are.

A second objection is the one I earlier tried to forestall by limiting the identity claim to basic acts for the time being. As Glanville Williams puts the objection: 'an act has been taken as a willed movement—but this does not reach the full meaning of the word'.[27] What is needed, Williams tells us, are the *circumstances* in which a bodily movement is done, and the *consequences* a bodily movement causes. These are parts of an act too. Hyman Gross gives the example of the act of breaking and entering.[28] Various bodily movements of smashing glass, easing through a window, etc. are not sufficient to make up an act of breaking and entering, Gross urges; for also needed is the absence of the circumstances that the movements were not consented to by the owner of the premises (as a test of his burglar alarm, for example).

Although I shall eventually argue that each complex action of breaking and entering is in fact no more than certain bodily movements, and that circumstances like absence of owner's consent are not parts of such an act, for now let us honour my earlier stipulation. The identity thesis itself only applies to basic acts, like moving one's arm toward the glass. By definition, there are

---

[27] Williams, *Criminal Law*, 16. Williams ultimately came to accept the identity thesis. See id., *Textbook of Criminal Law*, 2nd edn. (London, 1983), 147–8.
[28] Gross, *A Theory of Criminal Justice*, 54.

no other elements to such basic or simple acts. That is why they are so basic and simple.

A third objection was once a standard staple in the philosophy of action,[29] but it has resurfaced in legal theory in Hyman Gross's more recent book.[30] The objection is twofold. First, for any given act, there are many different bodily movements with which one can perform that act. Richard Peters's example: the act of signing a contract can be done through the movement of one's toes as well as by the movement of one's fingers, as long as you can grasp a pen with either.[31] Secondly, for any given bodily movement, it can give rise to many different acts. To use a somewhat Gross example: the identical rhythmic pelvic motions can constitute an act of fornication or an act of adultery.[32] The conclusion of both these points is supposed to be that actions therefore cannot be bodily movements.

There are two misconstruals of the identity thesis in this objection. The first is that just referred to, the failure to restrict the acts one considers to basic acts. It is less plausible to suppose that the basic act of moving one's fingers can be accomplished by toe movements than it is to suppose that a complex action like signing a contract can be accomplished by such diverse movements.[33] But the second misconstrual is even more drastic: Peters, Gross, and others take the identity asserted to be between *types* of acts. With what type of movement can the type of act that is contract-signing be identified? With what type of act may the types of movement involved in adultery be identified? 'None' is the only plausible answer, as the Peters and Gross counter-examples show. But that is not what the identity thesis asserts. The thesis is only: each instance of a contract-signing is an instance of *some* bodily movement, toe- or finger-moving, or some other; each instance of fornication, or of adultery, is an instance of

[29] The argument was a standard staple against the identity thesis in 1950s and 1960s philosophy of action. See David Hamlyn, 'Behaviour'; *Philosophy*, 28 (1953), 132–45; A. I. Melden, 'Action', *Philosophical Review*, 65 (1956), 529–41: 530–2; Richard Peters, *The Concept of Motivation* (London, 1958), 12; Melden, *Free Action*, 128–31; Annette Baier, 'The Search for Basic Actions', *American Philosophical Quarterly*, 8 (1971), 161–70.

[30] Gross, *A Theory of Criminal Justice*, 51–5.

[31] Peters, *The Concept of Motivation*, 13.

[32] Gross, *A Theory of Criminal Justice*, 51–2.

[33] See Don Locke, 'Action, Movement, and Neurophysiology', *Inquiry*, 17 (1974), 23–42: 32.

pelvic motions, or *some* motions (use your imagination). The identity of act-tokens with movement-tokens is in no way affected by the Peters–Gross kind of argument against type-identities.

A fourth objection requires some background to be understood. This is what I shall call the 'category mistake' objection. This objection is based on a general view about the power of ordinary language to determine questions of reference and identity. One looks to how ordinary speech uses terms like 'action', 'movement', and the like, to see if they appear in the same category of discourse. If they do not, then it is a 'category mistake' to use them in the same assertion. It is not meaningful to assert or deny, for example, that 'actions are identical to bodily movements', if 'action' and 'movements' are categorically different terms.[34]

Illustrative of this kind of objection to the Austin–Holmes identity thesis is A. I. Melden's linguistic observation that 'motive' and like words ('reasons', 'intentions', etc.) are only properly employed to explain actions.[35] It is linguistically deviant to speak of there being motives for bodily movements like arm-risings; rather, we speak of motives for actions like raising one's arm. From such linguistic facts Melden concludes that concepts of action have different 'logical features' that place them in a different category of discourse from concepts of bodily happening. There is thus a conceptual 'gap' between physical-movement descriptions and action descriptions, much like the gap many philosophers think exists between descriptive and normative statements. Such a gap prevents one from identifying raising one's arm with one's arm rising in the same way the is–ought 'gap' prevents one from identifying the goodness of a knife with its sharpness. In both cases, Melden concludes, either asserting or denying such an identity is meaningless—that kind of conceptual confusion called a category mistake.

In retrospect it is remarkable that an entire generation of analytic philosophers took this style of argument seriously. From the

---

[34] *The* influential expression of this view is Ryle's in ch. 1 of *The Concept of Mind.* For a sympathetic expression of this ordinary-language view applied to action, see Myles Brand (ed.), *The Nature of Human Action* (Glenview, Ill., 1970), 4.

[35] Melden, *Free Action*, 73–82.

Second World War to (roughly) 1970 this high-handed way of suspending questions of reference and identity was very much accepted. The influence of such former philosophical consensus is still felt in criminal-law theory. It is just this two-languages view that justified Herbert Hart's confidence that all he had to do to rebut the Austin–Holmes identity thesis was to show that our patterns of ordinary usage of action verbs did not, on its surface, coincide with the thesis.[36] The two-languages view is also evident in George Fletcher's rejection of the identity thesis as incompatible with the way we see human action: 'The difference between bodily movements and human acts is that when someone is acting, we can perceive a purpose in what he or she is doing.'[37] Such purpose is itself understood when we drop the 'methodology of the natural sciences', dealing with matter in motion, in favour of the 'inter-subjective understanding' needed to grasp goals and purposes.[38] Such linguistic dualism is also one of Hyman Gross's main motivations for his conclusion that 'acts and bodily movements are different kinds of things'. Gross reaching such a conclusion in part because of linguistic facts like the fact that 'the appropriate move to defend against a charge of doing harm is different for each'.[39] Similarly, at the root of Antony Duff's claim that 'persons and actions . . . cannot be explained by an analysis which seeks to reduce them to supposedly simpler elements' (such as bodily movements) is Duff's claim that 'the *meanings* of the concepts of person, action . . . are given in their ordinary usage; and ordinary usage shows that persons and intentional actions are directly observable' (which bodily movements are not).[40]

As we have noted before, no amount of linguistic observations about ordinary-usage patterns can suspend questions of reference and identity in this way. Consider the phrases, 'second planet from the sun', 'Evening Star', and 'Morning Star' in their ordinary usages before Babylonian astronomers discovered that such phrases in fact referred to the same thing, namely, the planet Venus. Presumably the first phrase was ordinarily used in 'planet talk', the latter two in 'star talk'; presumably the second phrase

[36] Hart, *Punishment and Responsibility*, 101.
[37] Fletcher, *Rethinking Criminal Law*, 436.   [38] Ibid.
[39] Gross, *A Theory of Criminal Justice*, 51.
[40] Duff, *Intention, Agency, and Criminal Liability*, 130.

was used in 'evening talk', the third, in 'morning talk'. If these are categories of discourse, presumably it is a category mistake to assert or deny the identity of the Evening Star with the Morning Star with the second planet from the sun. Since I take this conclusion to be absurd, I take the argument leading to it to be equally so. No amount of linguistic observation of the kind Melden engaged in should convince us that acts are not in fact identical to bodily movements. That such an identity should lead to some linguistic oddities should not be surprising if the identity itself has not been widely enough known, for a long enough period, to get reflected in ordinary-usage patterns.

A fifth objection that has crept into the legal literature is an epistemic objection: acts are known to those who perform them in a distinctive way, a way in which movements$_I$ are not known to those whose bodies so move. Such an objection is suggested by Wittgenstein's epistemic criterion for action. 'What is left over if I subtract the fact that my arm goes up from the fact that I raise my arm?', Wittgenstein famously asked.[41] His own answer was an epistemic one: the actor is not surprised by his own acts in the way he may be by movements of his body.

There is a simple-minded way to take this suggestion of Wittgenstein's: 'Acts like raising one's arm are known in that non-inferential, non-observational way characteristic of first-person mental experiences generally; bodily movements like arm-risings are not known in this way. Therefore, the two things are different.' What makes this use of Wittgenstein so simple-minded is that we have contrasted acts of arm-raisings with just those movements of arm-risings that are not even plausibly involved (in any way) with acts of arm-raisings. To object to the identification of the event referred to by 'Jones raised his arm at t' with the event referred to by 'Jones's arm rose at t', it is totally beside the point to compare both this act and this movement to some other movement, such as some reflex movement of Smith's arm. The identity thesis asserts there to be an identity between acts and *those movements that take place when acts take place*.

A less simple-minded way to form an objection to the identity thesis out of Wittgenstein's epistemic observation was stated by Herbert Hart: 'our *primary* awareness of our own actions is not

---

[41] Ludwig Wittgenstein, *Philosophical Investigations*, 3rd edn. (London, 1958), §621.

that of a physiologist: it does not include a knowledge of the muscle movements required'.[42] Hart's insight here was elaborated on by one of his former students, Eric D'Arcy: 'Austin's theory carries the further implication that only a physiologist could tell us the name of our acts. . . . I could not name or identify the muscles involved, and am not aware of the existence of most of them'.[43] Acts are not to be identified with the bodily movements that occur during them, on this view, because whatever else acts are, they are known to the actors who perform them in a certain way (*pace* Wittgenstein), and the accompanying bodily movements are not so known.

I have no quarrel with the Wittgensteinian premiss of this objection. Acts do characteristically, although perhaps not inevitably, have what Carl Ginet has called recently an 'actish phenomenal quality'. [44] If we notice the movements of our limbs at all (which Hart and D'Arcy deny we often do), we may not be aware of them in the same way as we are aware of our moving them. Yet even with these epistemic concessions, the negative conclusion about the identity thesis doesn't follow. How I come to discover that some thing exists is one question; what it is that I discover to exist is another. One cannot answer metaphysical questions of essence and identity solely by epistemological observations of the Wittgensteinian kind. The Morning Star was discovered in a somewhat different way from the Evening Star: one looked in the morning for the one, and in the evening for the other. My own intentions are usually known to me in a way different from how they are known to a third-person observer, the latter having to make behavioural inferences since he lacks my first-person experience. Yet these epsitemic facts should not convince us in either of these cases that there are two different stars, or two different mental states.[45] An elephant may be known in quite a few different ways by the proverbial blind men, each touching a different part, but it is still just one elephant.

I conclude that despite the widespread rejection of the identity thesis in contemporary criminal-law theory, there have been no

[42] Hart, *Punishment and Responsibility*, 103.   [43] D'Arcy, *Human Acts*, 8.
[44] Carl Ginet, *On Action* (Cambridge, 1990), 13. See also Lawrence Davis, *Theory of Action* (Englewood Cliffs, NJ, 1979), 2–3, and Katz, *Bad Acts and Guilty Minds*, 127, for further descriptions of the introspective 'feel' of acting.
[45] A point Hart himself used against Wittgenstein, in Hampshire and Hart, 'Decision, Intention, and Certainty'.

persuasive reasons advanced to justify such rejection. I next turn to the question of whether contemporary philosophy has advanced any better reasons.

1.3 Civil War between Volitionalists: Mental-Action Theorists versus the Identity Thesis

The most trenchant criticism of the identity thesis in contemporary philosophy appears to come from what are often called 'mental action' theorists. A mental-action theorist is close to the Austin–Holmes overall view about the nature of acts, because he, like Holmes and Austin, believes that there is a distinctive mental state of volition or willing that is an essential feature of acts. Unlike Holmes and Austin, however, the mental-action theorists believe that all there is to action is this mental act of willing. Our acts, on this view, never 'break the skin'; there are only our internal acts of willing, all else including bodily movements being the *effects* of acts.[46]

On such a view, bodily movements are not identical to acts precisely because bodily movements are the effects of acts (and nothing can be the cause of itself). The crucial premiss for the mental-action theorist in rejecting the identity thesis is thus: Any movement$_T$ *causes* the associated movement$_I$. The act of raising my arm on some occasion, for example, causes (and is thus not identical to) the bodily movement of my arm's rising on that occasion.

This is not a very intuitive premiss, for it does not square very well with the language we use to refer to acts. 'Moving one's hand' seems to involve one's hand moving as an essential feature. Still, as I myself shall argue at the close of this chapter as well as in Chapter 11, how we name or refer to actions can mislead us as to their nature. 'Killing' seems to refer to the state of affairs of a death occurring as a necessary feature, yet perhaps the complex description 'killing' only *refers to* the simple act of moving one's finger even though it picks out such an event by one of its effects, namely, the death of a person. Analogously, H. A. Prichard argued, 'when we refer to some instance of this activity, such as our having moved our finger . . . we refer to it thus not

---

[46] For a description of mental-action theorists, and a bibliography of who they are, see Brand, *Intending and Acting*, 6–15; Moya, *The Philosophy of Action*, 18–29.

because we think it was, or consisted in, the causing our finger to
move . . . but because we think it had a certain change of state
as an effect'.[47]

We must thus put aside the seeming linguistic oddity of the
premiss as much of an argument against it. Still, what considerations might induce us to accept it? Prichard argued for it by
rejecting the idea that a causal sequence between events can itself
be an event: 'though the causing a change may require an activity, it is not itself an activity'.[48] This rejects my earlier expressed
qualification to the identity thesis, to the effect that acts were
ultimately to be identified with the causal sequence volition-
cause-bodily movement, not just with bodily movements *simpliciter*. Such rejection forces those holding the Austin–Holmes
view to a choice: should acts be identified with bodily movements
simpliciter, or with the volitions that cause them? *That* choice
Prichard thinks to be easy. Not only do bodily movements *simpliciter* occupy a different spatio-temporal region from their associated acts, but, Prichard urges us, introspection reveals what is
more essential to action:

When, e.g., we think of ourselves as having moved our hand we are
thinking of ourselves as having performed an activity of a certain kind,
and, it almost goes without saying, a mental activity of a certain kind,
an activity of whose nature we were dimly aware in doing the action and
of which we can become more clearly aware by reflecting on it. . . . Con-
sequently, there seems to be no resisting the conclusion that where we
think of ourselves or of another as having done a certain action, the
activity of which we are thinking is that of willing . . .[49]

Phenomenal experience, like surface linguistic features, can
provide no more than clues to underlying natures. Even granting
Prichard all he says about the deliverances of introspection, it
still might be the case that acts are to be identified with bodily
movements and not with the willings that cause them. The main
reason for rejecting Prichard's analysis, however, does not lie
here but with his initial premiss: why cannot causal sequences
between events themselves be events? Is not 'breaking a rack' in

[47] Prichard, *Moral Obligation*, 190.
[48] Ibid. 189. Prichard's mental-action theory is summarized in Glen Langford,
Human Action (Garden City, NJ, 1971), 8–21; see particularly p. 15.
[49] Prichard, *Moral Obligation*, 189.

billiards an event, even though it consists of a causal sequence of numerous events (of each ball imparting its motion to the ball(s) next to it)?[50] If so, then we are not thrust to a choice between movements (as effects of willings) and willings (as causes of movements). We can have both: acts are willings-causing-movements, not just willings and not just movements.

A set of alternative arguments for the mental-action theory is given by Jennifer Hornsby. The first of these arguments aims to establish directly the crucial premiss of the mental-action theorist, namely, that basic acts *cause* their associated bodily movements (and thus cannot be either fully identical to them or constituted in part by them). According to Hornsby:

(1) It is a necessary truth that expressions of the form 'Rachel moved$_T$ her arm' entail expressions of the form 'Rachel caused her arm to move$_I$'.[51]

(2) The expression 'Rachel caused her arm to move$_I$' is equivalent to 'Rachel's deed caused her arm to move$_I$'.[52]

Since Rachel's deed just is Rachel's moving$_T$ her arm, then the basic act of moving$_T$ her arm caused the bodily movement of the arm moving$_I$. From this the conclusion follows that basic acts can neither be fully identical to bodily movements nor can they be constituted in part by such movements, for nothing is the cause of itself. (In addition to rejecting the identity thesis this premiss also makes the mental-action theory quite plausible, for the cause of such bodily movements surely is some kind of mental state or activity.)

There is nothing wrong with the first of Hornsby's two premisses. I myself will argue in Chapter 8 that all descriptions of actions, including complex ones such as 'Smith killed Jones', are replaceable by expressions like 'Smith caused the death of Jones'. Moreover, the second premiss has some plausibility to it as well, if we think of more complex descriptions of actions. 'Smith caused the death of Jones' is replaceable by 'Smith's act (of, for

---

[50] See Brand, *Intending and Acting*, 16, 55, 67. See also Thalberg, *Perception, Emotion and Action*, 85–112; Thomson, *Acts and Other Events*. That events have parts, and that those parts may themselves be events, is a plausible metaphysics of events. See the discussion in Ch. 14.

[51] Hornsby, *Actions*, 124–6.   [52] Ibid. 126.

example, moving his trigger finger) caused the death of Jones'.[53] But if we restrict our gaze to those states of affairs associated with basic descriptions of actions—bodily movements like Smith's trigger finger moving$_I$—then Hornsby's second premiss simply begs the question against those like me who hold that the moving$_I$ of the finger is part of the moving$_T$ of the finger. For on the latter theory there is no simpler or more basic act that Smith did in order to move his fingers—he simply moved his fingers. There is thus no act with which to replace 'Rachel' in Hornsby's example, even though we have all the more basic act descriptions to substitute for 'Smith' in my killing example.

Hornsby at one point recognizes that her second premiss 'might seem to beg the question against a certain important doctrine about agency, which holds that an agent's causing an event is not to be construed in terms of one event's causing another'.[54] Yet as stated Hornsby's second premiss also begs the question against a rather different doctrine from the doctrine that 'agent-causation' is primitive; the premiss also begs the question against the view here defended, that bodily movements$_I$ are *parts* of the most basic acts that a person can do (and thus cannot be the *effects* of some yet simpler or more basic action). Hornsby's second premiss does nothing to establish a rebuttal to this crucial point, and by itself does nothing to establish the falsity of the identity thesis.

Which acts *are* basic is the ultimate issue between the mental-action theorist and the volitional theory that I defend. To argue for the latter theory will be to defend the causing of certain bodily movements$_I$ as the simplest acts we can do. Before coming to that crucial issue, we should first examine two more negative arguments against the identity thesis.

Hornsby's second argument is built on the distinction I earlier mentioned, the distinction between Rachel's arm moving$_I$ and Rachel's muscles contracting$_I$. I concluded earlier that these are distinct events, and typically the second causes the first to occur. Hornsby assumes both of these points to be true, and argues:[55]

[53] Judy Thomson has questioned such reductions. Thomson, 'Causality and Rights: Some Preliminaries', *Chicago–Kent Law Review*, 63 (1987), 471–96. For my disagreement see Moore, 'Thomson's Preliminaries about Causation and Rights', 514–15.
[54] Hornsby, *Actions*, 15 n. 8.   [55] Ibid. 20–8.

(1) Rachel's moving$_T$ of her arm causes the contracting$_I$ of her arm muscles.

Yet suppose, according to the identity thesis, that:

(2) Rachel's moving$_T$ of her arm is identical to the complex event consisting of a volition-causing-Rachel's arm-moving$_I$.

Then:

(3) The complex event (consisting of a volition-causing-Rachel's arm-moving$_I$) itself causes the contracting$_I$ of her arm muscles.

The problems with concluding that (3) is true are two: First, it seems to require that causation work backwards through time. For part of the complex event that causes the contracting$_I$ occurs *after* the contracting$_I$ occurs. And effects are not supposed to precede the occurrence of their causes. Secondly, recall that muscle-contractings$_I$ typically cause arm-movings$_I$. That fact, conjoined with (3), gives rise to the unhappy implication that one event (a muscle contracting$_I$) is *both* the cause of part of a complex event and the effect of that same complex event.

I take the conclusion of this *reductio* to indeed be absurd for both of these reasons. Since the argument is valid, that leaves rejection of (1) or (2) to be the only way to avoid the absurd conclusion (3). Mental-action theorists reject (2), which is the identity thesis: 'Unless we allow that actions are distinct from movements$_I$, we cannot say how the latter are, but the former are not, caused by the contractions$_I$ of muscles.'[56] We must thus find a way to reject (1) if we are to salvage the identity thesis.

Fortunately we have good reason to reject (1). When I move$_T$ my arm, my arm muscles do indeed contract$_I$ but my action does not cause them to contract. Rather, the arm muscles contracting is one of the intermediate events that must occur as part of the *causing* of the arm moving$_I$ by the volition to move it. Certain brain events, certain nerve events, and certain muscle movements are all part of the causal mechanism that connects volitions to

---

[56] Ibid. 24.

bodily movements. Being contained in such actions as moving$_T$ my arm, of course such muscle contractions$_I$ are not caused by such actions.[57]

The mental-action theorists' puzzle that generated the view that movings$_T$ of arms cause contractings$_I$ of arm muscles is admittedly untouched by the above considerations,[58] so we should briefly advert to this puzzle. As stated by Hornsby:

> A man makes it happen that his muscles contract. How does he do this? 'By clenching his fist.' So what he does that makes it happen that the muscles contract is clench his fist. Is this not to say that they are actions of fist clenching which make it happen that the muscles contract, which, in other words, cause them to contract? This is the train of thought that has puzzled people.[59]

This familiar puzzle is generated only because it is idiomatic English to say things like 'He caused his muscles to contract by moving his arm' and 'What he did to make his muscles contract was to move his arm'. Given the admittedly idiomatic nature of such expressions, the assumption is then made that the 'by' in the first, and the 'make' in the second, refer to causal relations between actions of moving and events of muscle contractions.

Exactly what such words do refer to in these contexts is the subject of a surprisingly large literature.[60] Even a cursory scan of that literature shows that, if sometimes 'by', 'make', etc. do describe causal relations, as often they do not. The generic sense of such expressions is that they relate means to ends. They tell us, that is, what means an agent utilized in order to achieve some end he has adopted. In the examples given, a paraphrase would be: 'He ended [muscles contracting$_I$] by meansing [moving$_T$ his arm]'. Actions that are means to some end are often *causes* of the states that are the ends. This is why English allows 'caused' to be substituted for 'by', 'make', and the like. Yet causal relations are not always involved in using some act as the means to achieving some end. Imagine the golf instructor telling his client how to hit the ball squarely on the tee: 'Follow through on your swing.'[61] When we do concentrate on following

---

[57] See Thalberg, *Perception, Emotion and Action*, 68–71.
[58] Hornsby, *Actions*, 25.   [59] Ibid.
[60] See the helpful summary in Bennett, *Events and Their Names*, chs. 8–9.
[61] The example is Hornsby's own. See *Actions*, 76 n. 1.

through, we do hit the ball squarely. It is idiomatic English to describe this fact as: 'He caused the ball to be hit squarely by following through on his swing'; or 'What he did to make the club hit the ball squarely was to follow through with his swing'; or even 'His following through on his swing caused the ball to be hit squarely'. What we mean when we say any of these things, however, is not that the later action of following through with the swing caused the earlier event of a square hit on the ball. We mean only that the action of following through with the swing was adopted as a means to achieving the end of having a square hit on the ball. The oddity of having part of the action that is one's means occur *after* the end sought occurs is readily explained: this is the only way the golfer knows how to get his muscles to move correctly *before* the hit. The only causing of the square hit is done by those prior muscle movements. The relation of the later action of following through to the earlier hit is the relation of means to end, but not of causation. Such examples show that idiomatic English usage of 'by', 'make', and even 'cause' give us no reason to believe that Rachel's moving$_T$ of her arm causes the contracting$_I$ of her arm muscles. And without this premiss, Hornsby's second argument against the identity thesis collapses.

This brings us to the real issue between identity theorists and mental-action theorists. To see this issue, consider a third and last argument of Hornsby's.[62] Suppose that Rachel moves$_T$ her arm on some occasion. This action of hers involves her arm muscles contracting$_I$. Her moving$_T$ of her arm also causes the death of another named Jones. With this scenario in mind, we might think:

(1) Rachel's contracting$_T$ of her arm muscles, Rachel's moving$_T$ of her arm, and Rachel's killing of Jones, are all actions Rachel performed.
(2) These are not three actions she did, but only one.

Yet the supposition of the identity theorist is:

---

[62] Although she does not cleanly separate this third argument from the second, they are in fact distinct. At one point Hornsby appears to believe that she is only making the third argument, for she 'remains agnostic' on whether we ever contract our arm muscles by moving our arm (which was the insight that generated the first premiss of her second argument). See ibid. 75–6 n. 1.

(3) Rachel's arm-moving$_I$ is part of Rachel's moving$_T$ of her arm.

Since from (2) we may infer that whatever Rachel's moving$_T$ of her arm is composed of, is also what her contracting$_T$ of her muscles is composed of, we get from (3):

(4) Rachel's arm moving$_I$ is part of Rachel's contracting$_T$ of her arm muscles.

And this is absurd, because Rachel's contracting$_T$ her muscles is over before Rachel's arm-moving$_I$ is over.[63] Hornsby's conclusion: 'unless we push actions right back inside the body [by rejecting (3)], we cannot make good sense of talking about an action as a person's contracting$_T$ his muscles'.[64]

Because (4) is absurd and must be rejected, and because the 'coarse-grained theory' of act-token individuation asserted by (2) I shall myself defend in Chapter 11, we must reject (1) if the identity thesis (3) is to be preserved. We must deny, in other words, that there is any 'good sense' to be made 'of talking about an action as a person's contracting$_T$ his muscles'. In the ordinary case, where a person has not undergone special training, he does no contracting$_T$ of his muscles as an action of his; rather he moves$_T$ his arm, and there is no simpler or more basic action by which he performs such movement$_T$.

Whether this is so is thus the crux of the matter about this third argument of the mental-action theorist against the identity thesis. This issue is also the question begged by the first argument of Hornsby's, for there what was crucial was whether 'Rachel caused her arm to move$_I$' could be replaced with 'some deed of Rachels' caused her arm to move$_I$'; the only deed of Rachel's that could be more basic than Rachel moving$_T$ her arm, would be some deed like Rachel contracted$_T$ her arm muscles. This issue also determines whether muscle contractions$_I$ are parts of the causal mechanism by which volitions cause movings$_I$, as I contend, or whether such contractings$_I$ are effects of the actions of moving$_T$ one's arm, as the mental-action theorist contends in her second argument. In short, we have reached the crux of the issue between the identity theorist and the mental-action theorist. None of Hornsby's three arguments even touch this crucial issue.

---

[63] *Actions*, 23.   [64] Ibid. 24.

The issue is whether there are any actions we do in order to move$_T$ our bodies, which actions are therefore more basic than moving$_T$ our bodies. Are there such actions as the contracting$_T$ of our muscles, the sending$_T$ of a nerve impulse to such muscles, or the firing$_T$ of the right neurons in the brain to cause such nerve impulses to be sent? The identity thesis defends the intuitive answer to this question, which is in the negative. Although muscle movements, nerve impulses, and neuronal excitation are all events that do occur when we act, none of them are themselves acts we as persons perform. Persons only move their bodies.

There are three considerations to be raised in the defence of this intuitively plausible position. I shall call the first the epistemic step. This step is taken once we conclude that moving$_T$ our bodies is the simplest or most basic thing we *know* how to do. The simplest or most basic acts are those we just will because we don't know how to bring them about indirectly by willing something else as a means. I know how to move my arm—I try, and it moves. I do not know how to move my muscles, the impulses in my nerves, the ionization potentials of the neurons in my brain, except by moving my arm. Epistemically, moving my arm is the simplest or most basic thing I know how to do.

The second step I shall call the basic-action step. This step is taken once we conclude that what we as persons can do by way of a basic action is a function of what we know how to do directly. Even though other events assuredly take place within our bodies when we move them, those other events are not our basic actions for that reason. Sending a nerve impulse can be a basic action a person performs only if such sending is something a person knows how to will. Since we do not know how to will such things, their occurrence when we move our bodies is no basic act of ours.

The third step I shall call the complex-action step. This step is taken when we conclude that effects of *mental* actions are not themselves parts of (real, or physical) actions. Unless those physical effects are themselves willed or intended, we should conclude that there is no complex action here even though such effects are the physical events that go on in our bodies when actions go on. To test this, imagine a scenario in which the actor performs what is admittedly a mental action. For example: he tries to move a recently amputated leg (for which he has the phantom-limb experience). Suppose such mental action has physical effects: the effort causes the pupils of his eyes to contract, his facial muscles to twitch, and his lips to turn downward. Has he performed the actions of contracting$_T$ the pupils of his eyes, twitching$_T$ the muscles of his face, or curling$_T$ his lips downward? A negative answer is intuitive, and such an answer denies that causing physical effects by mental acts of willing or trying is sufficient to characterize such causing as (complex) actions of a person.

This three-step argument has as its conclusion that any physical effect V_I-ing must be caused by a willing, trying, or other mental act, and it must also be the object of such a mental act or mental state, in order for there to be a basic action of V_T-ing. Since only bodily movements are these kinds of physical effects, only bodily movements can be basic actions.

This argument may be attacked in any of its three steps. Consider the epistemic point first. There are two sorts of counter-examples to the point that the only physical effects we know how to bring about directly are the movements of our bodies. One is where we acquire the skill to will changes internal to our bodies. For example, we learn how to will the constricting of our capillaries in discrete locations within our body. So trained, suppose a pregnant woman who wishes to abort her fetus wills (and thus causes) the constriction of the blood-vessels leading to her uterus, thereby cutting off the blood supply to her fetus, causing it to die.[65] Is not such a constricting$_I$ part of an action of constricting$_T$ (which is also, given its further effect, also an action of killing)?

The second sort of counter-example stems from imagining that we can will physical effects external to our bodies. In a recent film, for example, a pilot operates a plane by use of a helmet that is wired to the plane's controls; the pilot has only to think, 'Left turn', to cause the controls to execute a left turn.[66] When he thinks such a thought, and it has its intended effect, has he not acted in 'turning the plane'?

The main answer to both of these counter-examples is disappointingly simple: we do not in fact have the capacity to do either of these things directly. If we ever do acquire the ability to

[65] I am indebted to Charlotte Crane for this speculation.
[66] The example is due to Mark Grady.

will the constriction of our blood-vessels, the contraction of our muscles, the sending of nerve impulses, etc., then the most basic acts we know how to do will not be bodily movements; likewise for the telekinetic-like powers of the fictional pilot. But as it stands, the levers by which we move the world are our own bodies and the movements one can will with them.[67]

This says something important about the identity thesis. That thesis is revealed to be a contingent corollary of a deeper truth about actions, namely, that actions are the causings of those effects that are themselves described by the objects of the volitions that cause them. This deeper truth says nothing about *what* those effects might be that are both caused by volitions and described by the propositional content of volitions. It turns out that as an empirical matter the only effects of willing that fit the bill are bodily movements. The identity thesis is thus born of this contingent fact about human beings and their psychology. If our capacities change dramatically in the future, then we would have to amend the identity thesis accordingly.

In the two cases supposed, the amendments would be minor enough. To accommodate the first, 'bodily movement' would have to include those inner movements that we can directly control, capillary contraction joining things like swallowing and breathing. To accommodate the second, bodily movement should include movements of artificial limbs, which is essentially what the hypothetical plane becomes when it is willed to move with the magic helmet. More extreme examples, such as outright telekinesis, might force us to abandon the identity thesis entirely; but if it turns out that we can so bypass our bodies as *the* means by which we change the world, such thesis should be abandoned.

The second step of the argument is not vulnerable to this empirical kind of attack where one trots out recherché thought experiments of imaginable capacities. The second step may seem vulnerable to a kind of conceptual objection, however. The conceptual objection would be that all those physical events that take place within the human body when actions take place must themselves be actions. If the moving₁ of the arm is part of the

---

[67] For a brief summary of the psychological literature on what it is that we can be trained to will directly, see Gregory Kimble and Lawrence Perlmuter, 'The Problem of Volition', *Psychological Review*, 77 (1970), 361–8 374–5 (heart rate, blood pressure, pilomotor response, rate of oxygen absorption, ear-wiggling).

action of moving$_T$ the arm, so the contracting$_I$ of the arm muscles must be part of the action of contracting$_T$ the arm muscles, the movement$_I$ of the nerve impulses must be part of the action of sending$_T$ the nerve impulses, etc. The objection is that all parts of the causal mechanisms necessary for the action of moving one's arm must themselves be actions too.

Yet this objection is confused. We sometimes do speak of subpersonal 'agencies' doing things like 'sending$_T$ a nerve impulse to the biceps'. But this anthropomorphic way of speaking is not to be taken seriously. Functionalists in the philosophy of mind, Freudians, cognitive psychologists, and workers in artificial intelligence have long thought it fruitful to subdivide the mental states and actions of whole persons into the little states and little actions of little people.[68] For example: 'The ganglion cells in the back of the retina pre-process the information received from the external world, sorting out and encoding what is worth sending up the optic nerve'. Whole persons do not send messages up their optic nerves, even though such goings on are necessary for whole persons to do such things as look over their shoulder. However one draws the line between the actions that persons do and the subpersonal routines that agencies within persons do, nerve-sendings and muscle-contractings remain on the subpersonal side of the line.

The objection to the third step of the argument is also a conceptual objection. This objection disagrees with the deeper truth that underlies the identity thesis. That truth is that the causal sequence volition–cause–physical effects, does not constitute an action unless the object of the volition was to cause just that physical effect. The third step of my argument relies on this deeper truth to conclude that facial tics, for example, are not actions even when they are caused by mental acts like the mental acts involved in trying to move a non-existent limb; in such cases the object of the willing (to move one's leg) is not the facial movement caused.

The beach-head for the conceptual objection to my view is established by shifting examples to those where the facial tic, pupil contraction, or other bodily movement$_I$ is intentionally

---

[68] I discuss functional subdivisions of self briefly in the next chapter, since my account of volitions is functionalist.

brought about to serve some further end.[69] For example, X wants to scare Y and realizes that an effective way to do so is to manifest the behavioural features of his trying to move his non-existent leg (for Y will believe X is darkly angry or mentally disturbed, since that is how X looks during such times). X accordingly tries to move his phantom limb while he faces Y. This effort of will causes the eye, lips, and facial movements earlier described; these in turn cause Y to be frightened. Has not X performed the action of frightening Y by moving$_T$ his facial muscles, lips, and pupils? If we think the answer to that question to be 'yes', the argument continues, then surely the very same movements are actions when caused by the very same mental acts (of trying to move the phantom limb) even when such movements are *not* adapted as means to some further end like frightening Y. *Ergo*, the desired conclusion: physical effects of willing to move one's arm—like muscle-contracting$_I$—need not be the objects of one's willing in order to be actions like contracting$_T$ one's arm muscles.

Identity theorists should allow the intentional-action beachhead, but they should defend the perimeter around it vigorously. Where X intends to frighten Y by means of his facial tic, that facial movement is X's action. But that is only because in such cases X's intention is to move his face in just this way. Since he does not know how to move his face in this way directly, he does so indirectly, by willing to move his phantom limb. There is nothing mysterious about concluding that such facial movements are actions that X performs. Such cases are no more problematic as instances of action than are cases where one *wiggles one's ears* indirectly by moving the ears with one's hands.

Drop out the intention to move one's facial muscles, however, and such movements cease to be actions. Where one is simply trying to move one's phantom limb, and such mental act of trying causes the eye and facial movements$_I$ described, one has not moved$_T$ one's facial muscles, contracted$_T$ one's pupils, or moved$_T$ one's lips. It is misleading to say that the movements in this unintentional case are caused by the very same mental act as they are in the intentional case, for there is this difference: in the latter case the willing of the movement of one's phantom leg is accompanied

---

[69] I am indebted to Leo Katz and Antony Duff for forcing me to respond to this objection. The objection is also glancingly suggested in Annas, 'How Basic Are Basic Actions?', 204–6.

by an intention to cause a movement of one's lips, facial muscles, and pupils, whereas in the former case it is not. The criterion that excludes contractions$_I$ from being contractings$_T$—the inclusion of the contracting$_I$ within the object of a willing or intention—is thus capable of distinguishing these two cases.

Why should this difference (in whether the bodily movements$_I$ are within the object of some willing or intention) make a difference in action theory? Because our bodies are the only means we have of effecting changes in the world, where the 'we' refers to our personal selves. Whether we choose to effect change in the world depends on whether some bodily movements are willed or intended by us. When they are not, they and their further effects are not events of which we are the authors; when they are, we are such authors. Although we are in some sense authors whenever we mentally act, and although we are in that same sense authors of those bodily events that we do not will or intend but that are caused by our mental actions, we have not chosen to have any effect in the world in such cases. Only when we will or intend such movements do we choose to begin some chain of events in the world beyond our head; only then do we act in the sense of 'act' relevant to criminal law, namely, physically act.

I thus conclude that the mental-action theorists have not made it plausible to us that our basic actions of moving$_T$ our bodies are in reality just our mental acts of willing our bodies to move. The movings$_I$ of our bodies are accordingly not simply the effects of our movings.$_T$. That leaves open the possibility that our movings$_T$ are partially identical with our movings$_I$. To say that there are no good reasons in philosophy or in legal theory to resist such identification is not of course to say that there are good reasons to make it. Yet is not the identification plausible on its face? All theorists of action concede there to be some intimate connection between my *act* of moving$_T$ my arm on some occasion and the movement$_I$ of my arm on that occasion. Thus, even mental-action theorists speak of 'associated movements$_I$' or 'doing-related events'.[70] Surely they are right about this—moving my arm *is* closely related to my arm-moving. The simplest thing to say is that that intimate relation is one of the greatest intimacy, namely, identity. If there is no good reason to resist such an identity, I take that to be a good reason to accept it.

[70] See Davis, *Theory of Action*, 79.

2. THE EXCLUSIVITY THESIS: ALL ACTS ARE IDENTICAL TO BODILY MOVEMENTS

Much of the criticism of the volitional theory of action within legal theory has focused on the exclusivity thesis. When Holmes said that acts were 'nothing more' than willed muscle contractions, and when Austin said that the only acts 'properly so called' were bodily movements, they were excluding the *circumstances* in which such acts were done, and the *consequences* such acts caused, from being part of the acts themselves. As Bentham put it earlier, 'an act . . . is confined to the person of the agent',[71] and as the Restatement of Torts put it later, 'The word "act" . . . does not include any of its results even the most direct, immediate, and intended.'[72] It is this exclusion of the death seemingly a part of killing, the lack of consent seemingly a part of rape, the ownership by another seemingly a part of theft and burglary, that so exercised the ordinary-language philosophers of action and those criminal-law theorists that they influenced.

The critics of the exclusivity thesis typically assume that Austin, Holmes, Bentham, *et al.* must be denying that killings, rapes, thefts, burglaries, etc. are really acts. For example, D'Arcy: 'Austin's doctrine . . . boldly proclaims that most of the names which we speak of as denoting acts do not do so at all: to kick or punch, to lift or carry, to speak, or strike a match, or sign a contract—none of these is an act.'[73]

Usually legal theorists defending the exclusivity thesis do so by treating it as a *useful* thesis to stipulate for criminal-law purposes. They eschew it as a metaphysical thesis that is *true*, because they can't imagine stomaching the implication D'Arcy references. If a rape or a killing is not an act, what is? By contrast, I shall defend the exclusivity thesis as a metaphysical truth about acts (and, because it is a metaphysical truth, it is also part of the definition of 'act' useful for criminal-law purposes). Such a defence is possibly only if we deny D'Arcy's implication of the exclusivity thesis, for I too think it crazy to conclude that killings, kickings, speakings, etc. are not acts.

[71] Bentham, *Introduction to the Principles of Morals and Legislation*, 74.
[72] American Law Institute, *Restatement of the Law of Tort* (St Paul, Minn., 1934), i. 6–8.
[73] D'Arcy, *Human Acts*, 9. See also ibid. 67.

The way around D'Arcy's implication is to identify every complex action of killing, raping, etc. with some basic act of moving one's body. Then one can say, with Austin and Holmes, that the only acts there are are bodily movements without also implying that killings etc. are not acts. That identification of all complex acts with some basic act is precisely what I take the exclusivity thesis in its most plausible form to assert.

As mentioned before, the identity here spoken of is an identity of act-tokens and not act-types. The exclusivity thesis cannot claim that killings as a *type* of action are identical to some discrete set of basic acts—for one person can kill another in any of an indefinitely large number of ways, probably so numerous that virtually any basic act could, in the right circumstances, be a killing. Rather, the claim is only that for each act of killing, there is a corresponding basic act of limb movement, and that these two nominally distinct acts are in reality one and the same event.

To make out such an identity will be to argue: (1) that neither a non-bodily movement (i.e. an omission or purely mental action) nor a non-willed bodily movement like a reflex can be a killing; and (2) that where there is a willed bodily movement 'involved' in a killing, that willed movement (not some death caused by it) *is* the killing. It is not convenient to argue for either proposition here, since we first need to have some greater understanding of the diverse structure of complex action descriptions before us. I shall thus defer the argument to Chapters 10 and 11 respectively, which we shall come to after we have charted how complex action descriptions are put together. Preliminarily, however, we can see the intuitive plausibility of each proposition from things we have already examined in the previous section.

With regard to part of the first proposition, recall Hornsby's assertion that 'Smith killed Jones' is equivalent to 'Smith caused the death of Jones'.[74] If Hornsby is correct about this, as I shall argue in Chapter 8 that she is, then one only need see that neither omissions nor mental acts (when unaccompanied by bodily movements) can be causes in order to see that bodily movements are necessary for killing. Only sticks and stones, and the bodily movements that move them, can break your bones; stillnesses, and simply willing that you be hurt, cannot harm you. Now of

[74] Hornsby, *Actions*, 13, 124–6.

course certain other items that are not willed bodily movements can cause harm, namely, natural events like falling human bodies and natural states like another person having a communicable disease. Yet causation of harm by the involuntary movements or states of a person's body are not the same as the *person* causing such harm. Otherwise we could perform the complex action of killing simply by being so ugly that we scare people to death.

With regard to the intuitive plausibility of the second proposition, recall the distinction Prichard made between the event that is referred to by a description, and the features a description may include as an aid to picking out that event.[75] If we apply that to 'killing', we may say that although the death of another is a feature of that description, that does not mean that the death of another is part of the event referred to. As Donald Davidson puts the point: it is a mistake to think 'that when the description of an event is made to include reference to a consequence, then the consequence itself is included in the described event'.[76] Something is a poison by virtue of the consequence it has for human beings when they ingest it; that does not mean that deaths or sickenings are parts of poisons. Analogously, an event is a poisoning if it is an event causing sickenings or death; that does not make the death/sickening state of affairs part of the poisoning; such death/sickening rather makes 'poisoning' one way to pick out the event in question, namely, by using one of the more dramatic effects of that event to pick out the event itself.

This isn't an argument *for* identifying an act of trigger finger movement with an act of killing, when death ensues. It is only the removal of a linguistic blinder to seeing how intuitively plausible is such an identification. Once the blinders are removed, we can appreciate the plausibility of the identification. If one didn't make such an identification, think of how many distinct acts we do when we do anything. When we move our finger, we also: move the trigger; shoot the gun; hit the man; kill the man; scare his neighbour; commit murder; etc. What is counter-intuitive is not that these are distinct *types* of acts that we do when we move our finger, for they are; rather, the sticky wicket is to say that these are seven different particular acts that we do at the very same time. This seems to make even the laziest of us very busy!

[75] Prichard, *Moral Obligation*, 190.
[76] Davidson, *Essays on Actions and Events*, 58.

If one finds these two propositions to be as intuitively plausible as do I, then the exclusivity thesis is equally plausible. For granting these propositions, we can say with Austin, Holmes, and Donald Davidson, 'mere movements of the body . . . are all the actions there are',[77] and *not* be committed to the absurdity of denying that killings, rapes, etc. are actions.

[77] *Essays on Actions and Events*, 59.

6

THE METAPHYSICS OF BASIC ACTS III: VOLITIONS AS THE ESSENTIAL SOURCE OF ACTIONS

1. CLARIFYING THE MENTAL-CAUSE THESIS

We now have to examine the 'mental' part of the Austin–Holmes theory of action. Acts are not plausibly identified with bodily movements *simpliciter*, but they should be identified with the complex event volitions-causing-movements. It is time we examined the volitional part of this identification.

As with the identity thesis, we will go some distance in defending this thesis if we first clarify what it does and does not say. To spell out this part of the theory will require examination of a considerable number of variables. To keep some semblance of organization, I shall discuss such variables around nine choices a volition theorist faces and about which she must make some decision.

1.1 Volitions as States or Agents?

The first choice is this: is a volition a datable state or event, as the name 'volition' suggests? Or is this a misleading way of referring to 'the will', a faculty within persons that has the power to cause change to occur in the world? This choice impacts on what sort of causal relationship we think to exist between volitions (or the will) and bodily movements: if 'volition' names datable mental states or events of a person, then the causal relation between volitions and movements should be of the ordinary, garden-variety kind that exists between natural events. If, however, 'volition' is taken to refer to a faculty of will that as an object causes bodily movements, then we must think that persons possess a kind of unique causal power. Such causal powers are often called 'causal agency' and are thought to be *sui generis* and distinct from the normal event-or-state-causation that characterizes natural events. Persons can just make things happen, on this view, and such exercise of unique causal powers is an unanalysable primitive.

Historically, these two conceptions go back to Hume and Locke, respectively. As a matter of history, it is pretty plain that the nineteenth-century proponents of the volitional theory of action chose Hume over Locke here. John Austin, for example, recognized that 'it is commonly supposed that there is a certain "will" which is the cause or author of both [volitions and movements]'.[1] But, said Austin, 'this same "will" is just nothing at all'.[2] Reciting straight David Hume on causation as merely the regular concurrence of events, Austin scorned 'fancied beings called "powers", of which the imaginary "will" is one'.[3]

History aside, there is good reason to eschew the Lockean version of the volitional theory. For we have seen a more thorough explication of the Lockean vision in our own time by Richard Taylor[4] and Roderick Chisholm,[5] a vision I too bought into in an earlier explication of action.[6] Austin was right: the vision is an empty one. It simply doesn't explain anything to posit the existence of a *sui generis* causal relation, and then to grace persons with a unique power to create such a relation. It reduces you to saying helpful things like 'A human action is something a person makes happen'.[7] As Donald Davidson put it, in rejecting the Lockean account: 'the concept of agent causation lacks these features [of ordinary causal explanations] entirely. What distinguishes agent causation from ordinary causation is that no expansion into a tale of two events is possible, and no law lurks. By the same token, nothing is explained.'[8] Indeed, how could one explain a datable event like an action with any notion of object- (rather than event-) causation? If one holds that 'a person causes . . .' is *not* elliptical for 'some event in the person causes . . .', how can a temporally located event like a killing be explained? As Carl Ginet has noted recently, 'if the cause of the mental

[1] Austin, *Lectures on Jurisprudence* i. 412.   [2] Ibid.   [3] Ibid.
[4] Taylor, *Action and Purpose.*
[5] Roderick Chisholm, 'Freedom and Action', in K. Lehrer (ed.), *Freedom and Determinism* (New York, 1966); id., 'The Agent as Cause', in M. Brand and D. Walton (eds.), *Action Theory* (Dordrecht, 1976).
[6] M. Moore, 'Responsibility and the Unconscious', *Southern California Law Review*, 53 (1980), 1563–1663: 1572.
[7] Ibid.   [8] Davidson, *Essays on Actions and Events*, 53.

occurrence is just me, just the enduring entity, and no event at all, then it cannot explain what it needs to explain'.[9]

I therefore put aside the 'person- (or will-)-as-primitive-cause' interpretation of the volitional theory. 'Volition' names a state or an event within the mind of an actor. When we talk of the faculty of will, or the activity of willing, or when we speak of wills or persons causing, I shall take it as elliptical for 'a volition caused . . .'.

1.2. Volitions as States or as Actions?

The second choice faced by the volitional theorist is between active and passive readings of 'volition'. Does 'volition' name an activity that a person performs in his mind? Or does it name a non-actional state that a person has in his mind but that he does not actively bring to mind? The distinction is between mental acts like trying to remember a name, and passively experienced mental states like the thoughts that just come to you.

Carl Ginet in his recent exposition of the volitional theory opts for the first of these alternatives. 'It is appropriate to characterize this volitional part of the experience of voluntary exertion as mental *action*.'[10] This is not to adopt the mental-action theory of act that we criticized in Chapter 5, as do mental-action theorists like Prichard, McCann, Hornsby, and Davis. Unlike these mental-action theories, Ginet's is a volitional theory because act is not identified solely with the mental activity of willing but, rather, with the causation by that activity of some non-actional, bodily events.

Still, there are problems that should lead us to construe the volitional theory away from Ginet's active interpretation. One is the problem Gilbert Ryle adverted to many years ago in his argument against mental-action theorists like Prichard: if every act requires a willing to be an act, and willing is itself an act, then mustn't there be a further act of willing the willing in order for that willing to be a act?[11] And if so, mustn't there be a third-order willing (to will to will) in order for the second-order willing to be an act? And so on. This once widely accepted objection of Ryle's became known in the trade as the infinite-regress objection.[12]

---

[9] Ginet, *On Action*, 13. [10] Ibid. 30.
[11] Ryle, *The Concept of Mind*, 67.
[12] This objection, for example, was repeated by Melden in his 'Willing', *Philosophical Review*, 69 (1960), 475–84: 476.

This is certainly not the knock-down winner of an objection that Ryle and his followers thought it was, for there is a variety of responses available to those in Ginet's position. One is to assert that acts of willing are unlike other acts in that, for this class of acts, there need be no willings causing them in order for them to be acts.[13] Another response is to identify willings to will with willings so that logically there can be no possibility of events of willing causing other events of willing.[14] The availability of such responses had led some mental-action theorists to think that Ryle's infinite-regress objection 'is now of historical interest only'.[15] Yet there is this much sting left to Ryle's objection: to define acts like moving one's hand in terms of duplicate mental acts of willing the movement seems an ontologically expensive way of gaining very little ground, for acting is still left unexplained. And if one goes on to explain the mental action of willing, to avoid Ryle's regress one will have to explain it in non-actional terms. Lawrence Davis, for example, gives a functional specification of acts of willing in terms of their causal roles.[16] And if one is willing to give such a non-actional account of the act of willing, why put in the extra act (i.e. of willing)? Why not just give that non-actional account of the act of moving one's hand to start with? I conclude that Ryle's objection has a sting to it quite removed from what he thought it had: the objection becomes an objection to the non-reductive nature of mental-action theories. Either such theories don't explain much (if acts of willing are not themselves further explained in non-actional terms); or such theories give such non-actional explanations too late in the sense that they posit a superfluous mental act between what are unproblematically parts of acts (like the moving₁ of one's arm) and the non-actional events that cause them. Better to take 'volition' to name those non-active mental states to begin with, eliminating the unneeded *acts* of will.

---

[13] See e.g. Hugh McCann, 'Volition and Basic Action', *Philosophical Review*, 83 (1974), 451–73: 472. As Myles Brand notes, this response has the appearance of being a bit *ad hoc*. Brand, *Intending and Acting*, 13.

[14] For this familiar response, see e.g. Gilbert Harman, 'Willing and Intending', in R. Grandy and R. Warner (eds.), *Rationality* (Oxford, 1986), 372.

[15] Davis, *Theory of Action*, 39.

[16] Ibid.

1.3. Volitions as Desires, Beliefs, Intentions, or Choice/Willings?

The third choice facing the volitional theorist is how he should conceive of these non-actional states named volitions. Are they species of more familiar kinds of mental states, such as beliefs, desires, or intentions? Or are they *sui generis* states unlike any of these more familiar inhabitants of our mind?

One can find volitionists occupying quite a variety of each of these positions. Some have urged that volitions are a combination of both beliefs and desires. Early in the last century Joseph Buchanan, for example, urged that volitions consisted of 'the association of a muscular action with the energic conception of that action and a predominating desire to perform it'.[17] Later in the century William James chose the belief part of Buchanan's definition in framing his own well-known 'ideo-motor' theory of action initiation; on this theory, volitions are simply beliefs or thoughts whose content (or 'image') is the motor movement they cause.[18] This construal of volitions as beliefs has continued to find favour in contemporary psychology, in which 'ideas' or 'images' of motor movements are taken to be the immediate mental causes of those movements.[19] Such construal also has many adherents in contemporary philosophy of action, where the state that initiates actions is conceived of as a belief of some kind.[20]

In contrast to any of these views, John Austin conceived of volitions as a kind of desire or wish.[21] Austin carefully distinguished the desires that were volitions from those that were not in terms of their objects: a volitional desire had as its object the immediate act of moving one's limbs, whereas desires that were not volitions had as their objects further consequences the actor

[17] Joseph Buchanan, *The Philosophy of Human Nature* (Richmond, Ky., 1812), 300.

[18] William James, *Principles of Psychology* (New York, 1890), vol. ii, ch. 26.

[19] See Alvin Goldman's summary, in 'The Volitional Theory Revisited', in M. Brand and D. Walton (eds.), *Action Theory* (Dordrecht, 1976), 77. See also Kimble and Perlmuter, 'The Problem of Volition'.

[20] I refer to those philosophers who think that intentions initiate actions, and then analyse intentions in terms of beliefs. Some urge that the relevant beliefs are predictive beliefs about one's own future behaviour (see Brand, *Intending and Acting*, 148–52, for a discussion); others such as Donald Davidson focus on beliefs about the desirability of an action as being the intention that initiates it. Davidson, *Essays on Actions and Events*, 100.

[21] Austin, *Lectures on Jurisprudence* i. 419.

sought to achieve by moving his limbs. Thus, my desire to move my arm was a volition for Austin; my desire to signal a left turn was not a volition because its object was not an immediate one but a removed object I sought to accomplish by moving my arm.

What Austin did not do is draw any distinction in the *nature* of the states that were volitions. Volitions were just like desires in terms of the kinds of states they were; they differed only in the kinds of *object* they took. Austin did think that volitions were accompanied by beliefs: every volition is accompanied by an expectation or belief that the bodily movement wished will immediately follow the wish.[22] Yet unlike James and Buchanan, for Austin the belief is neither identical to the volition nor a part of the volition.

Among contemporary volitionalists, Alvin Goldman appears to have signed on to Austin's conception of volitions as a kind of desire, want, or wish. For Goldman distinguishes 'occurrent wants' from 'standing wants', and then argues that only occurrent wants are the causes of bodily movements.[23] With some hesitation, Goldman then identifies volitions as occurrent wants whose propositional contents describe just the bodily movement that they cause.[24]

Many volitionalists conceive of volitions as being neither beliefs or desires but, rather, as being a kind of intention. Both Bentham and Mill speak of volitions as being intentions to do acts like moving one's fingers,[25] and in this they have been seconded by many more contemporary volitionalists.[26] As with

---

[22] *Lectures on Jurisprudence* i. 412.

[23] Goldman, *A Theory of Human Action*, 86–8.

[24] Goldman, 'The Volitional Theory Revisited', 68–77. Compare id., *A Theory of Human Action*, 88, 91–2, where Goldman wishes to distance his theory of occurrent wants from volitions.

[25] Bentham, *Introduction to the Principles of Morals and Legislation*, ch. 8, §2; Mill, *A System of Logic*, 35.

[26] See Wilfrid Sellars, 'Thought and Action', in K. Lehrer (ed.), *Freedom and Determinism* (New York, 1966), id., 'Volitions Reaffirmed', in M. Brand and D. Walton (eds.), *Action Theory* (Dordrecht, 1976), id., 'Actions and Events', where Sellars conceives of volitions as intentions whose contents form the conclusion of practical inferences and which themselves cause bodily motions; John Searle, *Intentionality* (Cambridge, 1983), 95–6, and George Wilson, *The Intentionality of Human Action*, 2nd edn. (Stanford, Calif., 1989), each of whom conceives of volitions as 'intentions in actions'; Michael Bratman, *Intention, Plans, and Practical Reason* (Cambridge, Mass., 1987), 30–134, who speaks of both 'present intentions' and 'endeavourings' as immediate causes of actions; Brand, *Intending and*

Austin's and Goldman's treatment of volitions as desires, those who conceive of volitions as a kind of intention must employ two distinctions. The first is that between what is sometimes called 'bare', 'future', 'present', or 'pure' intentions, on the one hand, and what are often called 'further intentions' or 'intentions-with-which', on the other.[27] A bare intention is an intention to do a future act; a further intention is identical with these instrumental desires that motivate actions, for this is the further end an actor has that motivates him to perform some action. It is my intent to go downtown that is a bare intention; the intent to get a haircut that motivates my going downtown is a further intent. It is the first sort of intention that is relevant to volitions, for volitions are conceived of as a kind of bare intention.

The second distinction is in terms of the propositional content of bare intentions, for volitional intentions are to be distinguished from other intentions in terms of their differing propositional contents. Volitional intentions have as their object a description of a bodily motion; non-volitional intentions are those intentions having as their object some complex action to which some bodily motion is a means. My intent to go downtown is a non-volitional intention; my intent to move my leg is a volitional intention.

Other volitionalists do not believe that volitions are happily conceived as species of desire, belief, or intention. The late Alan Donagan chose 'choosing' as the name for the mental states that uniquely initiate action, whereas Gilbert Harman, Paul Grice, and Brian O'Shaughnessy each talk of 'willings' as distinct from intendings as this cause of actions.[28] It is unclear how distinct such theorists believe willing or choosing to be from intending. A clearer example of this variant of the volitional theory is provided by the mental-action theorist H. A. Prichard, who held that 'this character [of willing] is *sui generis* and so incapable of

---

Acting, who uses the phrase 'immediate intention' to refer to volitional causes of actions.

[27] See G. E. M. Anscombe, *Intention*, 2nd edn. (Ithaca, NY, 1963), 1–3; Davidson, *Essays on Actions and Events*, 88.

[28] Alan Donagan, *Choice: The Essential Element in Human Action* (London, 1987); O'Shaughnessy, *The Will*; Harman, 'Willing and Intending'; H. P. Grice, 'Intention and Uncertainty', *Proceedings of the British Academy*, 57 (1971), 263–79.

being defined, i.e. of having its nature expressed in terms of the nature of other things.'[29]

The choice between these competing conceptions of volition is not a minor matter, for it determines the nature volitions can plausibly be thought to have. Beliefs are cognitive states, and if volitions are a kind of belief then they too share this nature; desires and further intentions are species of the 'pro attitudes'[30] that motivate actions, and if volitions are a kind of desire they would share this motivating nature; bare intentions are neither cognitive not motivational states, but operate as executors of our background cognitive and motivational states,[31] and if volitions are a species of bare intentions they will share this executory nature. Finally, if volitions are none of these three kinds of states but have their own unique nature, then we shall have to say what that nature might be and how such *sui generis* states are related to the more familiar states of belief, desire, and intention.

For reasons that will become fully apparent only later, the most plausible construal of the mental-cause thesis is to take 'volition' to name a species of intention. I make this construal because of the *im*plausibility of the alternative construals. As I argue in more detail in the succeeding major subsection of this chapter, it is not plausible to treat volitions as beliefs of any kind; for believing that I will do something, or believing it desirable to do it, or believing that a certain act-type exists, are all quite compatible with my not having yet decided to so act and with my not in fact so acting at all. It is also not plausible to treat volitions as wants of any kind; as we shall see the best reason to believe in the existence of volitions lies in the executory role such states play, executing our motivating wants and their accompanying beliefs into the actions that serve them. We describe this unique function of volitions better if we do not collapse them into the background motivational states of desire which they execute.

'Choice', 'decision', 'willing', and 'trying' capture this executory role rather well, and there is much in the work of those who

---

[29] Prichard, *Moral Obligation*, 189.

[30] Donald Davidson's phrase to cover all motivational states in practical reasoning. Davidson, *Essays on Action and Events*, 3–4.

[31] A point I shall argue for later. See generally Bratman, *Intention, Plans, and Practical Reason*.

so characterize volitions with which I agree. Indeed, in so far as these are taken to be synonymous ways of referring to that sub-class of bare intentions that executes our more general plans into discrete bodily movements, the disagreement is merely about the least misleading label with which to refer to a state whose nature is agreed upon. To the extent such labels purport to refer to a state other than the executory bare intention just described, how-ever, they lessen the plausibility of the mental-cause thesis. For then the thesis would require us to conceive of there being four types of Intentional states, not just those of desiring, believing, and intending. The worry here is not only one of ontological par-simony; it is also the worry about the characterless nature of this fourth type of state, 'conational' states of 'willing'. One of the old objections against volitions has been that they have no char-acter other than being 'the unique cause of actions', and thus that positing them is only a move of theoretical desperation.[32] Conceiving of volitions as a kind of intention helps to blunt this objection, for intentions are a familiar and well-accepted type of mental machinery.

Happily we have no need to resort to a fourth kind of mental state in order to think of volitions, for bare intentions fit the bill nicely. As we shall see, in order to execute even simple practical syllogisms, in order to resolve conflicts between our desires and between our beliefs, in order to account for the phenomenon known as weakness of will, and in order to save us from being 'Sartrean persons' (where everything is always up for grabs at all times), there must be states of bare intention. Such states execute our background motivational and cognitive states into actions, do so even when our desires or beliefs are in conflict, account for our failures to execute such beliefs or desires (as in cases of akrasia, or weakness of will), and project our resolutions in these matters into the future so that we need not redecide such matters again and again. Volitions fit into this executory role of bare intentions very neatly, for they are simply such bare intentions having as their objects the simplest bits of bodily motion that we know how to do. Volitions are simply the last executors both of our more general intentions and of the background states of desire and belief that those more general intentions themselves execute.

---

[32] See Taylor, *Action and Purpose*, 68–9; Ryle, *The Concept of Mind*, 64–5; Melden, 'Willing', 477–8.

Michael Bratman, who has described this executory role of intentions with exceptional clarity, none the less doubts whether volitions (which he calls 'endeavorings') are truly intentions in any but a weak sense of the word.[33] Bratman's worry is that volitions do not share with ordinary intentions two demands of rationality. The first might be called the demand for intention–belief consistency: I cannot rationally intend to hit a target while I believe it is impossible for me to hit the target. The second might be called the demand for agglomerativity of intentions: if I intend to hit target 1, and by the same shot also intend to hit target 2 (which stands behind target 1), then I must intend to hit targets 1 and 2.

Bratman's worry is that intentions in the strong sense satisfy these two principles, and the fact that volitions do not shows that they are intentions only in a weak sense. The sort of example Bratman constructs to show that endeavourings do not satisfy these principles is well-known.[34] Suppose my physiology is such that I can (with great difficulty and only occasional success) wiggle my left ear; I have the same capacity with respect to my right ear; it is, however, impossible for me to wiggle both ears simultaneously, although the effort to do so does not interfere with my chance of success in wiggling one of my ears; and I know all of this. Suppose further that I must wiggle an ear to give a prearranged signal, but it does not matter which ear I wiggle. I therefore decide to maximize my chances of success by trying to move each ear at the relevant time. If I succeed in wiggling my left ear, that ear-wiggling is an action of mine, and, on the volitional theory, I must have willed (or endeavoured) to move it; similarly, however, I must also have willed (or endeavoured) to move my right ear. Yet it cannot be true that I willed (or endeavoured) to move my left and my right ear, for I knew that to be impossible. Volitional states thus do not satisfy the demands of belief–intention consistency and agglomerativity of intentions.

Interesting as it is, Bratman's worry ultimately should not convince us that volitions are not to be conceived as a species of bare intention. For other sorts of bare intention do not satisfy

[33] Bratman, *Intention, Plans, and Practical Reason*, 130.

[34] Ibid. 113–19, 130–5. I have changed Bratman's example in order to talk of volitions where he talks of intentions and endeavourings with more remote objects than bodily movements.

these two demands of rationality either: I can intend to hit a target even though I know I can't hit it;[35] and I can rationally intend to hit target 1 and intend to hit target 2, even though I do not intend to hit both targets 1 and 2.[36] On the other hand, intentions are subject to a demand for intention consistency: I cannot rationally intend to hit a target and intend that I not hit the target. But volitional intentions are subject to this same demand of rationality: I cannot rationally will that my ears move, and also will that they do not move.

1.4. Volitions as Intentional or as Non-Intentional Mental States?

The fourth choice facing the volitional theorist is whether to characterize states of volition as Intentional states or not. A mental state is an Intentional state if it takes an object (or 'has content'). I do not just desire, believe, hope, or intend; I desire *that* something be the case, believe *that* something is the case, etc. Whatever comes after the 'that' is the object (or content) of Intentional states.

Not all mental states are Intentional, e.g. states of mood or sensation,[37] and it is possible that volitions are non-Intentional

---

[35] See generally Rogers Albritton, 'Freedom of Will and Freedom of Action', *Proceedings and Addresses of the American Philosophical Association*, 59 (1985), 244. In specific criticism of Bratman, see Wilson, *The Intentionality of Human Action*, 268; Hugh McCann, 'Rationality and the Range of Intention', *Midwest Studies in Philosophy*, 10 (1986), 191–211. I also discuss this point briefly in the next major section of the chapter.

[36] Bratman considers this possibility but rejects it because it flies in the face of the co-ordinating role intentions play in executing our larger plans. Bratman, *Intention, Plans, and Practical Reason*, 117. Yet we can maintain the co-ordinating role for intentions if they usually (but not in these recherché cases) agglomerate as the demand for agglomerativity requires. Bratman's scenarios are, after all, very contrived and rare so that the demand for agglomerativity can *almost always* hold.

[37] If I am wrong about this, so much the better for my Intentional construal of volitions. On whether Intentionality is the mark of the mental, see William Lycan, 'On Intentionality and the Psychological', *American Philosophical Quarterly*, 6 (1969), 305–11.

[38] Myles Brand splits the difference here, urging that some volitions are Intentional but that others are not. Brand's thought here is that some actions (like tugging on one's moustache) are not filtered through any high-level plans, and therefore the 'immediate intention' (or volition) does not execute any such plans; from this he infers that such volitions need themselves have no representational content. Brand, *Intending and Acting*, 241. By contrast, I would think that even these intrinsically motivated volitions need content in order to guide movements. Such volitions are, in such cases, very small plans.

too.[38] To take this line, however, would be to abandon the clues that lead us to posit the existence of volitions to start with. As we shall see, it is largely the inability of states of belief and desire to rationalize our acts that lead us to posit the executory state of volition. To fulfil this cognitive function, such states had better have cognitive content. Of course, we might believe that there is an immediate cause of actions, that is unique to action, without such phenomenal clues. But, absent such clues, why would we posit such a state? Intermediate causes, as Alan Donagan notes, are always ripe candidates for elimination because unneeded in parsimonious explanations.[39]

1.5. The Objects of Volitions: Propositions or Real-World Things?

If volitions are best construed to be Intentional states, we next need to enquire as to their Intentional objects. This involves us in three enquiries, pursued in this and the succeeding subsections. The first enquiry is the most general: when we say that Jones willed or intended to move his fingers, how are we to take the object of his intent? Is it a thing in the world, the event or action of the moving of the finger? Or is it a description or concept of this thing?

As with all Intentional states, statements about volitions usually create what are variously called 'opaque' or '*de dicto*' contexts. An opaque context is one where the reference is unclear (or 'opaque') rather than being clear (or 'transparent').[40] 'I believe that Jones is a spy', 'I desire that Smith not buy that hat', and 'I intend to win the award', all typically create opaque contexts; for the reference of 'Jones', 'Smith', 'that hat', and 'the award' is not simply or straightforwardly to Jones or Smith, or some particular hat or award. Words that occur in the objects of these mental-state constructions refer only 'indirectly' to their respective entities, but directly to some description or concept about these entities that the believer, desirer, or intender has in mind. One can see the indirectness of the reference by substituting other concepts or descriptions that refer to the same thing and then

---

[39] Donagan, *Choice*, 146.
[40] The classical discussion of this is in Willard Quine, *Word and Object*, 141–56. For a discussion of Quine's views on opacity, see M. Moore, 'Foreseeing Harm Opaquely', in J. Gardner, J. Horder, and S. Shute (eds.), *Action and Value in Criminal Law* (forthcoming, Oxford, 1993).

enquiring into the truth value of the overall expression thus created. Suppose, for example, the award in question is the award given to the slowest learner in the class, but I do not know that about the award, thinking it to be given to the fastest learner. I do intend to win the award, but I do not intend to win the award given to the slowest learner. I thus intend to win the award only under some descriptions of it but not others.[41]

Some have urged that statements about volitions may create 'transparent' (or '*de re*') contexts,[42] and in certain usages this is true. For example, 'The agent intended his action of *moving his index finger* to signal the firing of the cannon'. The italicized phrase 'moving his index finger' picks out an action, but it does so transparently; for we can substitute any other phrase referring to that action without changing the truth value of the overall expression in which it occurs. We can say, for example, that the agent intended his action of lifting the finger to the inside of his ring finger to signal the firing of the cannon, and not have changed the truth value of the sentence.

That such transparent constructions can be created with 'intent', 'will', or other volitional words should occasion little surprise. For we can also create such transparent usages of all mental-state verbs.[43]

1.6. *The Objects of Volitions: Actions or Movements?*

The second enquiry about the objects of volitions is whether or not the propositional content of a volition uses movement$_T$ descriptions (that is, action descriptions) or movement$_I$ descriptions (bodily-event descriptions that are non-committal about the event being an action). It is no good attempting to finesse this choice by objecting, 'But the particular movement$_I$ that is the object of some particular volition *is* a movement$_T$, so there is no choice to make'; because Intentional objects are descriptions (not

[41] On the opaqueness of the objects of intentions specifically, see M. Moore, 'Intentions and *Mens Rea*', in R. Gavison (ed.), *Issues in Contemporary Legal Philosophy* (Oxford, 1987).

[42] Davidson, *Essays on Actions and Events*, 84–5; Wilson, *The Intentionality of Human Action*, 11–13, 120.

[43] That fact does not alter in the least the fact that volitions take as their objects propositions. 'Jones intends (wills, chooses, etc.) to move his finger' is opaque, because the intending takes as its object the description 'move his finger', not the event and object themselves.

the real-world things to which they appear to refer), so that the descriptions 'movement$_T$' and 'movement$_I$' are different objects of a propositional attitude.

Austin did not decide this matter with any conscious thought. When he speaks of willing 'the movement', he does not choose which object he thinks volitions to have.[44] Interestingly, however, when he discusses particular examples, he chooses the non-actional event description, as in 'I wish [will] that my arm should rise'.[45] This is the right way to go here for the volitional theorist. For the alternative is to say that the objects of volitions are action descriptions, as in 'I will my raising of my arm'. And if the object of volitions were action descriptions, then the volitional theory would be guilty of epistemic circularity. For such a theory analyses 'action' as 'the volitional causing of bodily movement$_I$', yet if the volitional component is a volition *to do an action*, the account is viciously circular.[46]

1.7. The Objects of Volitions: Internal Bodily Events, Bodily Motions, or Remote Consequences?

The third and final variable concerning volition's object that requires clarification is what sort of non-actional description we should take to be volition's object. There are three possibilities here. We might think that willing, like (non-simple, or non-immediate) intending, can take any non-actional descriptions as its object. I might as easily will your death as the movement of my finger that causes is. Bentham used 'will' in this way,[47] and it is a perfectly idiomatic usage of the word.

The second possibility is that we restrict the objects of willings to those simple bodily movements$_I$ that correspond to basic-act descriptions. This was explicitly John Austin's view: 'A consequence of the act is never *willed*,' although such a consequence may be, in Austin's terminology, intended.[48] The only objects of willings for Austin were those immediate bodily motions by which I bring about further consequences like trigger movements,

---

[44] Austin, *Lectures on Jurisprudence* i. 412.   [45] Ibid. 413.

[46] As noted by Prichard, *Moral Obligation*, 190–2; Searle, *Intentionality*, 85; Brand, *Intending and Acting*, 11; Donagan, *Choice*, 87; and Davidson, *Essays on Actions and Events*, 78–80.

[47] Bentham *Introduction to the Principles of Morals and Legislation*, 82.

[48] Austin, *Lectures on Jurisprudence* i. 421.

bullets flying, and deaths.

The third possibility is to restrict the objects of willings to those internal events that are constituents or causes of bodily movements. According to a volitional theorist like Carl Ginet:

> I do not will to move my body. The content of volition is not concerned with movement, only with . . . directed force at the movement (a momentary force vector). . . . Movements take time and are properly the objects of intentions rather than volitions. Volitions do not plan ahead, not even very slightly. They do not plan at all; they execute.[49]

Of these three possibilities, we can put aside the third rather quickly on long-familiar grounds. We do not know enough about the internal workings of our bodies to will 'force vectors' or anything like the internal levers and pulleys that Ginet hypothesizes. A. I. Melden was perhaps the first of many ordinary-language philosophers to make this point. 'Surely I must know which muscles to move,' Melden argued, if I can properly be said to move them as an action (or, as I would say, will them to move).[50] Finding that we lack such knowledge, Melden concluded: 'We do not move our limbs by manipulating any sorts of interior levers or pulleys within the body.'[51]

The relevance of this old criticism is this: Melden's epistemic insight, while not by itself yielding his metaphysical conclusion, none the less provides the basis for an inference about the likely objects of volitions. What we know we can do—not by doing something else, but simply by doing it—is a good phenomenal clue to the objects of our willings. We could of course be mistaken about this; yet, as I have argued elsewhere, our first-person awareness generally provides evidence of the nature of the underlying states.[52] Such surface phenomenal (and linguistic) features do not *constitute* those states, as Melden and ordinary-language philosophy often assumed; but it is to be taken seriously as we search for metaphysical natures. What we are searching for here is where the mental operations involved in practical reasoning—belief–desire sets and the willings that execute them—'hook on to' the only physical means of changing the world that we have at our disposal, our bodies. What we are aware of directly willing

---

[49] Ginet, *On Action*, 33.   [50] Melden, *Free Action*, 59.
[51] Ibid. 63.   [52] Moore, 'Mind, Brain, and the Unconscious'.

is a promising stab at what objects our volitional states do have. As Arthur Danto points out, 'We all know, in a direct and intuitive way, that there are basic actions, and which actions are basic actions.'[53] We do not have a similar awareness of Ginet's sort of internal 'force vectors' or of any other sub-personal mechanisms involved in the physics of bodily motions.

Our choice is thus between the first two possibilities. What motivated Austin to take Bentham to task for 'misusing' the word 'will' so as to license willed consequences of movements as well as willed movements? Austin was certainly not accusing Bentham of this kind of scientific mistake: 'Bentham thinks that we can will the death of another just like we can will our arm rising, viz., we will it and it happens just like that.' Bentham did not think that our wills had such magical powers to 'act-at-a-distance', nor was that Austin's accusation. Rather, Austin thought that something in the nature of the state he called willing ('volition') dictated that its object be restricted to those items we non-instrumentally cause (bodily movements) and that that object not include those events we instrumentally cause by moving our bodies.

Austin's only stated reason for this restriction (of the permissible objects of willing) is that such an object restriction was his only way of distinguishing volitional desires from the more general desires that motivate them. Austin concluded that 'bodily movements . . . are the appropriate objects of volitions' because that gave him the 'true test' for how volitions differed from other desires: 'they are the only desires immediately followed by their appropriate or direct objects'.[54]

Our discussion in the next major section of this chapter will show that this isn't much of a reason for the restriction, since volitions can (and should) be distinguished from desires by the *nature* of the state, and not (only) by their differing *objects*. There are better reasons to abide by Austin's restriction.

One such reason stems from the kind of clue our phenomenal experience gives us. Just as we do not have the experience of willing any sub-personal physical components of bodily motions such as Ginet's 'force vectors', so we do not have any experience of willing deaths as such. True, we will movements that cause deaths, but that is to reinforce Austin's restriction, not to rebut it. And, as before, our phenomenal experience should be taken as

<hr>

[53] Danto, 'Basic Actions', 145.   [54] Austin, *Lectures on Jurisprudence* i. 423.

a good clue to underlying structure, giving us some reason to restrict the objects of willing to the movements of our bodies.

The argument from phenomenal experience my seem weaker here than it was against Ginet. The objection would be that most of the time we only experience the willing of some more complex state of affairs of which our bodily movements—many of them—will be components. Consider activities such as playing the piano or speaking an English sentence. The objection is that the last mental state of which we are aware that executes our more general desires—our willings—has a more general object than 'movement of this finger now'. Its object is 'keeping a melody' or 'expressing an idea'.

Undeniably we learn to string together various of our bodily movements into complex routines with such dexterity that, once we have mastered the routine, we can literally not pay attention to what we are doing (at the level of bodily movements).[55] I do not pay attention most of the time to how I move my fingers to play a piano sonata or to how I move my tongue in order to say 'The cat is on the mat'. I just play the one and say the other. Yet, the phenomenal clues we seek are not only to be sought at the level of trained, adult awareness. Rather, we should ask, how did the pianist and the English-speaker acquire their complex skills? What were the simplest objects of their attention then? And to learn how to play the piano is to learn (among other things) how to move one's fingers; to learn to speak English is, among other things, to learn how to move one's tongue.[56] Remembering that we seek phenomenal clues to the nature of those mental states that execute desires into action, a good place to look for such clues is at the mental states we had to acquire in order to acquire our various skills of action. The objects of those states were bodily motions. It is a reasonable (although far from inevitable) hypothesis that those same states exist to cause those

[55] See Richard Jung, 'Voluntary Intention and Conscious Selection in Complex Learned Action', *Behavioral and Brain Sciences*, 8 (1985), 544–5; Baier, 'The Search for Basic Actions', 166; J. Ripley, 'A Theory of Volition', *American Philosophical Quarterly*, 11 (1974), 141–7.

[56] Jung, 'Voluntary Intention and Conscious Selection in Complex Learned Action'; Wilfrid Sellars, 'Metaphysics and the Concept of a Person', in Karl Lambert (ed.), *The Logical Way of Doing Things* (New Haven, Conn., 1969), 244; Goldman, 'The Volitional Theory Revisited', 71–2. See also J. Anderson, *The Architecture of Cognition* (Cambridge, Mass., 1983), for a description of the mental states involved in learning how to knot a tie.

same motions when they occur later in life as part of speaking or playing, even though awareness of those states has receded.[57]

1.8. Volitions as Functional States, Physical States, Behavioural States, or Irreducibly Mental States?

The eighth choice faced by the volitional theorist is about the ultimate nature of volitions: are they (ultimately) brain states? some irreducibly mental states, existing in time but not in space? dispositional states, where the disposition is to engage in certain behaviours? some functional state of the brain but not (essentially) itself a brain state? How one answers this question is fully dictated by the nature one thinks mental states in general (not just volitions) possess. Respectively, physicalism, dualism, behaviourism, and functionalism are the names for the general views just stated.

Implicit in my foregoing exposition of volitional states is a supposition that mental states generally are functional states. A functional state is a state whose essential nature is specified by the functional roles such a state plays in causing, and being caused by, other states and events.[58] 'Pain', on this view, names a state specified by its role in causing certain behaviour, being caused by certain stimuli, and entering into a variety of relationships to other mental states such as beliefs and intentions.

Such functional states are physically realized in the physical system that is our brain. Pain, for example, may for us have the structure of C-fibre stimulation in the central cortex of the human brain. Other creatures may have pain, however, and have

---

[57] The consideration from phemenology is not the only reason one might have for construing the object of volitions to be more remote consequences, not bodily motions. Michael Bratman, for example, shows in detail how we need interlocking intentions for the future ('plans') in order to execute our background belief–desire sets into actions. He also urges that such future intentions are executing states not reducible to beliefs or desires, and that further executing states he calls 'present intentions' are needed. Significantly for the present issue, Bratman urges that present intentions cannot be restricted in their objects to the bodily motions we are now bringing about because that makes them insufficiently rational to fit with our more general future intentions. Bratman, *Intention, Plans, and Practical Reason*, 134–5. Yet we could grant all of this, and still think that there must be yet another executory state, a volition, that does have such a restricted object.

[58] The description of functionalism that follows in the text is a condensed version of Moore, 'Mind, Brain, and the Unconscious', where more details and references to the literature are available.

such functional state realized in quite different structural features. It is this variation in physical realization, but invariance of functional role, that makes such states functional (and not straightforwardly physical) states.

Depth metaphysical functionalism—the kind of functionalism which I generally think to be true—holds that the nature of such functional states is subject to further and further scientific refinement. It is not, in other words, a matter of linguistic or conceptual discovery that volitions have certain functional roles. Neither phenomenal experience of such states, nor well-established usage of their names, can yield necessary or conceptual truths about the essence of such states. Such roles are a matter of scientific discovery as we discover, among other things, more and more about possible functional roles given what the physical system's structure can do.

In general terms, volitions are specified by the role they play in proximately causing bodily motions and in being the effects of both our more general intentions and the belief–desire sets the latter execute. Volitions are mediating states, and what they mediate between are out motivations and our intentions, on the one hand, and our actions, on the other.

Functionalism is not only a metaphysical position about the nature of mental states. It is also a programme for research into the nature of mind. The functionalist's research strategy for seeking to discover the physical realization of states like volitions is what is called 'homuncular functionalism'. The strategy works like this: one starts with mental states we do experience, as I have argued we experience volitions; one then hypothesizes how such states themselves might be subdivided into simpler operations of specialized functions that together can do what a volition (as one whole state) can do. One might, for example, divide willing into an initiating function and a 'gating' function, the latter being a last-minute control function.[59]

The hope eventually is to subdivide by functions until one reaches a level of simplicity in functioning for each 'homunculus' that is so simple that the relevant physical equipment can do it. A 'homuncular switchman', for example, would have as 'his' task

---

[59] Suggested in Benjamin Libet, 'Unconscious Cerebral Initiative and the Role of Conscious Will in Voluntary Action', *Behavioral and Brain Sciences*, 8 (1985), 529–39.

simply turning on and off a two-valued switching mechanism. A physical structure like the brain—with the two-valued (positive, negative) ionization potential in its cells—can do such simple tasks and we can at this level thus eliminate any metaphorical reference to a 'switchman'.

The trick in such a research strategy is to keep approaching the nature of a mental state like volition both functionally and structurally: the more we know of the physical structure of the brain, the more we know about possible ways in which its parts might be specialized in their functions; the more we know about the functional subroutines that underlie our experience, the more we know what to look for in the physical features of the brain. Approaching the nature of volitions from physical structure is often called a 'bottom up' approach; starting with our experience and subdividing it functionally is often called the 'top down' approach.

All of this is a familiar and well-regarded position in the philosophy of mind. I raise it not only to flesh out a bit the nature of volitions, but also to defuse some large-scale objections to volitions in post Second World War philosophy. As we shall see in the last major section of this chapter, a functional view of volitions cuts the ground from under several large objections that convinced an entire generation of philosophers (and two generations of legal theorists) to have no truck with volitions, or 'mysterious acts of will'. I shall defer defusing these objections until after we have outlined why we might believe there are such mental states as volitions.

1.9. Volitions as States of a Whole Person, or as States of Subpersonal Homunculi

Before coming to the arguments for and against the existence of volitions, we have one more interpretive choice to make. This choice can be seen only if we attend to a distinction familiar to homuncular functionalists. This is the distinction between the personal versus the sub-personal levels of explanation.[60] At the personal level the mental states of belief, desire, and intention

---

[60] For an introduction to the person/sub-personal distinction, see Dennett, *Content and Consciousness*, 93–6. In light of Dennett's more recent pragmatic construal of mental states, he would hardly agree with my statement of the distinction in the text.

that we refer to in explaining human action are to be taken literally, which is to say, realistically. The states involved, in other words, are thought to really be states of desire, belief, and intention that persons really have. At the subpersonal level, by contrast, our talk of desires, beliefs, or intentions of subagencies like dream censors, information pre-processors, or translators, has a cast of metaphor to it. We don't really think that dream censors can *intend* to deceive us or that information pre-processors have *beliefs* about what is relevant, but we may think it useful to pretend that they do. Such pretence is dropped when we can cash out the metaphor in terms of a literal description of the physical structures and their functions that make it possible for us whole persons to have states of belief, desire, and intention.

It may be tempting to construe volitions as sub-personal states, for doing so relieves us of any obligation to present the normal evidence (from phenomenology and behaviour) we need to verify the existence of true mental states.[61] Yet such a construal would render the volitional theory trivial. It surely is the case that there is some state (or disjunctive set of states) that triggers motor movement, and it is almost as obviously likely that such states differs when the motor movement is an action from when it is not. The interesting (if more controversial) volitional thesis is that there are such voluntary motor-initiating states that are truly states of mind of the person. Such states of volition then have to be argued for at the same level at which we argue for the existence of belief, desires, and (other) intentions. This puts volitions in the company of those other states involved in practical inference, states that have been studied since the Greeks as part of the common-sense (or 'folk') psychology of rational persons.

2. THE ARGUMENT FOR VOLITIONS

2.1. Introduction

With the mental cause thesis clarified, we now have to enquire into what reasons we might have for believing it to be true. What

---

[61] Myles Brand appears to succumb to the temptation referenced in the text in so far as he construes what he calls the 'cognitive component' (content?) of what he calls 'immediate intention' (volition) to be inaccessible to conscious functioning and, for that reason, not to be capturable by a folk-psychological concept. Brand, *Intending and Acting*, 154.

reason is there to believe that there are mental states of volition? Like all ontological questions, this one too obliges us to identify other items that we are relatively certain exist and to ask, does the best explanation of their existence also require that volitions exist? This is the 'non-foundationalist' or 'inference to the best explanation' strategy for seeking to reveal and to justify our ontological commitments that we utilized briefly in Chapter 4 (where I argued that events exist despite their supervenient status).

It is very plain from the literature on volitions that the basic thing that gives rise to the inference (that there must be volitions) is the existence of human actions as a natural kind. Our sense has long been that 'something' is indeed 'left over' if we subtract the fact that our arm rises from the fact that we raised our arm, to paraphrase Wittgenstein's famous observation.[62] That something is the volitional causation of the bodily-movement$_I$ event of the arm rising.

What is and what is not a natural kind is not a matter that we can know a priori. Whether jade is a natural kind, two natural kinds, or no natural kinds, is a matter of contingent fact about which we must develop a theory. Just because we have a *concept* of human action that differentiates it from our concepts of other events, therefore, is no guarantee that there is a natural kind that corresponds to that concept.

The conclusion that human actions form a natural kind is a plausible one. We have a first-person awareness of active control when we act that we do not have when we are merely 'acted' upon by forces beyond our control. This is the 'actish phenomenal quality' noted by Carl Ginet recently, and by many others in the past.[63] We also readily sense the difference between the actions of other persons, like their arm being raised by them, and

---

[62] Wittgenstein, *Philosophical Investigations*, §621. For discussions of the volitional theory beginning with this intuitive insight of Wittgenstein's, see e.g. G. N. A. Vesey, 'Volition', *Philosophy*, 36 (1961), 352–65; Melden, 'Willing'. As Alvin Goldman notes, 'the primary aim of the theory [of volitions] is to distinguish voluntary from non-voluntary events'. 'The Volitional Theory Revisited', 67–8.

[63] Ginet, *On Action*, 11–13. The famous neurologist Penfield induced movements in various of his subjects by placing electrodes on their cerebral cortices. Despite the subjects' not being aware of the stimulation, but only of their movements$_I$, Penfield observed that 'the patient will often say in the end, "you made me do that." He recognizes the fact that it is not a voluntary movement of his own.'

other events involving their bodies, such as their arm going up as a reflex. In the familiar hyperbole of Oliver Wendell Holmes: 'Even a dog distinguishes between being stumbled over and being kicked'.[64]

Not only does there seem to be phenomenologically experienced difference between movements$_T$ that are actions and those movements$_I$ that are not actions, we have embedded this experienced difference in our morality, our law, and our sense of what a self or a person is. One of the keystones to thinking of ourselves metaphysically, morally, and legally is to conceive of personal agency as distinct from mere physical involvement of our bodies in the chain of causation of some harm or benefit.

These are strong clues suggestive of the fact that human actions do form a natural kind whose nature thus demands a theory. At the very least, they put the burden on those who 'believe that the two forms of behaviour are in fact identical to account for the fact that, casually, they seem so different'.[65] On the other hand, such clues put a burden on those who hypothesize that human actions do form a natural kind, to formulate a plausible theory of what the nature of that kind is. The ultimate proof that there is such a kind is to have a theory showing there to be a unified nature shared by all members of the kind—in the old language, an 'essence'.

In the previous section of this chapter we previewed a number of such theories, all of them purporting to explain why we experience and categorize human actions differently from the way we categorize and experience non-actional bodily events. One explanation is that persons have unique causal powers and it is the exercise of those powers that we sense when we sense that we have acted. Yet this very short account in terms of some supposed primitive agency of persons we put aside as being too short: it doesn't explain anything to say that actions are those results we bring about through the exercise of our primitively unanalysable causal powers.

Another non-starter is to 'explain' why we experience actions differently from other bodily events in terms of goals, purposes,

[64] Holmes, *The Common Law*, 7.

[65] Kimble and Perlmuter, 'The Problem of Volition', 363. Kimble and Perlmuter go on to review some of the older psychological literature questioning whether voluntary and involuntary movements are not in fact the same.

motives, intentions, willings, beliefs, or desires, when all such 'explanations' are deemed to be non-causal explanations. This is the interpretivist view I introduced briefly in Chapter 4, according to which for an event to be a human action is for it to be interpreted to be a human action by some observer–interpreter, not necessarily the actor herself. This sceptical view explains nothing. Even if we were to grant that there is some point to seeing human behaviour as fitting a certain pattern,[66] which pattern is then described in the non-physical, non-causal language of goals, purposes, etc., such interpretive 'understanding' hardly amounts to explanation.[67] No explanation competitive with the volitional theory is given by such interpretivists for the simple but sufficient reason that no explanation at all is given by such approaches.

If one is to frame explanations truly competitive with the volitional theory in order to better argue for that theory, one should start with Donald Davidson's kind of causal account of human action.[68] After dispensing with the ordinary-language-philosophy interpretivists of action, Davidson proposed that actions are those bodily-movement events that are caused by belief–desire sets. An act, on this view, is a bodily movement caused by a belief–desire set.

Not just any belief–desire set, of course, will do here. The content of the belief and of the desire must fit the form of what Aristotle called a 'valid practical syllogism' (and what Davidson calls a 'rationalizing' relation). Take the ordinary case of someone, X, going to a barber's shop because he wants to get a haircut. In the first person, as he reasons them through, the content of his belief–desire set is:

(1) I desire that I get a haircut.

[66] On when there is a point to purely interpretive (as opposed to explanatory) exercises, see Moore, 'The Interpretive Turn in Modern Theory'.
[67] As some interpretivists will concede. See G. H. Von Wright, *Explanation and Understanding* (Ithaca, NY, 1971).
[68] I refer to the account Davidson developed in his 1963 paper 'Actions, Reasons, and Causes', reprinted in his *Essays on Actions and Events*. As we shall see, Davidson himself has changed his theory in several ways, in part in response to some of the criticisms we shall consider. See his 'How Is Weakness of the Will Possible?', and his 'Intending', also reprinted in his *Essays on Actions and Events*. See also id., 'Reply to Michael Bratman', and id., 'Reply to Bruce Vermazen', both in B. Vermazen and M. Hintikka (eds.), *Essays on Davidson: Actions and Events* (Oxford, 1985).

(2) I believe that, if I go to that barber shop, I will get a hair-cut.
Therefore:
(3) I go to that barber shop.

On the non-volitional, causal theory of action here considered, X's going to the barber's shop is an action just because it was caused by a belief–desire set of the right ('rationalizing', 'valid') form. Such a form requires that the content of the desire match the consequent clause of the content of the means–end belief; further, the antecedent clause of the content of the means–end belief must also match a description of the action done that he himself has or would adopt.[69]

Because I take Davidson's to be the most plausible alternative to the volitionalist's explanation of why actions seem so distinctive from other bodily events, to argue against the former account will be to argue in favour of the latter account. This argument best proceeds in two stages. In the first stage what needs establishing is that the Davidsonian account leaves out something that it cannot afford to leave out, namely, what I earlier called 'bare' or 'future' intentions. In the second stage of the argument, I will seek to show that among the bare intentions an actor must have in order to have his motive move him to action are those kinds of bare intentions we call volitions.

2.2. The Need for Intentions

Coming to the first stage of this argument for the existence of volitions: what is wrong with the belief–desire theory can be seen by attending to the conclusion of practical inferences such as the one just schematized. Aristotle said that the conclusion of a practical syllogism is an action. You want something sweet, believe the thing in front of you is sweet, and 'straight away' you eat that thing.[70] Aristotle's notion that action follows 'straight away'

---

[69] I choose the sloppy word 'match' advisedly, because just what exactly must be the two relations between propositions is a matter of much dispute and complexity.

[70] Aristotle, De Motu Animalium, 701^{a}7–15. For modern doubts as to whether even Aristotle meant what he said—that the conclusion of a practical syllogism is an action—see Robert Audi, Practical Reasoning (New York, 1989), 18–19, 192–3 nn. 6, 14; Anthony Kenny, Aristotle's Theory of the Will (New Haven, Conn., 1979), 142–3.

from a belief–desire set seems to leave something out: don't I have to form some propositional attitude having as its contents that I go into the barber's shop, before I do that act? Shouldn't we say that the conclusion of a practical syllogism is not an action—for how can an event *be* a conclusion[71]—but, rather, the propositional content of some mental state? Taking this latter tack allows us to preserve practical reasoning as a true form of reasoning, because the conclusion can be as propositional as are the premisses.

Aristotle at the very least seems to be committed to conceiving of the conclusion of a practical syllogism as being a description of an action, not the action itself, for it is only in this way that there can be a match between the antecedent clause of the content of the means–end belief and the conclusion. (And such match cannot be dispensed with without coming up with some other idea of what the *validity* of a practical syllogism comes to.) Further, not just *any* description of an action can be the conclusion of a valid practical syllogism. The description must be in some sense the agent's own. The only clear sense we have of the latter notion is when an agent has some propositional attitude having that description of his action as its content.

So at the very least we need to supplement the belief–desire account of action with some sort of propositional attitude, one having as its content a description of the action done that matches the antecedent clause of the content of the actor's belief. More, such concluding propositional attitude, whatever its nature, must be seen as a more direct cause of those bodily movements that are parts of actions than belief–desire sets; what belief–desire sets causes are this attitude, and what it causes are such movements. Belief–desire sets exercise their causation, as it were, through this concluding propositional attitude.

Now I wish to argue that we must conceptualize this concluding propositional attitude as a bare intention, not as a belief (say, that one will go downtown) and not as a desire (say, to go downtown). If successful, this will give us the conclusion of the first stage of the argument for volitions, namely, that there are bare intentions that intervene as effects of belief–desire sets and as the

---

[71] A point Davidson eventually came to concede, having initially adopted Aristotle's view that the conclusion of practical inference was the action itself. See Davidson, 'Reply to Bruce Vermazen', 221.

more immediate causes of those bodily movements₁ that are parts of actions.

There are a number of considerations that argue in favour of the view that it is intentions that immediately cause movements, not beliefs or desires. The first of these was introduced by Davidson himself. As Davidson noted, desires for something sweet, to get a haircut, etc. never generate or support all-things-considered judgements about the desirability of the appropriate actions.[72] Unless I am an obsessional neurotic about keeping my hair trimmed (to a degree never observed even in mental wards), I desire not only a haircut; I desire many other things as well, the attainment of which can conflict with my getting a haircut on any given occasion. My desire for a haircut is only in this sense prima facie. This prima-facie character of my general desire for a haircut will carry over to any specific desire I might form about the particular act that I believe will satisfy the object of my general desire. I will not desire to go to that barber's shop all-things-considered; from the contents of my desire and belief, I can only infer that going into that barber's shop is a prima facie desirable thing to do. It is not, in other words, a desirable act to perform *tout cour* but only desirable because it has the property (I believe) of causing my hair to be cut. If I aim to conclude my practical reasoning in a way that can justify my going into that barber's shop, I need a stronger judgement: I need to conclude that that act is, given *all* its properties and all the other things I desire, the thing to do now.

It may be tempting to think that one can dispense with the need for some *all-out* concluding propositional attitude in the following way. Each belief–desire set of an actor is accompanied by a more particular desire to do the act that realizes the object of the more general, motivating desire. A motivating desire to get a haircut, in the example given, is accompanied by a desire to go downtown. Likewise, a motivating desire to get some writing done is accompanied by a desire to stay at home (in light of the actor's belief that that is the only way in which he can get any writing done). Two such motivating desires may indeed conflict in their realizability on a given occasion, and actors obviously do do one or the other of the two possible actions. Yet there is no

---

[72] Davidson, 'Intending', in *Essays on Actions and Events*.

need, this account concludes, to posit *all-out* propositional attitudes to account for such resolutions; it is sufficient to say that the strongest prima-facie desires won out, where 'won out' means caused the action. One can make do, in other words, with (prima-facie) motivating desires, means–end beliefs, and (prima-facie) desires to do the action. There is no need to posit all-out intentions or all-out anything else in order to account for our ability to resolve conflicts between our desires.

The problem with this attempt to conceptualize concluding propositional attitudes as being the strongest competing prima-facie desires to do an act is that it leaves out the resolving function of such concluding propositional attitudes when there is (as there always is) conflict. In the face of conflict between prima-facie desires, there seems to be a resolution when the actor decides which of the alternative courses of action he is going to pursue.

There are four reasons to think that there must be such a 'resolving' or 'all-out' propositional attitude. One is our phenomenology; we seem to experience a change of state between being undecided in the face of conflict, and having resolved what it is we are going to do. Yet on the 'strongest-prima-facie desire' account, nothing has changed between these two times. Our prima-facie desires are just that at both times, merely prima-facie desires. Thus the concepts of choice, decision, and intention are born as labels for a state we sense to exist when we have resolved our conflicts. Such concepts label a state whose function is to act as a kind of barrier between the losing prima-facie desires and actions.

Secondly, there are some choices we seem to make that go against our strongest desire. I refer to cases of akrasia, or weakness of will. As the latter name with perfect accuracy suggests, sometimes our resolution of conflicting desires goes against what we most want. It then becomes impossible to explain our conflict resolution (and thus our action) by the strongest-desire account. For we would have to say, either that the act was caused by the *weaker* desire to do that act, or that the act was caused by the stronger desire to do some other act inconsistent with the act done. It is more plausible to think that what immediately caused the act done was neither the weaker nor the stronger desire, but a different state, a choice or intention. This allows us a very familiar description of akratics; they intentionally do acts that flout their strongest desires.

Thirdly, there are some choices we seem to make where there are no strongest desires. Michael Bratman calls these Buridan cases, after the story (attributed to Jean Buridan) of the burro that starved to death because it was equally positioned between two equally desirable stacks of hay and could think of no reason to choose one stack over the other.[73] In Buridan cases there are no strongest desires, yet unlike the burro we resolve the conflict between two equally strong desires all the time. In such cases what causes us to act cannot be merely the strongest of two competing prima-facie desires; rather we must resolve the conflict with some third and distinct mental state, one that is not prima facie in its content but reflects the all-out desirability of doing one action over another.

Fourthly, there is what I shall call the 'Sartrean function' of all-out or resolving propositional attitudes. Jean-Paul Sartre once opined that it was bad faith not to regard all choices as equally open to us at all times.[74] Sartrean persons are free, free even of the inertia or precedential effect of their past decisions and resolutions. Thus, if I desire an academic career, and I desire a career in professional sports, and it is clear to me that I cannot do both, this conflict of prima-facie desires continues even after I choose to pursue advanced degrees, accept a teaching position, etc. The strongest prima-facie desire simply continues to win, but the conflict remains as intense as are the comparative strengths of the conflicting desires.

Fortunately, none of us are Sartrean persons, for to regard everything as up for grabs at each instant of time would be to paralyse oneself. Recognizably human people are non-Sartrean: they regard their decisions, resolutions, choices, etc. as fixing matters that do not again need recalculation.[75] When two prima-facie desires conflict not just in their simultaneous realizability on a given occasion, but more generally (as is the case with long-term career plans), our resolving of the conflict at one time seems to carry forward throughout time. This suggests that we forge

[73] Bratman, *Intention, Plans, and Practical Reason*, 11, 22–4.

[74] Jean-Paul Sartre, *Being and Nothingness*, trans. Hazel Barnes (New York, 1956).

[75] A facet of practical reason very nicely explored in Joseph Raz, *Practical Reason and Norms* (Oxford, 1975), and Bratman, *Intention, Plans, and Practical Reason*.

some new mental state out of the prima-facie desires in conflict, because once we have resolved the conflict with a long-term decision we do not in fact experience the conflict the same way. On the conflicting-desire account this experience of carrying resolutions through time is difficult to account for, because on the conflicting-desires account nothing has changed over time except possibly the strength of the desires. On the conflicting-desires account, the action potential of the losing prima-facie desires remains as it was before.

These four considerations militate against a strongest prima-facie desires account of practical reasoning. They argue for there being some all-out propositional attitude that resolves our conflicts, both at a time, and over time. If one is convinced thus far, there remains the question of what sort of all-out propositional attitude there is that resolves conflicts and mediates between our motivating desires and our actions. One might think that this all-out, mediating state is either an all-out *desire* to do the act in question, or a *belief* that one will do the action, or an impression or image of the act in question, or a *belief* that it is, all things considered, desirable to do the action. Why must we introduce a third sort of mental state—of intention, choice, or willing—that is neither a type of desire nor a type of belief?

Consider desires first. John Austin thought that volitions were desires of a certain kind,[76] seconded in contemporary philosophy by Alvin Goldman, who at one time at least conceived of volitions as 'occurrent wants'.[77] According to Austin, the conclusion of my earlier practical syllogism is: 'Therefore, I *desire* that I go into that barber's shop'. The only things peculiar about such volitional desires, Austin might say, are: first, that its object is a particular act that is 'immediate' in the sense of simply described; secondly, that the desire is all-things-considered and not merely prima facie; and thirdly, that the desire is unaccompanied by any means–end belief but only requires a belief that the act I am about to do is the act all-out desired.

I have no quarrel with the first of these differences, for desires may easily differ in their objects and still be desires. The *nature* of a propositional attitude does not have to be different just because the *object* of that attitude is different. The third differ-

[76] Austin, *Lectures on Jurisprudence* i. 411–12.
[77] Goldman, *A Theory of Human Action*, 88.

ence between Austinian 'volition-desires' and regular desires is a real difference. To see the difference, we need to repair to cases of intrinsically motivated actions and see whether we are prepared to judge these to be cases of *desire*-causing-actions. Suppose I move my forefinger inward once, for no reason. (Granted, I rarely and perhaps never do this. '*Act gratui*' that are basic acts may exist only in existential fiction. Still, it is surely not inconceivable that there could be such unmotivated basic acts.) The only practical reasoning I could do for such an act that would fit within the two-premised schema earlier introduced would be:

(1) I desire that I move my forefinger inward.
(2) I believe that if I move my forefinger inward, I will move my forefinger inward.

Therefore:

(3) I desire that I move my forefinger inward.

This is a pretty a silly bit of imaginary practical reasoning. For the belief is so trivial that we should doubt that it ever performs any motivating work for this class of actions. A more accurate schema is simply: I desire that I move my forefinger inward.

Such cases of intrinsically motivated basic actions thus present an argument against the belief–desire causal theory of action that is independent of the arguments thus for presented; for intrinsically motivated basic acts are acts, yet there is no belief–desire set causing them once we reject the above schema as silly.[78]

Our present concern with intrinsically motivated basic actions is not to form an independent argument against the belief–desire causal theory. We have already amended that theory so that it admits the need for some concluding, all-out propositional attitude to be caused by belief–desire sets that itself then causes actions. The present question is whether such a mediating attitude can accurately be described as an all-out *desire* to do the act.

The relevance of intrinsically motivated basic actions to this question is this: what if these cases themselves cannot be described as cases where *desires* cause the acts in question? What

---

[78] John Searle relies on such intrinsically motivated actions (what he calls 'spontaneous actions') to argue for the existence of 'intentions-in-actions' that are not identical to either beliefs or desires. Searle, *Intentionality*, 84.

if we think that only intentions are the causes of intrinsically motivated basic actions? Then I take it we have good reason to think that intentions also operate as the immediate causes of action when there is extrinsic motivation; at least there is no reason of parsimony to conclude that they do not, since we have already admitted intentions as a third and distinct kind of mental state into our ontology.

We do recognize cases where desires cause behaviour without any accompanying beliefs. Consider this case: a prisoner rattles the bars in his cage *because* he wants out; he doesn't rattle the bars *in order* to get out, because he does not believe for a moment that his act will get him out. His desires simply cause his action. Similarly, my desire to beat Bobby Fischer in chess can *cause* my heart rate to go up or, potentially, any number of non-actional types of behaviour of or within my body. These are genuine instances where desires without beliefs do cause behaviour, but they are very different from supposing that desires can cause intrinsically motivated basic actions. For there is no intelligence in these behaviours (some of which are not even actions). There is no guiding of the behaviour by the propositional content of the desires, as there is in the instrinsically motivated cases. This gives us some reason to think that this intelligent, guiding function is performed by a state other than desire. Bare intention to do the act is a good candidate for this kind of guiding state.

In any case, the main objection to Austin's conceptualization of volitions as desires lies with his second difference with normal desires: such volitional desires would have to be all-out and not merely prima facie. They would thus have to serve what I called the resolving function of concluding propositional attitudes, and this function too seems to mark a difference in kind and not merely of degree.[79] Background states of positive or 'pro' attitude—what I have been calling desires—seem to perform a different, inclining function in our practical reasoning than do concluding states of resolution, guidance, and execution. On a functionalist view of mind, we do well to take this difference of function to be a difference of kind. Since 'intention' has long been the name for this concluding, resolving, guiding, executory mental state, I see no reason not to retain the label.

---

[79] See Davidson, *Essays on Actions and Events*, 102.

Now let us consider predictive beliefs. Mightn't the concluding, all-out propositional attitude be one of belief rather than of some third kind of mental state ('intention', 'will', etc.)? This is not the much debated question of whether an intention to do A is necessarily accompanied by a belief that one will do A. Rather, the question is whether what we might call an intention to do A *is*, in reality, a belief that one will do A. It is this stronger identity claim that the belief–desire theorist needs if he is to exclude intentions from being the concluding propositional attitude that immediately causes actions. Still, if one can defeat the weaker, accompaniment claim, we will have also defeated the stronger identity claim. That is how I shall proceed.

Consider three variations of this purported accompaniment of intention with predictive belief.[80] The first is that to intend A is to believe that one will do A. This surely is too strong, for I may intend to do something very difficult even though I believe I will not succeed. For example, I can intend to lose weight without believing that I will succeed; I can intend to wiggle my ears (say in a biofeedback laboratory) without believing that I will succeed.

The second and weaker variation is that to intend A is to believe that there is some possibility, however slight, that one will succeed in A-ing. Again, however, even this is too strong. I may be so weak of will, and know that I am so, that although I intend to lose weight I do not believe there is any chance that I will. Likewise, I can intend to wiggle my ears without believing that I have any chance of success; I may, for example, simply get paid in some psychology experiment for intending to move my ears whether I succeed or not, so that I am quite willing to try to wiggle my ears even while I am quite certain that I will fail.

This suggests a third and even weaker variation: to intend to A is to believe that one will try to do A.[81] Yet this variation won't work either. Now we need to separate the accompaniment claim from the identity claim. It may well be true that a belief that one will try to do A accompanies an intention to do A, but an inten-

[80] The three variations are discussed in Brand, *Intending and Acting*, 148–51. On this much discussed topic, see also Grice, 'Intention and Uncertainty'; Davidson, *Essays on Actions and Events*, 91–6; D. F. Pears, 'Intention and Belief', in B. Vermazen and M. Hintikka (eds.), *Essays on Davidson, 'Actions and Events'* (Oxford, 1985).
[81] Suggested by Hampshire and Hart, 'Decision, Intention, and Certainty'.

tion to do A cannot be identical to a belief that one will try to do A without generating an infinite regress of ineliminable intentions. For trying to do A is, at least in part, intending to do A. If I try to rob a bank, I do some act in execution of an intention to rob the bank. Moreover, with respect to basic actions, to try to wiggle one's ears is to intend to move one's ears. For basic acts there is no more basic act one can do in order to do them, so there are no other parts to trying to do such acts than intending to do them.

Once this connection of trying to intending is seen, it is of no use to analyse 'X intends to A' as 'X believes he will try to A', because that is the same as saying 'X believes he will intend to A'. 'Intend to A' within the content of such a predictive belief of X's must also then be analysed in terms of a further predictive belief, and so on *ad infinitum*.

These difficulties in reducing intentions to predictive beliefs may lead one to attempt a reduction of intention to certain non-predictive beliefs. Consider two possibilities, William James's 'ideo-motor' theory of action initiation, and Donald Davidson's more recent views on what sort of beliefs immediately cause actions. James held that what immediately causes us to act is an impression, image, or representation of the action performed.[82] Although James didn't call these beliefs, we should. Otherwise we again would face the possibility of there being representations in the mind that are not the content of any propositional attitude. To avoid that possibility, I would construe James's ideo-motor theory to be one according to which the immediate cause of actions is the occurrent belief that a certain type of act exists. The problem with such a theory illustrates a generic problem for all belief theories of intention, including those we have considered as well as Davidson's yet to be considered. This is the problem that the phenomenology of belief doesn't come close to capturing the phenomenal feel of acting and intending to act. This is particularly clear of James's ideo-motor theory. According

[82] James, *Principles of Psychology*. For contemporary explications of the ideo-motor theory that are quite sympathetic to it, see Kimble and Perlmuter, 'The Problem of Volition'; Anthony Greenwald, 'Sensory Feedback Mechanisms in Performance Control: With Special Reference to the Ideo-Motor Mechanism', *Psychological Review*, 77 (1970), 73–101. More critical contemporary accounts are in Vesey, 'Volition'; Goldman, 'The Volitional Theory Revisited'; and Brand, *Intending and Acting*, 175–81.

to James:

> every representation of a movement awakens in some degree the actual movement which is its object . . . We do not have a sensation or a thought and then have to add something dynamic to it to get a movement . . . Try to feel as if you were crooking your finger, whilst keeping it straight. In a minute, it will fairly tingle with the imaginary change of position . . .[83]

Accept James's invitation: is his description of the phenomenology accurate, or is it more accurate to conclude, as have many of his critics in psychology, that:

> It is clearly possible to think of such a movement as raising the hand, either as a movement to be made or as a movement not to be made. And I can find no truth in the assertion that, when I think of such a movement, my limb fairly tingles with the tendency to move, and that it is necessary to exert some inhibitory power in order to prevent its movement. To merely think of a movement and to intend or to will a movement are entirely distinct . . .[84]

The experimental evidence bears out this phenomenology. Ideas of movement do not themselves produce movement without something else being done by the actor whose body it is.[85]

The evidence is equally against the predictive-belief accounts of intention we earlier considered. Predicting that I will do something doesn't seem at all the same as resolving, deciding, choosing, willing, or intending to do it. I am an observer of my own future behaviour when I predict what I will do; that reflective attitude seems far removed from the first-person, committed attitude of one who has settled on a course of action.

Even Donald Davidson's mature theory of intention, which was formulated to avoid a number of the pitfalls just discussed, succumbs to this last objection. Davidson identifies intentions with all-out judgements: 'a judgement that something I think I

---

[83] James, *Principles of Psychology*, ii. 526–7.
[84] William McDougall, *Outline of Psychology* (New York, 1923), 291. See also Brand, *Intending and Acting*, 176: 'James' version of the theory is too strong . . . it is a deep insight of folk psychology that action is initiated only by events with noncognitive, motivational features.'
[85] See Vesey, 'Volition', 359; Kimble and Perlmuter, 'The Problem of Volition', 372–3.

can do, and that I think I see my way clear to doing, a judgement that such an action is desirable not only for one or another reason but in the light of all my reasons . . . *is* an intention'.[86] One construal of this is that the judgements that are concluding propositional attitudes are themselves beliefs: an intention to A is a belief that, all things considered, A is desirable and that the world is such that there is no impossibility to my A-ing.[87]

Such a belief (in all-out desirability) no more satisfies our sense of intending than do the predictive beliefs or the Jamesian beliefs earlier examined. The phenomenology of believing—even believing that an action is all-out desirable—does not match the phenomenology of deciding, planning, choosing, or intending to do it.

Moreover, Davidson's proposed identification of intention with believed all-out desirability is subject to two of the objections we interposed to the reduction of intentions to the strongest desire.[88] One of those was Buridan cases. In such cases there is no *belief* that one option is all-out more desirable than the other, just as there is no strongest *desire* for one option as opposed to the desire for the other. In such cases there thus must be some other sort of state that causes our actions, since there is no strongest prima-facie desire, nor any all-out belief in desirability, to do the work.[89]

---

[86] Davidson, *Essays on Actions and Events*, 101.

[87] Some interpretation of Davidson is required because of his own unwillingness to say exactly what an all-out *judgement* might be in terms of familiar propositional attitudes. See Davidson, *Essays on Actions and Events*, 97 n. 7, 101–2. Bratman thinks Davidson identifies intentions with 'the acceptance of an all-out evaluation' that the action is desirable. Michael Bratman, 'Davidson's Theory of Intention', in B. Vermazen and M. Hintikka (eds.), *Essays on Davidson: Action and Events* (Oxford, 1985), 15. To me, this acceptance, or 'judgement', sounds like a belief, as it did to the late Alan Donagan. *Choice*, 152.

[88] These are both nicely explored in Donagan,*Choice*, 152–6.

[89] Davidson attempts to defuse this objection as follows: 'if there is reason to reach some decision, and there are no obvious or intrinsic grounds for decision, we find extrinsic grounds. Perhaps I flip a coin to decide. My need to choose has caused me to prefer the alternative indicated by the toss; a trivial ground for preference, but a good enough one in the absence of others'. Davidson, 'Reply to Michael Bratman', 200. Davidson seems to be saying that in Buridan cases there is a belief about all-out desirability, namely, a belief that doing what the toss of the coin has indicated is now more desirable, all-out. Yet can this be so? That I desire to do either of the two opposed actions, rather than forgo both, does cause me to choose a decision procedure that is perfectly arbitrary (in terms of reflecting any increment of desirability). Yet when the decision procedure generates an outcome, how can I—given its known arbitrariness—now believe that the option

Secondly, the objection stemming from weakness of the will is as potent against Davidson's reduction of intentions to beliefs as it was against the reduction of intentions to (strongest) desires. For it is as plausible to think that there are cases where we act against our own belief about what, all-out, it is desirable to do, as it is to think that there are cases where we act against our strongest prima-facie desire. (Indeed, these may well be nearly identical classes of cases.) If this is so, it is hard to account for such cases on Davidson's account of action, for the immediate causes of such akratic actions is unlikely to be the belief that some action other than the one done was, all things considered, the desirable action to perform. Needed is some state that has as its content a description of the action done by the akratic. Intention fits the bill, allowing one to say the akratic intended to A even though he believed some action other than A was the best thing to do in the circumstances.

The upshot is that we have good reason to think that the concluding propositional attitude that mediates between our background motivations and those of our movements that are our actions is not a desire or a belief of any kind. We need to posit a third kind of state, an intention to do the action done, as the executor of our desires and beliefs. What we should next see is if such executory intentions include what we have defined as volitions.

2.3 The Need for Volitions

It may seem that once we have satisfied ourselves, first, that there must be intentions to act that operate both as the firm conclusions of our practical inferences and as the immediate causes of our actional movements, and secondly, that such intentions are not themselves a kind of desire or a kind of belief, we have done all we need to do to satisfy ourselves about the existence of volitions. For are not volitions just such executory intentions to act? They are, but volitions are only one kind of such bare intentions to act, namely, intentions whose object is that bodily movement$_1$ that is part of the action. Concluding in general that there are bare intentions *to act* does not conclude the question whether there are bare intentions having these non-actional, bodily-

generated is more desirable? Surely my only belief is still that either is more desirable than neither, and I must choose; but neither can be believed by me to be more desirable.

movement$_I$ descriptions as their objects. I should conclude, for example, that I must have an intention to go downtown from the argument and examples of the previous subsection; that does not conclude the question of whether I also must have an intention that my left foot move$_I$, then my right, as the volitions which execute more general intentions to go downtown.

The argument that there must be these more immediate sorts of bare intentions begins with the insight that there must be a causal relation between our more general intentions (or plans) and the most discrete of the bodily movements that execute such plans. After all, that is what makes us actors able to execute our plans with various bits of complicated, skilful behaviour. Many of the movements of the fingers, arm, torso, and eyes of the pitcher who throws a curve ball, in so far as they cause a good curve ball to be pitched, are themselves caused by the pitcher's intention to throw a curve ball. That much is not problematic if we make the common-sense assumption that the intention to throw a curve ball causes the curve ball to be thrown.

Neither is it very problematic to think that there must be both initiating causes for each of such discrete movements, and also monitoring, feedback, and correcting causes for the continuation of each such movement. Some sort of causal mechanisms must exist in order for our intention to throw a curve ball to cause just those movements that themselves cause a curve ball to be thrown. Given the complexity of skilled routines like throwing a good curve ball, mechanisms that monitor and correct the beginnings of mistaken bodily movements are as necessary as are mechanisms that initiate the sequence of movements.[90]

The real question is whether any of these causal mechanisms are not only causal mechanisms but are also described by the object of an intention. If we form an intention to throw a curve ball, and that causes us to throw one, are the discrete movements of our fingers across the lacing on the ball something we also intend? Or are those movements simply events that are caused by our intent to throw a curve ball and that themselves cause the ball to move in a curved flight, but that are not themselves intended?

---

[90] For a summary of the psychological literature on feedback loops and other mechanisms in conforming bodily movements to representations of those movements, see Brand, *Intending and Acting*, chs. 7–8.

Before we can answer this question we first must ask and answer another: how do we tell what someone intends? More specifically, how do we draw the line between what a *person* intends, and what some functionally defined homunculus within him 'intends' to 'do' in order to achieve what the whole person intends?[91] Myles Brand, for example, urges that we 'immediately intend' (Brand's phrase for volitions) the specific bodily movements that execute our more general (what Brand calls 'prospective') intentions and yet, Brand thinks, 'the monitoring and guidance of motor activity is in general [not] consciously accessible.'[92] Brand obviously has some criteria for when a *person* immediately intends that is not dependent on consciousness.

We in fact cannot be this cavalier in dismissing accessibility to consciousness as the touchstone of the personal, at least with regard to the agency and intentions of persons. To see this, recall that stones, trees, rivers, and human bodies all have agency in some sense of the word. Rivers *flow* to the sea, *explode* with force at the bottom of a fall, *erode* rocks, etc. Yet the agency and doings of rivers, homunculi, and other non-personal things are not like the agency and doings of persons. Persons are aware of their agency and its exercise, aware of the actish phenomenal feel of human actions. Indeed, Locke rightly held that human beings become persons only through the possession and the exercise of the ability to have conscious experience.[93] Nowhere is this more true than for the conscious experience of acting, for this experience is one of the crucial experiences needed to be a person at all.

Since it is so easy to get this issue confused, it is worth pausing to say what is here meant by 'consciousness' and by 'accessibility to consciousness'. By 'consciousness', I refer to the kind of awareness we have as an experience. By 'accessibility to consciousness', I have two things in mind. One is what Freud called the 'pre-conscious', by which he meant that some content could easily be summoned into conscious awareness even though such content was not at present the object of such awareness. This is the sort of 'awareness' we have of our movements when we have

---

[91] This is to recall for service the functionalist's distinction, introduced earlier, between the 'personal' and the 'subpersonal' levels of explanation.

[92] Brand, *Intending and Acting*, 154.

[93] David Wiggins discusses this aspect of Locke in his *Sameness and Substance* (Cambridge, Mass., 19890), 149–52. See also Moore, *Law and Psychiatry*, 348–9; G. E. Scott, *Moral Personhood* (Albany, NY, 1990), 27.

mastered a common routine but our mind (our consciousness, in this idiom) is 'somewhere else'. If the routine becomes non-routine and difficult, our attention becomes directed upon its execution, but absent such attention-focusing difficulty our scripts for such routines are often only pre-conscious.

The second kind of accessibility to consciousness is that which Freud called 'unconscious', or more precisely, 'dynamically unconscious'. The striking Freudian claim here is of extended memory: we can literally remember very hard to recapture ('repressed') experiences, if we go through the right memory-jogging processes (free association, hypnosis, transference and re-experience, etc.). Although such experiences are difficult to recapture, and perhaps cannot be recaptured, if they could in principle be recaptured, that is sufficient to make them the *person's* mental states and experiences.[94] It is this in-principle recapturability that separates such personal states from the sub-personal states Brand posits to guide motor movement, because these sub-personal states are not, even in principle, accessible to consciousness.[95]

We are now in a position to ask how far down (in detailed motor movement programmes) volitions of persons go. When I do walk downtown because I intended to go downtown, how much of the detail of *how* I walked downtown formed the object of one of my volitions? Did I only intend to walk, leaving all other guidance of my body to sub-personal routines and mechanisms? Or did I have thousands of intentions, to move my left foot, then my right to avoid the kerb, my arms to swing forward, my toes to push off, etc.?

At the level of consciousness, I do not pay attention to most of these details of motor movement. And it is a good thing, for if I had to focus on each of these details, I'd never be able to focus on my larger goals, muse about philosophy, etc. As Brand points out: 'Consciousness has limited capacity. If that capacity were occupied with the monitoring and guidance of ongoing motor activity, there would be none left for the high-level

---

[94] Explored by me in much greater detail in Moore, *Law and Psychiatry*, 126–40.

[95] The difference between the non-personal 'unconscious' of cognitive psychology and the personal 'unconscious' of dynamic psychiatry are explored by me in Moore, 'Mind, Brain, and the Unconscious'.

prospective thinking necessary for the continued survival of the organism'.[96]

Conscious awareness, in other words, has better things to do than be squandered on the details of motor movement that were mastered in infancy. Yet the implication of this is not at all that, therefore, descriptions of these motor movements are not the objects of a person's intentions or volitions. For the accessibility to consciousness of these movements makes their control something a person does even if their doing has become so routinized that they can be done 'without (conscious) thinking'.[97]

If we examine the learning of complex motor routines, we much more often see objects of a person's volitions being quite consciously formed to describe the discrete motor movements that are being mastered:

Learning to play a piece of music, and, in general, learning to play the piano, involves the building of behavioural elements into patterns which can be intended as wholes. It is only the beginner who has to think out each step as he goes along. The point is familiar to anyone who has learned to ride a bicycle or to swim.[98]

Consider learning how to move your hands and fingers in a way that fulfils a more general intention of knotting your tie. As described by Myles Brand:

The first stage in learning to knot a tie involves declarative knowledge. . . . A person learns a list of propositions that describes a step-wise procedure; he then encodes these propositions and recites them correctly. 'Put the tie around one's neck; leave the wide side longer by one-third; and so on.'[99]

This is a familiar observation to psychologists who study the learning of complex motor skills.[100] Equally familiar is the obser-

[96] Brand, *Intending and Acting*, 154.
[97] As the law recognizes in categorizing such behaviour as 'habitual' behaviour for which the actor is fully responsible. See Model Penal Code, §2.01. Compare Kimble and Perlmuter, 'The Problem of Volition', 373–7, who conclude that 'highly practiced acts tend to recede from consciousness, to become routinized and automatic and, in that sense, involuntary'.
[98] Sellars, 'Metaphysics and the Concept of a Person', 244.
[99] Brand, *Intending and Acting*, 257. Brand is summarizing the psychological work of Anderson, *The Architecture of Cognition*.
[100] See Kimble and Perlmuter, 'The Problem of Volition', 376 ('the child performing a motor act frequently verbalizes the sequence of steps').

vation that, as skills increase, conscious awareness of directing each movement recedes. Indeed, as skills are mastered it is often the case that conscious attention to discrete motor movements hinders rather than helps the performance of these movements in a skilful manner. The present question is how far such descriptions recede from consciousness. Myles Brand holds that 'this knowledge [of how to tie a knot] it not accessible in propositional form . . . it is stored in the form of a list of procedures or rules for tie-knotting'.[101]

In terms of explicit instructions about how to move one's fingers, etc., in the doing of actions like knotting a tie, playing a piano, or pitching a baseball, this is perhaps sometimes true. Perhaps, as Stanley Fish has recently observed about Dennis Martinez, the pitcher, all he can now say in the way of explicit instructions to himself or to others is to 'throw strikes and keep 'em off the bases'.[102] Yet even in such a case, isn't there an awareness of many of one's bodily motions conforming to one's own script or representation as to how one throws a strike? Compare Brand's and Fish's phenomenology to that of Alvin Goldman:

Suppose you are aiming to throw a baseball to a certain spot, say, home plate. You are concentrating not on your arm movement, but on the spot to which the ball was thrown. Are you not dimly thinking of your arm movement as well? Suppose you are perfectly ambidextrous, so that you could throw the ball with either arm. Would there not be a difference between the thought of throwing to home plate with your right arm, and the thought of throwing with your left? Even if the focus of your attention in both cases is on home plate, or the catcher's mitt, your intention or volition would be colored by the thought of throwing the ball with one kind of movement—e.g., a right-handed movement—or by the thought of throwing with another kind of movement—a left-handed movement. . . . The thought of the arm-movement, or bodily movement, is peripheral or marginal, but it is not entirely absent.[103]

I would add that even when this dim awareness of the movement is absent, it is none the less accessible to consciousness. It is preconscious in the sense of easily called to mind if attention is focused on it, and so remains part of a person's mental states.

[101] Brand, *Intending and Acting*, 257.
[102] Stanley Fish, 'Dennis Martinez and the Uses of Theory', *Yale Law Journal*, 96 (1987), 1773–1800.
[103] Goldman, 'The Volitional Theory Revisited', 74.

There are surely limits as to how finely grained we focus our attention even as we learn complex bodily routines. Those limits perhaps contract to some extent as we have fully mastered the skill as well, so that some of what was once accessible is no longer so.[104] At some point as we descend in our descriptions of ever smaller bits of bodily movements, we reach the sub-personal level where the scripts that guide movements are not accessible to our consciousness. This argument about volitions is thus only an argument about degree: do the contents of our bare intentions reach the fineness of detail of discrete bodily movements like moving one's arm, or are they limited to larger goals like throwing a strike or walking downtown? If we take our learning experiences into account, conjoined with the accessibility of the scripts we learned even later in adult life when we have mastered many complex motor skills, the evidence strongly favours the former answer even for seemingly 'automatic' or habitual actions like throwing a baseball pitch, saying a sentence, or playing the piano. Motor skills that are not nearly so routinized, such as moving one's finger on a trigger so as to execute one's intention to kill another person, are *a fortiori* cases for supposing volitions with such finely grained propositional objects to exist.

3. THE ARGUMENTS AGAINST VOLITIONS

While often one can argue *for* some philosophical proposition with the best positive arguments one has and then pretty much leave it at that, the topic of volitions has become so encrusted with stock objections that some brief indication of why these are not cogent objections needs to be made. This is particularly true in light of the historical influence these objections have had in keeping *legal* theorists away from volitions.

The first of these was an objection to the causal relation asserted to exist between volitions, a mental state, and bodily movements$_I$, a physical event. The objection is perfectly general: how can a mental thing causally interact with a physical thing? Gilbert Ryle, for one, thought the volitional theory to be 'an

[104] As some volitionalists concede. See Ripley, 'A Theory of Volition', 145; W. F. R. Hardie, 'Willing and Acting', *Philosophical Quarterly*, 21 (1971), 193–206: 200–4.

absurd hypothesis' in part because of the inconceivability of causal relations existing between two such different sorts of things:

> According to the theory, the workings of the body are motions of matter in space. The causes of these motions must then be either other motions of matter in space, or, in the privileged case of human beings, thrusts of another kind [i.e. volitions]. In some way which must forever remain a mystery, mental thrusts, which are not movements of matter in space, can cause muscles to contract.[105]

Such a causal hypothesis violates laws on the conservation of mass and energy.

Ryle's indictment is conclusive, but not against the culprit against whom he lodged it. It *is* absurd to think of things that exist in time but not in space, having no mass and no energy, causing other things to occur, when those latter events do have spatio-temporal location and do involve objects with mass and energy. One of the beauties of a functionalist interpretation of volitions is that it avoids the horrors of interactionist dualism that Ryle so persuasively depicted. On the functionalist interpretation, volitions are not 'things that exist in time but not in space, having no mass and no energy'. True, the essential nature of volitions is given by their functional role and not by their physical structure; true, causal laws about them may well be formulable in terms of concepts using such functional (not structural) criteria; but there is nothing in the functional view that commits one to thinking that such functional states exist in some non-physical realm, or that there is a 'functional substance' different from any physical substance. Rather, a functionalist holds that each particular volition that a person possesses has a structural realization in the brain of that person. Each such volition thus has a physical nature that may non-anomalously interact with other physical objects. Such physically realized functional states of volition are not 'non-physical episodes which constitute

---

[105] Ryle, *The Concept of Mind*, 63–4. For the continued appeal of this old worry about metaphysical dualism being presupposed by volitional theories of action, see Duff, *Intention, Agency, and Criminal Liability*, 117–19; Kevin Saunders, 'Voluntary Acts and the Criminal Law: Justifying Culpability Based on the Existence of Volition', *University of Pittsburgh Law Review*, 49 (1988), 443–76: 461–9.

the shadow-drama on the ghostly boards of the mental stage', as Ryle thought they had to be.[106] They are no more problematic or puzzling than is the causal relation between two billiard balls. Sometimes the followers of Ryle both in philosophy and in law have raised a second causal objection. Beginning with the Humean observation that no two events or states can be causally related to one another if the names for them are logically related, the objection is that volitions are logically related to bodily movements and so cannot cause them. As restated recently by Alan White:

> There is the difficulty of characterizing the relation between the volition and the bodily movement which follows it, for this seems to be at the same time both logical and causal. It is logical in that the volition which is alleged to cause, for example, a movement of the index finger is necessarily the volition to do exactly that and it is causal in that it is the cause of the movement. But a cause of something ought to be logically independent of it.[107]

White's worry is that the object of my volition—say, that my index finger move$_I$—is identical to that which the volition causes—my index finger moving$_I$. Yet this is a non-worry. All Intentional mental states that are successfully realized in action have this 'logical relation' to the action that they cause. If I desire to walk downtown, and I do walk downtown *because* of this desire, White would have to deny that this was a causal relationship between my desire and my action. Not only does this make the meaning of 'because' in such statements a real puzzle,[108] but the 'logical relation' worry betrays a fundamental ignorance about Intentional states. The nature of such states is given by their functional roles; their Intentional content is one of their features by virtue of which such states may play their functional roles. Such content is a *description*—'moving$_I$ of my index finger', or 'walking downtown'. Such descriptions are not identical to the things to which such descriptions may refer, namely, the movements and actions. My moving my index finger is one

---

[106] Ryle, *The Concept of Mind*, 64.

[107] White, *Grounds of Liability*, 31. For like views, see Melden, *Free Action*, 53; Langford, *Human Action*, 19; Duff, *Intention, Agency, and Criminal Liability*, 128.

[108] See Davidson, *Essays on Actions and Events*, 11; Moore, *Law and Psychiatry*, 20–3.

thing; the description 'moving my index finger' is quite another, even if the second is often used to pick out the first in communication.

It is because the Intentional states (volition, desire, intention, etc.) that cause actions have certain descriptions as their objects, and because such descriptions typically refer to some movements or acts rather than others, that we may plausibly suppose such states to indeed be causes of certain movements/actions, and not others. If such states did not cause behaviour in this way, we would have to view persons as fundamentally non-rational creatures.

The third objection to volitions, rather than saddling the volitionist with the dualist metaphysics of Descartes, saddles him with the incompatibilist metaphysics of Locke. The objection stems from the (incompatibilist) assumption that no one can either be morally responsible or have acted unless he was *free*, where 'free' means 'uncaused'. All causal theories of action, the volitional theory included, are necessarily false: for the concept of action (as free) guarantees that there can be no caused actions. Volitions may exist, and they may be causes, but they cannot—conceptually cannot—be the causes of *actions* or parts of actions. Needed to account for action, this objection concludes, is some philosophically primitive notion of agency or of agent-causation.[109] Otherwise, the person-as-agent gets left out of a story that talks of volitional states causing those movements that are parts of actions.

The functionalist interpretation of volitions cuts the ground from under this objection by flatly rejecting its incompatibilist assumption. There is no reason whatsoever to suppose that physically realized functional states of volition do not themselves have causes. Indeed, part of the functional specification of such states is that they are caused by background states of belief, desire, and (future) intention. Those background states are themselves physically realized functional states, and part of their functional specification is that they are caused by other mental states or by features in the external world. Functionalism is a theory about what mental states like volitions *are*. The volitional theory of

---

[109] See e.g. Roderick Chisholm, *Person and Object*; Malcolm, 'The Inconceivability of Mechanism'; Melden, *Free Action*; Taylor, *Action and Purpose*; Moya, *The Philosophy of Action*.

action is thus a theory of what actions *are*—they are the sequence of volition–cause–movement. Such a theory may be false, but no *conceptual* argument can show it to be false. For what grounds such supposed conceptual truths as that action must be free but certain linguistic facts about patterns of ordinary usage? The nature of action as essentially free can no more be guaranteed by such surface linguistic facts than can the Evening Star be conceptually guaranteed to be different from the Morning Star.[110]

A fourth and better objection than any of these is the objection stemming from what is known as 'deviant causal chains'.[111] Suppose that I will that my arm rise, and an event occurs as an effect of that willing that meets that description, namely, my arm does rise. The problem is that such arm-rising need not be part of an act of raising my arm. Suppose someone else raised it for me because that person, wishing that my every volition come true, arranged it so that he would raise my arm whenever he saw that I wished to but couldn't. Or suppose that I so very much wanted to lower my leg (after a long time in traction when I could not move it) that when my leg is released from traction my excitement causes my leg to slip and fall downwards. In such cases, there are bodily movements caused by my willing those movements to occur—and yet we know intuitively that I did not perform the actions of raising my arm or lowering my leg.

This problem seems more serious than it is for the mental-cause thesis. There are two ways of avoiding the problem. One is to adjust the object volitions must have to be criteria for actions: one must will, not just that one's arm goes up, but that its doing so is explained by your willing that it go up. Donagan calls this the 'explanatorily self-referential' character of the objects of volitions, and shows how such objects avoid both epistemic circularity and deviant causal-chain counter-examples.[112] The other way round such counter-examples is to keep the object of volitions simple—that one's arm rises, for example—but adjust the causal

[110] I also think the linguistic facts to be against the incompatibilist here, for we talk of causing voluntary action all the time. See Ch. 4, *supra*.

[111] See Roderick Chisholm, 'The Descriptive Element in the Concept of Action', *Journal of Philosophy*, 61 (1964), 613–25; Taylor, *Action and Purpose*, 248–9; Davidson, *Essays on Action and Events*, 77–81.

[112] Donagan, *Choice*, 88. See also Searle, *Intentionality*, 85; Harman, 'Willing and Intending', 373.

requirement for action (that a volition *cause* a movement₁) by saying that volitions must *directly* (or proximately) cause such movements.[113]

My own route around such counter-examples is the second. While the first is ingenious, and avoids all the problems of spelling out a notion of a 'proximate' or 'direct' causal requirement, it requires more complexity to volitions than they seem to have. Volitions, as direct initiators of action, should have a simple nature. Attributing self-referential content to such simple action initiators seem more like an *ad hoc* philosophical reconstruction than a true psychological description. Moreover, as lawyers well know, some notion of 'proximateness' is required whenever we refer to causation, so there is no special problem in using such a notion as part of one's criteria for action.[114]

Alvin Goldman, in his well-known book *A Theory of Human Action*, attempted to suggest how to spell out the notion of 'direct' or 'proximate' causal connection between mental states and movements. Such connection must be one that operates in the 'characteristic manner in which beliefs and desires flow into intentional acts'.[115] As Goldman later noted, 'I assumed that this "characteristic" causal process is linked with the feeling of voluntariness, as opposed to pure reflex or autonomic activity.'[116]

Along with many others, I criticized Goldman's causal criterion of action as empty.[117] It either was completely circular, referring to intentional action in its analysis of action, or collapsed into my own epistemic criterion for action ('the feeling of voluntariness'). I now think the criticism to be misplaced,[118] for I no longer believe it fruitful to seek a *conceptual* essence to action. The sort of causal relation to which Goldman was referring is not to be discovered as a conceptual truth or by conceptual analysis. If actions form a natural kind, then we must seek a better and better *theory* as to the nature of that kind. Part of such a

[113] As in Goldman, *A Theory of Human Action*, 61–2; Brand, *Intending and Acting*, 17–23.
[114] The parallel between the two kinds of proximate-cause problems is nicely displayed in Brand, *Intending and Acting*, 17–23.
[115] Goldman, *A Theory of Human Action*, 62.
[116] Goldman, 'The Volitional Theory Revisited', 82.
[117] Moore, *Law and Psychiatry*, 72–3. For a modern, sophisticated repetition of the objection, see John Bishop, *Natural Agency* (Cambridge, 1989), 142–5.
[118] As do some of my critics. See Robert Schopp, *Automatism, Insanity, and the Psychology of Criminal Responsibility* (Cambridge, 1991), 108–20.

theory must spell out in increasing detail more about the sorts of causal connections that must exist (between representations of movements, and movements) if actions are to exist. The more detailed and plausible we can make such a theory, the more plausible we make the volitional theory as a whole. But that there plainly are instances of volitions (deviantly) causing movements₁, where those volitionally caused movements₁ are not actions, is not the knock-down objection to the volitional theory many have thought it to be.

A fifth objection to volitions is the objection that stems from phenomenology. Gilbert Ryle originated this objection as well:

when a champion of the doctrine [of volitions] is himself asked how long ago he executed his last volition, or how many acts of will he executes in, say, reciting 'Little Miss Muffet' backwards, he is apt to confess finding difficulties in giving the answer . . . ordinary men never report the occurrence of these acts . . . [yet] according to the theory, they should be encountered vastly more frequently than headaches, or feelings of boredom . . . it is fair to conclude that their existence is not asserted on empirical grounds . . . [but is] a special hypothesis the acceptance of which rests not on the discovery but on the postulation, of these ghostly thrusts.[119]

Ryle's phenomenological objection to volitions was most prominently elaborated on and repeated by A. I. Melden[120] and Richard Taylor[121] in philosophy, and the objection was carried over into legal theory by Hart [122] and those whom he influenced.[123]

Once again, a functional interpretation of volitions undercuts any claim to final authority that might be made on introspection's behalf. There is no necessary link of subjective experience to functional states, since the nature of such states is given by their functional roles and not by their accessibility to consciousness.[124] If one has good theoretical reasons to hypothesize a state that mediates between our general preferences and our particular

[119] Ryle, *The Concept of Mind*, 65.   [120] Melden, *Free Action*.
[121] Taylor, *Action and Purpose*.   [122] Hart, *Punishment and Responsibility*.
[123] White, *Grounds of Liability*, 31; Duff, *Intention, Agency, and Criminal Liability*, 44–7, 127.
[124] On the limited role of subjective experience in specifying functional states, see Moore, 'Mind, Brain, and the Unconscious', and, more generally, J. Biro and R. Shahan (eds.), *Mind, Brain, and Function* (Norman, Okla., 1982).

actions, the lack of phenomenological evidence will not be critical. In addition, as I earlier argued, Ryle, Melden, Taylor, *et al.* are wrong about the phenomenological evidence: if you include learning experiences in one's evidential base, and if you make the reasonable assumption that the states needed for such experiences continue after learning has occurred even though such states are no longer noticed, then the phenomenological evidence is pretty good for the existence of volitions.

A sixth objection, one that builds on the supposed introspective 'characterlessness' of volitions, was once voiced by Richard Taylor:

Quite apart from any introspective scrutiny of one's own mental life, one ought to suspect that volitions are fictional just from the way they are referred to . . . They are always referred to and described in terms of their effects, never in terms of themselves, leading one to suspect that perhaps they have no inherent characteristics by which they even can be identified.[125]

A functionalist interpretation of volitions short-circuits this objection very quickly. Functional states are customarily referred to by mention of their effects because the essential nature of such states is given in part by the causal relation between them and those effects (i.e. bodily movements). True, if we could not eventually discover a physical nature to such states, we would have reason to disbelieve in their existence. But we often refer to things in partial ignorance of the complete nature of those things, and this should particularly be expected of functional states.

These six objections comprise pretty much of what might be called the 'old case' against volitions. They are the artefacts of a philosophy of mind and a philosophical method that prevailed in the Anglo-American philosophy of the 1950s and 1960s but is now very much on the wane. For, as we have seen, each of them is undercut by a functionalist philosophy of mind and by a philosophical method that seeks real essences (though scientific theory) rather than nominal essences (through conceptual analysis). What I have said in response to each is undoubtedly too little to convince those enamoured of the older philosophy of mind and its style of philosophy, even if it is too much for those with different philosophical predilections.

[125] Taylor, *Action and Purpose*, 68.

In any case, newer worries about volitions come from quite different quarters. One of these stems from recent studies of the cerebral processes that initiate voluntary or willed motor movements. Benjamin Libet and his colleagues in physiology have recently set out to clock: (1) the onset of persons' awareness of a decision or willing to move the fingers or wrist of their right hand, and (2) the onset of those cerebral processes known to precede motor movements of this kind.[126] The former variable was measured by the introspective reports of subjects matching a revolving spot on a clock face. The latter variable was measured by the 'readiness potential', a scalp-recorded slow negative shift in electrical potential generated by the brain and which regularly appears up to a second or more before voluntary motor acts.

Libet's findings were that the cerebral processes in the brain regularly precede the actor's awareness of deciding (willing) to initiate a spontaneous movement of his finger or wrist. From the finding, Libet concludes that volitions (willings, decidings, intendings, etc.) do not initiate bodily movements; cerebral processes do that. At most, volitions may veto movements already initiated by unconscious cerebral processes.

Yet, as much of the more philosophical peer commentary on Libet's work also has noted,[127] Libet's findings show no such thing. Surely volitions, on a functionalist view, are themselves in each instance a physical state of the brain. Why should we think that those physical processes that underlie a change of those physical states should not precede an awareness of the onset of the volition? Perhaps the readiness potential measures just those physical processes! In either case, volitions retain their causal role as action initiators. In either case, Libet has not shown that

---

[126] Libet, 'Unconscious Cerebral Initiative and the Role of Conscious Will in Voluntary Action'. This article summarizes the findings in: B. Libet, E. W. Wright, and C. A. Gleason, 'Readiness-Potentials Preceding Unrestricted "Spontaneous" vs. Pre-planned Voluntary Acts', *Electroencephalography and Clinical Neurophysiology*, 54 (1982), 322–35; id., 'Preparation or Intention to Act in Relation to Pre-event Potentials Recorded at the Vertex', *Electroencephalography and Clinical Neurophysiology*, 56 (1983), 367–72; B. Libet, C. A. Gleason, E. W. Wright, and D. K. Pearl, 'Time of Conscious Intention to Act in Relation to Onset of Cerebral Activities (Readiness-Potential): The Unconscious Initiation of a Freely Voluntary Act', *Brain*, 106 (1983), 623–42.

[127] See particularly Arthur Danto, 'Consciousness and Motor Control', *The Behavioral and Brain Sciences*, 8 (1985), 540–1.

actions are not caused by states of volition—only a bit more about the physical workings of those states.

There is another worry about volitions that a functionalist interpretation not only doesn't answer, but reveals. If one looks at functionalist accounts of volitions extant in the literature,[128] the result is not very reassuring. There are not the subtle and persuasive functionally defined subroutines of willing as there are, for example, in the cognitive psychology of perceptual belief. Indeed, it would not be unfair to say that with regard to the will we have not progressed much beyond Plato's ancient functional division of the soul. Plato's three—will, passion, and reason—are not, in Dan Dennett's colourful phrase,[129] a whole army of very stupid homunculi performing very simple functions but, rather, three smart generals unwilling to delegate any tasks. Cognitive psychology has done much to replace one of these generals (reason) with an army of dumber subordinates; another general (passion) at least has been found to have lots of lieutenants; but the remaining general (will) has seemingly remained indispensably one 'field marshal' who utters monotonously similar, specific commands of movement.

This suggests one of three conclusions about the will: either that the functionalist approach to the will is still in its infancy and that this explains the dearth of fruitful subdivisions; or that there really is no such thing as a volition, despite my earlier arguments; or that Plato said pretty much what there was to be said about the will so long as one restricts oneself to a top-down approach, and that progress about the nature of will is to be found by pursuing the bottom-up strategy instead. Perhaps we know enough when we say there is a mediating cause of action called volitions to seek their physical realizations. Physiologists have begun the search for cerebral processes involved in conscious initiation of simple bodily movements, when those processes are clearly distinguished from deliberative, motiva-

---

[128] See Davis, *Theory of Action*; Brand, *Intending and Acting*.

[129] Dennett, *Brainstorms*, 80–1. Dennett's most recent work, *Consciousness Explained* (Boston, 1991), ch. 8, urges that a properly done, functional subdivision of the mental processes initiating action should not seek a serial or one-way process whereby our representations are first fully formed and only then executed in action.

tional, and other background processes, and such bottom-up studies may well yield fruit.[130]

Whether volitions exist is thus very much an open, scientific question. In this chapter I have tried to indicate how, if we allow Austin the most plausible interpretation of the nature of volitions in the dimensions I have examined, his volitional theory is our current best bet about the nature of action. Even if one dissents from this conclusion, the volitional theory is surely not to be dismissed as 'really nothing more than an outdated fiction—a piece of eighteenth-century psychology which has no real application to human conduct'—[131] as did Herbert Hart and the present generation of legal theorists whom Hart influenced.

---

[130] See Gary Goldberg, 'Supplementary Motor Area Structure and Function: Review and Hypotheses', *The Behavioral and Brain Sciences*, 8 (1985), 567–88, for a helpful summary of thirty years of research into the area of the brain most likely to be involved in action initiation.

[131] Hart, *Punishment and Responsibility*, 101.

PART II

COMPLEX ACTION DESCRIPTIONS AND THE *ACTUS REUS* REQUIREMENT

7

THE DOCTRINAL BASIS OF THE *ACTUS REUS* REQUIREMENT

As is well known, the statutes that make up the special part of any jurisdiction's criminal code do not use basic act descriptions like 'Don't move your fingers'. Such statutes use *complex* descriptions such as 'Don't kill', 'Don't rape', 'Don't break and enter the dwelling house of another at night', etc. Such descriptions are complex in the sense that they ascribe properties to acts in addition to the properties of being a bodily movement caused by a volition. For an act of finger-moving to be a *killing*, for example, death must be an effect of the finger-moving; for an act of tongue-moving to be a *saying of an English sentence*, that act must both cause certain sounds to be emitted, those sounds must conform to the syntactical and semantical rules of English, and the speaker must be following those rules as she directs her tongue movements; for an act of knee-flexing to be a *concealing*, such knee-flexing must both result in the actor being out of sight and be motivated by an intention to get out of sight; for an act of pelvic thrusting to be a *rape*, such movements must both cause a penetration of a woman and be done in the circumstances that the woman has not consented; etc. Because the truth conditions of expressions using descriptions such as 'killing', 'saying', 'concealing', and 'raping' include the existence of such extra properties, we are entitled to think of such descriptions as *complex*. A *basic* description, by contrast, ascribes only the property of bodily-movement-caused-by-a-volition, which is essential to the act being an act of any kind at all.

The *actus reus* requirement of the criminal law can now be defined as follows: the act for which an accused is punished must be an act (bodily-movement-caused-by-a-volition) that has the properties required by some complex act description contained in some valid source of criminal law. As we shall see in Chapter 9, the principle of legality accepted in Anglo-American criminal law restricts the sources of that law to statutes of prospective application only. Anticipating the results of that chapter, we may reformulate the Anglo-American *actus reus* requirement as requiring that the accused must do an act that has the properties required by some complex act description contained in some statute in force at the time the act was done.

We can now see how distinct the *actus reus* requirement is from the act requirement earlier discussed. There are many cases where the accused will plainly have willed the movement of his limbs (and thus have satisfied the act requirement of the criminal law) but where his act will not possess one or more of the properties required for it to satisfy the complex descriptions of the *actus reus* requirement. A willed movement of one's body that causes a night-time breaking and entering of a dwelling house is still not an action of burglary if the actor owns the house in question. Criminal law plainly has two distinct requirements, the act requirement of its general part and the *actus reus* requirement of its special part.

In examining the *actus reus* requirement I shall pursue the same organizational strategy as was followed in discussing the act requirement. That is, I shall first (i.e. in this chapter) discuss certain scepticisms about there being an *actus reus* requirement in Anglo-American criminal law. This requires a by-and-large doctrinal analysis to ascertain whether the *actus reus* condition can meaningfully be separated from the *mens rea*, excuse, justification, and causation conditions. Secondly, in Chapter 8 I shall examine the question of whether or not there is one *actus reus* requirement, as opposed to there being as many *actus reus* requirements as there are criminal prohibitions in Anglo-American criminal law's special part. This is a matter of describing Anglo-American criminal doctrine and the complex descriptions of action that doctrine employs, asking whether all complex descriptions have any properties in common. Thirdly, in Chapter 9 I shall ask the normative question of whether criminal law should have an *actus reus* requirement, enquiring into the good(s) served by having such a requirement. Fourthly, in Chapters 10 and 11 I ask whether the things to which complex action descriptions purport to refer—complex actions—in fact exist. It is in these chapters that I shall seek to redeem the promissory notes issued in the discussion of the exclusivity thesis in Chapter 5, for it is here that I shall defend the proposition that every complex action is in reality nothing but a bodily movement caused by a volition.

1. THE PROBLEM OF BORDERS

One scepticism about the criminal law's *actus reus* requirement is the thought I shall call the 'scepticism about borders'. This is the thought that no sensible or useful distinction can be drawn between the *actus reus* requirement, on the one hand, and the *mens rea*, the causation, the (absence of) excuse, and the (absence of) justification requirements, on the other hand. I shall consider this scepticism by considering each of these lines in turn.

1.1. Actus Reus *versus* Mens Rea

By the '*mens rea* requirement' for criminal liability, I refer to what Sanford Kadish once aptly called the 'special *mens rea*' requirement.[1] Unlike the 'general' *mens rea* requirement, which includes absence of excuses like insanity, involuntary intoxication, etc., the special *mens rea* requirement focuses only on those mental states of an accused that have as their object the various elements of the *actus reus* of some particular prohibition.[2] The California Privacy Act, for example, prohibits (that is, has as its *actus reus*) the recording of any confidential communication of another without that other's consent when done by means of any electronic amplifying or recording device.[3] The special *mens rea* requirement of the statute is that one must do this action *intentionally* in order to be convicted.

Note that to specify the *mens rea* requirement for a crime such as the crime of surreptitious recording, one must use the *actus*

---

[1] S. Kadish, 'The Decline of Innocence', in S. Kadish, *Blame and Punishment* (New York, 1987). See also id., 'Act and Omission, *Mens Rea*, and Complicity'; S. Morse, 'The Guilty Mind: *Mens Rea*', in D. K. Kagehiro and W. S. Laufer (eds.), *Handbook of Psychology and Law* (New York, 1991).

[2] In the case of specific-intent crimes, the object of the prohibited *means rea* differs from the various elements of the *actus reus*. In the context of the present worry about how to distinguish *mens rea* from *actus reus*, I ignore the more easily distinguishable specific-intent crimes.

[3] California Penal Code, §632 (prior to 1967, §653j).

reus description. One cannot say what must be *intended* to violate the statute until one can say what must be done to violate the statute. This might lead one to think that the two requirements are not distinct. This is the first of several interpretations I shall make of Jerome Hall's generally sceptical conclusion that 'the separation of *mens rea* from the external aspect of a crime [i.e. *actus reus*] obscures the *necessary* connection between them'.[4]

We should indeed concede that there is often *this* kind of connection between the *actus reus* and the *mens rea* requirements, because such concession goes no distance towards the conclusion that the requirements are in any sense not distinct. That the same language is used to describe the object of a prohibited mental state as is used to describe a prohibited action in no way makes the two prohibitions the same. I may easily violate the *actus reus* requirement of the California Privacy Act by in fact recording a communication between others that is in fact confidential, without their consent and uses non-radio, electronic means, without *intentionally* doing so—e.g. I am merely testing my recording equipment, having no idea that a confidential communication is being recorded.[5]

So no breakdown of the line between the *actus reus* and the *mens rea* requirements is to be sought in the linguistic economy often used to express both requirements in some criminal statutes. A second and slightly more plausible scepticism about this line was once expressed by Hall in the following terms:

what is the meaning of the definition of 'actus reus' in terms of the entire external situation excluding mens rea but including 'a mental element in so far as that is contained in the definition of an act?' An 'act,' in penal theory, implies a mens rea. But this, by definition, must be excluded.[6]

Hall's query was how *mens rea* can be separated from *actus reus* when the latter, via the act requirement, has its own mental-state

---

[4] Hall, *General Principles of Criminal Law*, 321.

[5] See e.g. *People* v. *Superior Court*, 70 Cal. 2d 123, 74 Cal. Rptr. 294, 449 P. 2d 230 (1969).

[6] Hall, *General Principles of Criminal Law*, 227. It is for this reason that Douglas Husak too is sceptical of the *actus reus* requirement: 'If all acts are voluntary, proof of *actus reus* would require evidence of the inner, mental state (the "will") of the defendant; no longer would the scope of *actus reus* be confined to the external and physical. Hence definitions of *actus reus* that include voluntariness become confusing at best and useless at worst.' Husak, *Philosophy of the Criminal Law*, 91.

requirement, namely, that the accused *will* his movements? *Actus reus* includes mental states, so how can it be defined as if it excluded mental states?

The answer lies in the difference between the various mental states required by the *mens rea* requirement and the single mental state required by the act requirement. For there to be an act at all (and thus for there to be the *actus reus* for any crime), there must indeed be a mental state of volition causing the relevant bodily movements. As we have seen, many volitional theorists wish to call such mental states 'intentions', and volitions are a kind of intention. But notice that the intention required to act at all—the intention to move one's limbs—is not the same in its object as the intention described by the *mens rea* requirement. The latter intention has as its object complex act descriptions like 'killing', 'disfiguring', or 'recording a confidential communication'; it does not have basic act descriptions like 'moving my fingers' as its object.[7] We should thus describe the difference between the mental states prohibited by the *actus reus* requirement and the mental states prohibited by the *mens rea* requirement as at least a difference in the objects of the respective mental states: bodily-movement₁ descriptions are the objects of the mental-states part of the *actus reus* requirement, whereas complex act descriptions are the objects of the mental states of the *mens rea* requirement.

So long as the difference between bodily-movement and complex action descriptions is plain enough, then so is the distinction between the *actus reus* and the *mens rea* requirements. Or so it seems. What about statutes that prohibit *concealing* a weapon? Or *lying* to a friend? Or *attempting, abetting,* or *conspiring* to kill another? About such crimes as these, wasn't Bracton right when he said that 'your meaning imposes a name upon your act'?[8] If so, isn't the *mens rea* prohibition—do not intend to conceal, do not misspeak knowing that you are doing so, do not intend the death of another—a necessary part of the respective *actus reus* prohibitions? For how can one conceal something without

---

[7] Statutes that only require intentions of this latter sort—intentions to move one's limbs—are classified as strict-liability statutes, that is, statutes that have no *mens rea* requirement at all. This is the meaning of the common doctrine about strict-liability crimes, namely, that such crimes do require a 'voluntary act' (i.e. an intent to do a basic act).

[8] Bracton, 2 *De Legibus*, ed. Sir Travers Twiss (London, 1879), F. 101b, 127.

intending to conceal it, lie without knowing the falsity of what is said, attempt to kill another without intending the death of another? Etc.

For this class of examples, there is indeed a blending of the *actus reus* and *mens rea* requirements in the sense that one cannot have satisfied the *actus reus* requirement for such crimes without also having satisfied the *mens rea* requirement of such crimes. The mistake is to generalize from this class of examples to conclude, as Jerome Hall apparently did, that it is always the case that '*actus reus* implies *mens rea*'.[9] Rather, this kind of merger of the two requirements takes place only when criminal statutes utilize complex action descriptions that are, as I shall call them, 'intentionally complex'. As Elizabeth Anscombe has noted in her study of intentions, 'many of our descriptions of events effected by human beings are *formally* descriptions of executed intentions'.[10] 'Concealing' and 'attempting' are of this class. But many more complex action descriptions are not intentionally complex. 'Killing', 'maiming', 'burning', 'hitting', 'scaring', for example, are complex descriptions that do not presuppose that the actor intended to kill, intended to maim, etc.[11] As we shall see in Chapter 8, most of the verbs of action to be found in the special part of Anglo-American criminal law are of this latter kind, making Hall's sceptical generalization much too wide.

Even for that minority of action verbs that are intentionally complex, there is no reason that we cannot impose the *actus reus/mens rea* distinction on them. Consider the offence of

[9] Hall, *General Principles of Criminal Law*, 230.

[10] Anscombe, *Intention*, 87. For a more recent discussion of what Jonathan Bennett calls 'intention-drenched verbs', see Bennett, *Events and Their Names*, 205–6. A. C. E. Lynch uses Anscombe's insight to taxonomize various of the intentionally complex action descriptions contained in English criminal law. Lynch, 'The Mental Element in the *Actus Reus*', *Law Quarterly Review*, 98 (1982), 109–42.

[11] Compare Langford, *Human Action*, 29: 'it is the intention with which a person acts that provides the criterion of identity of an action. It is because Brutus intended to kill Caesar when he stabbed him that his action is identified as killing Caesar, and not merely as moving his arm.' If this were true of 'killing', then all verbs would have hidden mental-state requirements (in addition to the requirement of a willed movement). But Langford's observation is surely and obviously false. Brutus *killed* Caesar because the willed movements of his body caused Caesar to die. Brutus's intent is surely relevant to his culpability, but not to whether he did kill Caesar. He could, after all, have killed Caesar by accident, i.e. not intending to do so.

'impersonating another who is entitled to vote'.[12] The verb 'to impersonate' both in ordinary and in legal usage requires that one not only do things that might lead a registrar of voters to believe that one is someone else (who is a registered voter), but also that one do those things with the intent of so deceiving the registrar of voters. In such case, the *actus reus* of the offence is the objective part: the volitionally caused bodily movement(s) that themselves cause the state of likely belief by the registrar. The *mens rea* is the intent to so deceive the registrar as to one's identity.

Not only *can* we impose the *actus reus*/*mens rea* distinction on such intentionally complex verbs, we have good reason to impose such a legal distinction on these ordinary verbs. With respect to accomplice liability, for example, in applying the maxim that there can be no accomplice liability without a guilty principal, the *guilt* required of the principal is only that he do the prohibited act; with regard to the crime of 'impersonating', this is best construed to be the objective part, namely, that he do those things that in fact make it likely the registrar will be deceived as to his identity, not that he intend to so deceive the registrar.[13] More generally, wherever we need to distinguish wrongdoing (bad action) from culpability (bad actor) in the criminal law, we will need to impose the *actus reus*/*mens rea* distinction on these intentionally complex action verbs. Fortunately we can do so by again distinguishing volitions from all other mental states.

A fourth and final interpretation of the supposed necessary connection between *actus reus* and *mens rea* comes from the troublesome idea of criminal negligence. Negligence does not name a state of mind—such as an attitude of indifference or a conscious awareness of a risk—nor even the absence of a state of mind, such as failing to advert to a risk. This is one of the meanings of the common saying that negligence is an 'objective' requirement. Yet despite this, negligence is one form of the *mens rea* requirement for some crimes. The objection that stems from these facts is the mirror image of the objection that some mental-state

[12] This offence is discussed briefly in this context in Lynch, 'The Mental Element in the *Actus Reus*', 119.

[13] On the question of whether one should take into account the culpability of the principal in ascertaining the liability of the accomplice, compare J. C. Smith, 'Comment', *Criminal Law Review* (1987), 480–5, with Sanford Kadish, *Blame and Punishment* (New York, 1987), 181–6.

requirement is built into the *actus reus* requirement: now the objection is that a non-mental-state conduct requirement constitutes the *mens rea* requirement for some crimes, so how can one distinguish the *actus reus* from the *mens rea* of such crimes?

Consider negligent homicide. Killing another human being is usually thought to be the *actus reus* of this crime, and doing so negligently is its *mens rea*. The objection is that such a separation is artificial, for 'Do not kill another human being negligently' is only a shorthand designation for more specific prohibitions, like: 'Don't shoot at a target without adequate bullet-absorption material behind it', 'Don't construct or operate basement nightclubs without fire exits adequate to their patronage', etc. The objection would conclude that once one sees the true *actus reus* requirements of negligent homicide, there is nothing else required for liability that can be called the *mens rea* requirement.

This objection would have merit against use of the distinction for crimes of negligence if the special part of Anglo-American criminal law contained but one prohibition using the negligence concept: 'Don't act negligently'. About such a prohibition, it would make little sense to separate an *actus reus* requirement from a *mens rea* requirement, for the reason referenced by the objection: there is no type of action prohibited except for *negligent* actions, and, once these are found, there is no further *mens rea* requirement to be satisfied.

Unlike torts, the concept of negligence is not so used in Anglo-American criminal law. Rather, our codes prohibit: negligent killing of another human being; intercourse with a woman when one negligently fails to realize that she has not consented; etc. Such prohibitions use complex action descriptions like 'killing another human being', or 'having intercourse with a woman without her consent'. Because of this, it makes good sense to enquire into an accused's liability in two steps: first, was the particular act he performed an instantiation of the type of action prohibited by the statute? Secondly, did that act, in addition to having the properties required of it to be the type of act that is prohibited, also possess the property of negligence? This is precisely the division we impose for crimes of intention or recklessness, and such division is fruitful because it separates the question of what the accused did from the question of what he thought he was doing. In the case of negligence, the second question of course is not one of mental state, but it is usefully separated none the less. Whether an accused in fact killed another is one question, and whether it was reasonable to do so (either in terms of the objective facts of the situation, or in light of the facts reasonably available to the actor, given his capacities) is another. Even when *mens rea* does not name a true mental state, as it does not in negligent crimes, it is usefully distinguished from *actus reus*.

1.2. Actus Reus *versus Causation*

About the distinction between the *actus reus* and the causation requirements, we can be much more brief: there is no distinction here because there is no 'causation requirement' that is not fully incorporated into the *actus reus* requirement. For statutes eschewing the use of complex action verbs like 'killing' or 'abusing' in favour of the locutions 'causing the death of another' or 'causing the abuse of an infant', the *actus reus* just is such causing of some state of affairs.[14] For statutes framed in terms of the more idiomatic complex verbs of action, there is no causal question to be asked for liability beyond that asked in applying such verbs to start with.[15] Once a court has found that the defendant by his voluntary act caused the death of another—found that the defendant thus *killed* another—there is no further causal question to be asked. The *actus reus* requirement of either kind of statute requires, either explicitly or implicitly, that a causal issue be resolved, but, once that issue is resolved, there is no separate causal issue to be addressed.

1.3. Actus Reus *versus Absence of Excuse or Justification*

As we have seen, Meir Dan-Cohen has urged a residual definition of '*actus reus*': 'the *actus reus* designates all the elements of the offence except the *mens rea*'.[16] As Glanville Williams recognizes (and endorses), so defining '*actus reus*' means that it 'includes . . . not merely the whole objective situation that has to

---

[14] See e.g. American Law Institute, *Model Penal Code* (Philadelphia, 1962), §210.1 (one is guilty of criminal homicide if one 'causes the death of another human being'); Md. Stats. Ann., Art. 27, §35A (one who 'causes abuse' to a minor child is guilty of child abuse).

[15] This is to anticipate the result of the discussion that follows in Ch. 8.

[16] Dan-Cohen, '*Actus Reus*', 15. For a similar usage of the term *actus reus*, see Lynch, 'The Mental Element in the *Actus Reus*', 110.

be proved by the prosecution, but also the absence of any ground of justification or excuse'.[17] Since Williams and Dan-Cohen are right about including the absence of justification as part of the *actus reus*, wrong about so including the absence of excuse, I shall begin by distinguishing justifications from excuses before defining the *actus reus* requirement in light of the distinction between them.

In morality, justification has to do with the moral wrongness of some action. An action that is normally wrong, like killing, is not on this occasion wrong if it is justified (say, by the saving of the innocent lives being threatened by the person killed). Excuse, by contrast, has nothing to do with the wrongness of the act done even if it does have to do with the moral culpability of the wrongdoer. The killing done by an insane person is no less wrongful, even if he is not culpable in the doing of it because he is excused by reason of his insanity.

This moral distinction impacts upon the *actus reus* requirement in the following way. As we shall examine in Chapter 9, one rationale for there being some *actus reus* prerequisite to criminal liability is captured by the ideal of legality. Part of that ideal is that citizens can figure out what conduct is criminal, if they wish to, by consulting the criminal code. The law that defines the prima-facie elements of the various offences is certainly part of what they need to know in order to plan their behaviour in such a way as to avoid criminal sanctions, but so is the law that defines the justifications.[18] Citizens need to know that self-defence justifies a homicide otherwise criminal as much as they need to know that killing itself is prohibited. Thus, although for-

---

[17] Williams, *Criminal Law*, 20.

[18] This point may seem to be blunted by the presence of open-ended justificatory provisions of criminal codes, like Model Penal Code §3.02, which exempts from punishment any otherwise criminal act whenever 'the harm or evil sought to be avoided by such conduct is greater than that sought to be prevented by the law defining the offense charged'. Such open-ended justificatory provisions do offend against the values defining the ideal of legality; yet a paramount value—not punishing those who have done no wrong—intervenes to justify such provisions, for the possible circumstances of justification outrun the ability of any code to describe them. On the latter point, see Moore, 'Torture and the Balance of Evils', 342–3. While the possible circumstances in which moral wrongs are done also outruns the ability of any code to describe them, we rightly care more about *not* punishing the morally innocent than we care about punishing the morally guilty, so here it is the ideal of legality that is paramount.

mally one can separate justifications from the rest of *actus reus* (by the burden of proof), there is little reason to do so in light of the common ideal served by having both sets of doctrines clearly spelled out in advance.

This conclusion is reinforced if we consider the reason why[19] we have the substantive prohibitions we do. As I argue later in Chapter 9 the reason we prohibit murder, cruelty to animals, abusing a corpse, rape, etc. is that such acts are morally wrong. The justifications operate as exceptions to the rules describing such acts as wrong, and for that reason, too, the justifications are an integral part of the requirement of legally prohibited wrong-doing (*actus reus*).

Aside from this functional test of integration between *actus reus* and lack of justification, the unity of these two aspects of criminal liability can be seen by grasping how purely accidental formal, or procedural is the distinction between them. Take the federal offence of 'obstructing the passage of the mails' by way of example. In a well-known case,[20] a federal mail carrier was literally obstructed from carrying the federal mail on a riverboat on which he was travelling. The obstructing defendant was a state sheriff who stopped the riverboat (and thus the mail) in order to arrest the federal mail carrier under a valid state bench warrant for murder. The defendant's justification for obstructing the federal mails was that it was necessary in order to achieve a higher good, namely, getting one murderous individual off the streets (or the river). Does it matter to anything other than procedural issues—who has the burden of raising the issue, of producing evidence on it, of risking non-persuasion of the jury on it—whether we cast such a defendant's claim as one of justificatory defence or of no *actus reus* (that is, that a *literal* obstructing is not a *legal* obstructing)? I can't see that it does, making the *actus reus*/justification distinction simply a function of procedural consequence in an adversary system, but not a distinction of substance.

Or consider the matter from the point of view of a legislative draftsperson: are there any reasons other than procedure for preferring a homicide statute that reads 'Maliciously killing another

---

[19] See, more generally, Moore, 'The Limits of Legislation', *USC Law* (Fall, 1985), 23–32; id., 'A Theory of Criminal Law Theories'; id., 'Sandelian Antiliberalism'.

[20] *United States* v. *Kirby*, 74 U.S. (7 Wall.) 482 (1868).

human being is murder, except in self-defence', to one that reads 'Maliciously killing another human being is murder', where the second statute is accompanied by a set of defences to murder that include self-defence?

Yet another way of seeing the unity of *actus reus* and lack of justification has to do with the *actus reus/mens rea* distinction. Imagine that some defendant, D, believes that someone, X, needs defence from the aggressions of another, V. D shoots V under that belief. The objective fact of the matter is that V was *not* about to shoot X, although it could appear that he was. If absence of justification is part of what makes murder wrong, then the criminal law should apply the *mens rea* requirements of homicide to that element no less than to those elements that are part of the prima-facie case (here, the *actus reus* of killing another human being). That the criminal law, with some hesitations, does apply the same *mens rea* requirements to the absence of justification as to the killing[21] shows that they are equally good candidates to be considered part of the *actus reus* requirement for homicide.

There is one implication of treating the elements of *actus reus* and absence of justification as one integral whole that might be troubling to some. It is certainly an implication that goes against the prevailing case-law on the subject. The implication arises in what are called problems of 'justificatory intent'. Imagine again that an innocent party, X, is about to be shot by another, V, in circumstances where, objectively, defendant D would be entitled to defend X with deadly force. D, however, does not know that V was about to kill X, but none the less (for malicious motives of his own) shoots V dead just in time to prevent V from shooting X. Is D guilty of murder, or may he avail himself of the defence of defence-of-others so that at most he is guilty of attempted murder? If one accepts, as do I, the unity of justification and *actus reus*, the correct answer is that D has done no wrong. He is culpable because he tried to, but it is for that culpability for which he should be punished. D no more murdered

---

[21] See American Law Institute, *Model Penal Code*, §§2.02(4), 1.13(9)(10). The common law more hesitantly applied its *mens rea* requirements to the elements of justification in that mistake of fact as to justificatory matters had to be reasonable to negate the *mens rea* required. See e.g. Joshua Dressler, *Understanding Criminal Law* (New York, 1987), 186.

anyone than does the person who tries to shoot another to death but whose bullet misses. Both are very culpable, but neither has done the wrong of murder because neither has killed a human being who was *not* about to kill an innocent party.[22]

This is not for me a counter-intuitive conclusion, although it appears to be to many other criminal-law theorists and courts. Yet if one adopts the conventional view (that D in the last hypothetical is guilty of murder and not just attempted murder), the onus is on him to explain how justifications differ in any morally relevant way from exceptions contained in the definition of the *actus reus* proper. He has to say that a statute prohibiting 'malicious killing of a human being, except in self-defence' *does* differ from a pair of statutes, one of which prohibits 'malicious killing of a human being' and the other of which grants as a defence that it is no crime to kill in self-defence. The first sort of criminal code, he might say, fails to map on to morality's absolutist norms of obligation which are short and snappy like "Thou shalt not kill'. The second sort of code is preferable, on this view, because the *actus reus* of the homicide statute matches the absolute and exceptionless norms of morality. The justifications of self-defence or defence of others, on this view, do not make killings morally right, even in the weak sense of morally permissible; rather, such justifications would have to be viewed like a Hohfeldian liberty or privilege to do the wrong thing (i.e. kill in self-defence or in defence of others). These justifications, in other words, would be like the view of some regarding the liberty of women to abort a fetus: woman have such a liberty such that no one has a right that they not abort their fetuses, but it is still wrong for them to abort their fetuses.

This is the best I can do to make sense of how one has to see justifications like self-defence or defence of others in order to defend the view that these defences are in any non-procedural way distinguishable from the definition of the *actus reus* that is part of the prosecutor's prima-facie case. And in truth, it makes no sense to me. It is not (morally or legally) *wrong* to defend oneself or others with deadly force; these justifications operate

---

[22] On this view of justification, see Paul Robinson, 'A Theory of Justification: Societal Harm as a Prerequisite for Criminal Liability', *University of California Los Angeles Law Review*, 23 (1975), 266–92. Compare Fletcher, *Rethinking Criminal Law*, 559–69.

just like exceptions or any other feature of moral and legal norms excluding certain kinds of killing from being wrong. With regard to defence of others, sometimes it is even obligatory (not just permissible) to kill an attacker about to kill innocents; that is, it is right (not-wrong) in the strongest sense. I accordingly see no moral basis for the received wisdom that defendants like D are guilty of murder rather than attempted murder.

All of this changes if we move from justification to excuse. The formal ideal (of legality) behind the *actus reus* requirement is not served by regarding the excuses to be part of the *actus reus* of crime. An insane person—or one who is very young, involuntarily intoxicated, acting under duress, etc.—does not need to know that insanity will be an excuse to homicide if he commits it. Potentially excused persons may *want* to know this, but the ideal of legality is not enhanced if they do know it. Potentially excused persons only need to know when killing is wrongful and when it is not, in order to try as hard as they can to avoid wrongful killings. If they perpetrate a wrongful killing, and they are insane, we will excuse them, but the availability of that excuse is not something they need to know in planning their behaviour.

The substantive ideal behind the *actus reus* requirement is also not served by including the absence of excuse as part of the *actus reus*. The legal prohibition of serious moral wrongs is not affected by whether or not there are doctrines of excuse. Without including the justifications within the *actus reus* of offences, a code would be seriously deficient because it would not be punishing only acts that are moral wrongs. The same can hardly be said for the excuses, since they have nothing to do with the wrongfulness of conduct but only with the culpability of those whose conduct it is.

Assuming that there is good reason to draw a line between the *actus reus* (including absence of justification) requirement, and the requirement that conduct not be excused (the 'excuse requirement'), it remains to be shown that there is a conceptually coherent place at which to draw the line. This worry finds its main expression in the confusion surrounding the distinction between justifications and excuses. Kent Greenawalt has recently explored what he calls the perplexing borders of justification and excuse.[23]

---

[23] Kent Greenawalt, 'The Perplexing Borders of Justification and Excuse', *Columbia Law Review*, 84 (1984), 1897–1927.

In my first hypothetical (of D mistakenly but reasonably believing that killing V was necessary to saving X), for example, Greenawalt urges that D would commonly be said to be *justified* in his killing of V despite his mistake. Yet D's exculpatory ground looks similar to one of the classic grounds of excuse, namely, ignorance or mistake. So haven't we something like excuse built into justification (and, thus, into *actus reus*)?

Greenawalt's mistake is with his initial premiss. In the hypothetical, D objectively did a wrongful act. D may idiomatically be said to have been 'justified' in doing what he did because he was epistemically justified in believing what he believed. But D's *action* is not justified; it was wrongful. D's action is thus only excused. So long as one classifies mistaken beliefs about elements of normally justificatory defence to be excuses, there is no 'perplexing border' here. Only the absence of those factors that show an act otherwise wrongful not to be wrongful on this occasion are part of the *actus reus* requirement, and, by that criterion, a mistake about one's justification for acting is not itself a justification but only an excuse.

Although Greenawalt elegantly details a variety of other examples in his nicely written article, all are premised on the assumption that justification is a *subjective* matter, that is, that it is governed by what the accused reasonably believes and not by the objective facts of the matter. Change that assumption, as I have urged we must, and none of such examples present any conceptual problem for the line between the *actus reus* (including absence of justification) and the excuse requirements.

2. INTERPRETIVIST SCEPTICISM

As we have seen, interpretivists are sceptical about the act requirement in two dimensions. First, at the level of legal doctrine, they are sceptical about the bite the act requirement can have in light of judges' alleged ability to find the requirement satisfied or not, depending on how broadly or narrowly they 'time-frame' their search for an act. Secondly, at the level of metaphysics, they are sceptical about the reality of the distinction between bodily movements that are acts and those that are not:

they say that it is the interpretive stance that we as observers take towards such events that fully determines what they are.

Interpretivists are also sceptical about the *actus reus* requirement in two analogous dimensions. First, at the level of legal doctrine, they deny that the rules making up the prohibitions of the special part of the criminal law have any determinate content. Like the Legal Realists before them, such rules (as those prohibiting murder, arson, burglary, etc.) are only 'pretty play things',[24] so that the *actus reus* requirement is a requirement in name only. Really, it is a licence to judges to declare conduct they don't like criminal. Secondly, at the level of metaphysics, they deny that there are any natural kinds of actions. Even when they concede the reality of particular acts, they like the nominalists deny the reality of any natural groupings of those particulars into kinds. On such a view, it is only arbitrary human convention that carves up human behaviour so that killing is one kind of act and robbery is another.

These two scepticisms support one another. It is easier to believe that legal standards are indeterminate if the things to which such standards purportedly refer do not exist as natural kinds; and if the standards that purport to refer to natural kinds of acts fail to refer to such kinds because they don't exist, at least one form of objective interpretation of legal standards is impossible.

Taking the second scepticism first: the root intuition here is the same doubt about the metaphysics of events that we encountered before, in Chapter 4. Events are somehow more suspect as things than are material objects like gold or whales. Here, the scepticism is not to deny that particular events exist, but, rather, to deny that *kinds* of events exist in any sense other than as purely nominal kinds. We are all nominalist sceptics about some universals, such as the universal 'Figuoroa Streetness'. Although I think that there are bits of pavement running north–south in Los Angeles, that each bit partakes of 'Figuoroa Streetness' I take to be purely a function of human convention: we aggregated those bits, and not others, into Figuoroa Street, and it is only that conventional aggregation that makes Figuoroa Streetness a kind.

Whales, gold, water, intentions, anger, etc. are not conventional kinds like Figuoroa Streetness. Such kinds would exist

---

[24] Karl Llewellyn's deprecating description of legal rules. Llewellyn, *The Bramble Bush*, 3rd edn. (New York, 1960), 14.

even if there were no human conventions aggregating particular objects as being objects of a certain kind, and even if there were no name for such kinds. Conceding this to be true of whales, water, etc., the sceptic here considered none the less likens kinds of events to nominal kinds and not to these natural kinds. Stephen Munzer, for example, concedes that whales and water are natural kinds but none the less has doubts about whether such status applies to kinds of events like death.[25]

Yet if we repair to the reasons given to justify our beliefs in the reality of whales, water, etc. as natural kinds, those reasons are as applicable to event-kinds. The reason justifying my belief in the reality of kinds like gold, water, whales, starts with certain features of our linguistic behaviour. English speakers have words for whales, water, gold, and so do speakers in many other languages. A natural explanation for why there are such categories is because we were caused to invent them by the existence of such kinds. This explanation has to compete with the alternative explanation of the nominalist: that we have such categories only because we found it useful to invent them for one reason or another. The realist explanation seems to be the best of these competing explanations because it can account for the following fact about our linguistic behaviour: We guide our usage of words like 'whale', 'gold', 'water', not by the conventionally accepted indicators of their idiomatic usage, but by the real natures of the kinds referred to, as best we can grasp that nature at any point in time. Such linguistic behaviour makes sense if it is causally responsive to the hidden nature of natural kinds; it does not make sense if such behaviour is causally responsive only to human needs and interests (which the conventions in place at any given time may well be serving perfectly adequately).[26]

This 'inference to the best explanation' strategy for answering ontological questions applies as easily to events as it does to substances and material objects. As I have elsewhere argued in detail about death as a natural kind of event,[27] we have the same lin-

---

[25] Stephen Munzer, 'Realistic Limits on Realist Interpretation', *Southern California Law Review*, 58 (1985), 459–75: 468–9.

[26] This argument for realist interpretation of the words by which we refer to natural kinds is laid out at considerably greater length in Moore, 'A Natural Law Theory of Interpretation', 291–301, 322–38, 340–1. For an application of this interpretive view to criminal law, see Leo Katz, *Bad Acts and Guilty Minds*, 82–8.

[27] Moore, 'A Natural Law Theory of Interpretation', 291–301, 322–38.

guistic tendencies about 'death' as we do about 'whale', etc. The best explanation for these tendencies is again the realist one, not the nominalist one. On that basis, we should infer that events form natural kinds every bit as much as do material objects.

Thus, particular events can be instances of natural kinds, and some such events surely are. What of those kinds of event we call human actions? Are the kinds of human action natural kinds, so that there is a hidden nature guiding their identity and their individuation? Or are the kinds of which particular acts are instances only nominal kinds, kinds constituted by the conventions that define them? Action as such I take to be a natural kind (the nature of which was the subject of Chapters 4–6 of this book), but that is not my present question. The question is not, is human action a natural kind of event; it is, rather, are the kinds of human action there are (killings, hittings, etc.) natural or conventional?

I shall argue in Chapter 8 that action-kinds like killing, hitting, etc. are not natural kinds in that their essential nature is fixed by conventional stipulation or custom. Just as 'bachelorhood' is conventionally assigned to be the name of the class of things that shares three properties (personhood, maleness, and single marital status), so 'killing' is conventionally assigned to be the name of the class of things that share the properties volitionally caused bodily movement, death, and a proximate causal connection between the movement and the death. Conceptually (that is, true by the conventions that give the word its meaning) an action cannot be a killing where there is no death any more than a married person can be a bachelor.[28]

The important point here is to avoid any sceptical slide into nominalism about action-kinds. For even though the essence of such kinds is fixed by convention, and not by nature, such kinds still have a unitary essence. 'Killing' in this respect differs from

---

[28] Notice that nominal kinds like 'killing' or 'bachelor' may be defined in terms of words that themselves name natural kinds. Males, persons, deaths, causings, and bodily movings$_T$ may all themselves be natural kinds without an aggregation of them being a natural kind. The conventionalist reading of 'malice aforethought' (as it is used in the law of homicide), for example, says that 'malice' is the name for the disjunction of four kinds of mental state. (See Moore, 'A Natural Law Theory of Interpretation', 333–8, for a discussion of the conventionalist reading of 'malice' in the law of homicide.) That at least some of these four are natural kinds in no way makes *malice* a natural kind.

'Figuoroa Streetness'. Each act (that is a killing) is a killing because it has the property, causing death; it is not a killing simply because our society has called that particular action 'a killing', as each bit of pavement partakes of Figuoroa Streetness only if a label with that name has been attached to it. To say that the nature (or essence) of the kind killing is conventional is not to say that the kind has no nature. It does, and an act either is a killing or it isn't, irrespective of how a majority of native English-speakers happen to label that act.

I come, then, to the other kind of interpretivist scepticism here: that which focuses on the alleged indeterminacy of legal rules using words like 'kill', 'hit', 'abuse', and the like. It may seem that if we answer the nominalist scepticism about action-kinds, we will also have answered the doctrinal scepticism of the interpretivist. For if killings as a kind have a nature (albeit one that is conventionally fixed), mustn't the word 'killing' as used in the homicide statutes be taken to refer just to those particular events that are killings? If so, that seems to answer the interpretivist sceptic quickly, for then criminal statutes have a very determinate content to their *actus reus* prohibitions.

Any such fast answer to the interpretivist sceptic here runs head-on into the standard arguments and counter-examples to all forms of literalist interpretive schemes.[29] Consider but one such counter-example that Lon Fuller[30] advanced against Herbert Hart's own (quite different) form of literalism.[31] Imagine some statute makes it criminal to *sleep* in a railroad station. Sleep I take to be a natural kind of event, having a certain physiological and phenomenological nature. Fuller's query: should we interpret the statute to punish the person waiting for the train who nods off for a moment, because he is literally sleeping? And should we not punish the homeless person who spreads out his possessions for the night in preparation for sleep, but who is not yet literally sleeping? However one resolves Fuller's two cases, the point they

---

[29] Surveyed in M. Moore, 'The Semantics of Judging', *Southern California Law Review*, 54 (1981), 151–295.

[30] Lon Fuller, 'Positivism and Fidelity to Law', *Harvard Law Review*, 71 (1958), 630–72.

[31] H. L. A. Hart, 'Positivism and the Separation of Law and Morals', *Harvard Law Review*, 71 (1958), 593–629. Friends of Herbert Hart rarely think that he was guilty of any form of literalism in this essay, although what he said can be taken in that direction.

make is surely a good one: it is not a good interpretive methodology to interpret the *actus reus* provisions of criminal statutes
solely by the nature of the events normally referred to by the
words used in such provisions.

This opens the door for the interpretivist sceptic, for it may
seem that legal interpretations become wholly indeterminate once
we admit that a judge in every criminal case must use her own
value judgements (namely, to check the literal meaning of words
used in statutes to see if that meaning leads to results disserving
the values such a statute ought to serve). This opening of the
door to the sceptic is only illusory, however, as I have sought to
show in detail elsewhere.[32] Even if legal interpretation is a blend
of a variety of ingredients—things like literal meanings, statutory
purposes, precedent, and all-things-considered value judgements—there is no reason to think that such blends do not have
right answers. Interpretive judgements are in this respect no different from judgements resolving other moral dilemmas between
competing moral principles.[33]

[32] Moore, 'A Natural Law Theory of Interpretation'.
[33] Compare Richard Fallon, 'A Constructivist-Coherence Theory of Constitutional Interpretation', *Harvard Law Review*, 100 (1987), 1189–1286 (criticizing
Moore for the indeterminacy resulting from having four different ingredients in
interpretation), with Moore, 'Moral Reality', *Wisconsin Law Review* (1982),
1061–1156: 1151–2 (conflict of moral principles no reason to be sceptical of determinate resolution); id., 'Law, Authority, and Razian Reasons', *Southern California Law Review*, 62 (1989), 827–96: 845–8 (apparent conflict in morality's reasons
is no basis for scepticism about determinate resolution of the conflict).

8

UNITY IN COMPLEX ACTION DESCRIPTION AND IN THE *ACTUS REUS* REQUIREMENT

With regard to the act requirement, I treated the position that there were as many act requirements as there were verbs of action used in the various prohibitions of the special part as a sceptical position. The analogous position about the *actus reus* requirement need not be a sceptical position at all. For, in a sense, the *actus reus* requirement *is* as diverse as are the complex actions forbidden by criminal law. The *actus reus* requirement is that no penal liability may attach to any individual unless that individual has done an act that instantiates a type of action prohibited by some statute in force at the time the act was done. The content of such a requirement is of course as diverse as the types of action that the statutes of the special part prohibit—actions like killing, hitting, disfiguring, etc. In that sense, the *actus reus* requirement is indeed as many requirements as there are types of action prohibited by the diverse statutes of the criminal law.

Despite this, it would be disturbing if the numerous complex actions of killing, telephoning, driving, signalling, etc. shared no common features beyond the feature that made them one and all *actions* (namely, their being or including volitionally caused bodily movements). For then it would be cumbersome to describe (say in a criminal code) what must be true for there to be a killing, a maiming, a signalling, etc., beyond simply the doing of a basic act. Each complex description of action would have its own unique requirements. There would then be nothing general to say about the *actus reus* requirement beyond what was already said about the act requirement.

As we have seen, the Bentham–Austin theory of action sought greater unity to action description than this. Bentham thought that all complex action descriptions are constructed out of only three elements: basic acts, the consequences such acts cause, and the circumstances in which they are done.[1] As Bentham put it:

> Together with this [basic] act, under the notion of the same offence, are included certain circumstances: which circumstances enter into the essence of the offence, contribute by their conjunct influence to the production of its consequences, and in conjunction with the act are brought into view by the name by which it stands distinguished.[2]

This tripartite discussion of complex descriptions of actions into three parts was echoed by John Austin, who held that 'every act is followed by consequences; and is also attended by *concomitants*, which are styled its *circumstances*'.[3] This threefold division has become very much a staple of orthodox criminal-law theory, being echoed by Cook, Salmond, and others.[4] The American Law Institute in its Model Penal Code divides all offences into elements that are either acts, circumstances, or results,[5] and a like division has long been a staple of English criminal-law theory.[6]

The upshot is that the orthodox view holds not only that every complex description of an act requires that there be a basic act, but that what else a complex description requires is that certain

---

[1] Bentham, *Introduction to the Principles of Morals and Legislation*, 71, 77.
[2] Ibid. 80.   [3] Austin, *Lectures on Jurisprudence*, i. 433.
[4] Cook, 'Act, Intention, and Motive'; Salmond, *Jurisprudence*, 398–402; Kadish, 'Act and Omission, *Mens Rea*, and Complicity'.
[5] American Law Institute, *Model Penal Code*, §1.13(9). The Code divides up 'offence' (*actus reus*) into three categories: conduct, circumstances, and results. 'Conduct' is then defined as an act or a series of acts (§1.13(5)0, while 'act' in turn is defined as a bodily movement (§1.13(2)). It is this definitional matrix that justifies me in attributing to the Code the orthodox Bentham–Austin view: that all complex descriptions of action are equivalent to a description of a basic act plus one or more circumstances or consequences. However, when the Code actually uses its threefold distinction it alters one of the three terms: conduct now has a 'nature' that seems richer than mere bodily movement and yet is still distinct from either circumstances or results. See e.g. 2.02(2)(a)(i) and 2.02(2)(b)(i). This latter division is not the same as the Bentham–Austin view adopted in §1.13; it also is a *worse* division, introducing unnecessary fuzziness in the nature/result and nature/circumstance lines. See e.g. Kadish, 'Act and Omission, *Mens Rea*, and Complicity', 80–1; Paul Robinson and J. Grall, 'Element Analysis in Defining Criminal Liability: The Model Penal Code and Beyond', *Stanford Law Review*, 35 (1983), 681–762: 705–19. A like fuzzying of the Bentham–Austin threefold distinction has taken place in English criminal-law theory. See Richard Buxton, 'Circumstances, Consequences, and Attempted Rape', *Criminal Law Review* (1984), 25–34: 28–32.
[6] See the discussion in Buxton, 'Circumstances, Consequences, and Attempted Rape'.

consequences occur in the presence of certain circumstances. All complex action descriptions can be replaced by an equivalent description of: (1) a basic act; (2) a set of circumstances in the presence of which the doing of that basic act is forbidden; and (3) a set of consequences which, if they follow from the basic act, make the doing of that basic act forbidden. Statutes prohibiting the complex action of rape, for example, equivalently prohibit: (1) any volitionally caused bodily movement, (2) that causes penetration of the female to take place, (3) in the circumstance when that consequence is not consented to by the woman.

This version of the orthodox view is commonly called the 'elements approach' to *actus reus*.[7] The elements approach commits one to believing that there are *two* commonalities to the *actus reus* requirements of all statutes: first, as noted before that there be a basic act; and secondly, that, although the circumstances or consequences prohibited by different statutes differ, they are all either consequences or circumstances. This last commonality reduces the bountiful diversity of complex descriptions of actions to being complex in only two possible dimensions: circumstantially complex and causally complex. It is this equivalence of complex action descriptions to descriptions of basic acts being done in certain circumstances or producing certain consequences that we shall look at in the subsections that follow.

We shall thus be examining the fourth of the theses that make up the orthodox theory of action in criminal law, what I called in Chapter 3 the equivalence thesis. The equivalence thesis asserts that any complex action description used in a criminal statute is equivalent to a description of the canonical form 'In circumstances C basic act A causes consequences E.' The statute prohibiting 'killing a police officer while on duty', for example, by the thesis equivalently prohibits 'performing some basic act that causes the death of a person in the circumstances that that person was a police officer and that such killing took place while the latter was on duty'.

I shall approach the equivalence thesis in four steps. First we need to be clear about what the thesis does and does not assert. This is a matter of keeping in mind the differences between: descriptive and other uses of language, attributive and referential

---

[7] Robinson and Grall, 'Element Analysis in Defining Criminal Liability'.

uses of descriptions, and action-types and action-tokens. Secondly, I shall examine a variety of schemes for taxonomizing the various dimensions in which an action description may be complex. I shall here seek to defend the orthodox view that, for criminal-law purposes, only two dimensions of complexity need be distinguished. Thirdly, I shall then ask whether, conceding the existence of circumstantial complexity, all complex descriptions of action used in the criminal law are *at least* causally complex. Fourthly, I shall examine what I take to be the main objection to the equivalence thesis: that the relation between Jones and Smith's death is different in 'Jones killed Smith' from what it is in 'Jones caused Smith's death'. The general form of the objection is that causally complex action descriptions ('Jones killed Smith') require more than is required by descriptions either of a person's causal agency ('Jones caused Smith's death') or of basic act causality ('Jones's movement of his finger caused Smith's death').

1. DESCRIBING ACTIONS

As I discussed in Chapter 4, a once-held thesis of Herbert Hart's was that action sentences don't *describe* anything; they *ascribe* responsibility to a person for a result. We have seen reason to reject that thesis when applied to our usage of action sentences to talk about individual actions. There is an additional reason to reject ascriptivism when applied to statutory usages of the verbs of action. Ascriptivism renders incomprehensible what a legislature would be doing when it passed a statute saying 'No one shall kill a police officer'. At least the emotivist cousin of the ascriptivist could analyse such statutory utterances as expressing a negative attitude towards the type of action there described;[8] but, according to the more sceptical ascriptivist, such expressions do not describe anything, so there is nothing towards which the legislature could be expressing a negative attitude. Moreover,

---

[8] Emotivism in the philosophy of language is a close cousin of ascriptivism because both views were motivated by a common insight: language may be used to do various things besides describe how the world is. An emotivist believes that certain words in their typical uses do not describe anything; they express emotion or prescribe to others that they too should feel an emotion.

there is no class of persons towards which one could be ascribing responsibility (as there is a person to be the subject of responsibility ascriptions in act-token references like 'Smith killed Jones'). Given the prohibition of bills of attainder, criminal statutes pick out a class of persons only by the type of action that they have done. Once the ascriptivist says that no type of action is picked out by usage of action verbs, then no class of persons (to whom responsibility may be ascribed) is picked out either.

I conclude that the apparent legislative speech-act is the real speech-act here: the statutes that make up the special part of the criminal law *describe* something with phrases like 'kill a police officer'.[9] Next question: what do they describe? As the response to ascriptivism makes clear, not particular act-tokens. 'Smith killed Jones' refers to a particular action; 'No one shall kill a police officer' refers to a type of action of which Smith's may be one. Criminal statutes are always universally quantified statements (applying to *all* persons in a certain class), so the actions they describe are always *types* of action.

What type of type? In the last chapter I mentioned the distinction between natural kinds and nominal kinds, urging that there was no good reason to suppose that events do not come in natural kinds just as do some physical objects. Despite this fact, almost all of the action-types prohibited by the criminal law are not natural kinds of kinds. Kinds of acts that are natural kinds are acts of bodily movement. Criminal statutes never prohibit these kinds of kinds; rather, they prohibit 'breaking and entering a dwelling house of another at night', 'killing a police officer', 'hitting another', and the like. These descriptions refer to nominal kinds of acts, kinds whose essential nature is conventionally fixed. Such kinds do have a nature beyond simply the affixing of a common label to an otherwise disparate class of particulars. But the nature is fixed by conventional stipulation. A killing, for example, is *any* natural kind of act that has the property causing the death of another. That such a causal property is essential to any (particular) act being an act of killing is wholly a matter of

---

[9] Whether statutes are really speech-acts by a legislature is not a question we need here to decide. For a persuasive argument that they are not, see Heidi Hurd, 'Sovereignty in Silence', *Yale Law Journal*, 99 (1990), 945–1028. If passing a statute is not a speech-act, that is even more reason to take its *actus reus* prohibitions as descriptive and not ascriptive.

linguistic convention. Because of our interests in consequences like deaths of persons, we created a causally complex action verb to pick out particular acts by virtue of this one property. We do not have causally complex action verbs for all possible consequences of actions. For example, I can 'cause an avalanche'; I can't 'avalanche'. I can 'cause smoke to rise' from a fire; I cannot 'smoke' (in that sense), nor can I 'raise the smoke'. I can 'cause another to smile'; I cannot 'smile' (in that sense). The impossibilities here are matters of etymological accident: we have no word to pick out the acts that cause these consequences, although we could. There is no nature to killings, drownings, frightenings, hittings, etc. that is not also had by my invented avalanchings, smokings, and smilings.[10] We simply haven't cared enough about consequences like avalanches, rising smoke, or smiles to form verbs picking out acts by their causing of such consequences

Suggestive as it is, this is not a conclusive argument against a natural-kind view of killings, hittings, etc. For there surely are many natural kinds that exist in the world even though no English-language user has named them. More conclusive is the following thought experiment: can an act be a killing without causing death? Can an act be a hitting without causing physical contact with the body of another? Can an act be a rape without causing penetration? Etc. These properties seem as obviously essential to their respective types as the property of being unmarried seems to the type bachelor. Whereas if 'killing', 'hitting', and 'raping' named natural kinds, these questions should not seem conceptually closed. They should seem more like the question of whether an item can be a tiger even if it has no stripes, does not live in Asia, and feeds only on vegetables. Even if there are no albino, African, vegetarian tigers, it does not seem conceptually guaranteed that there could not be, as part of the very meaning of 'tiger'.[11]

Usually this is put as the difference between referential versus attributive uses of descriptions like 'killing'.[12] To use a descrip-

---

[10] Judy Thomson coins such new verbs in her *Acts and Other Events*, 129.

[11] See generally Moore, 'A Natural Law Theory of Interpretation'. The example is from Hilary Putnam, 'The Meaning of "Meaning"', in H. Putnam, *Mind, Language, and Reality* (Cambridge, 1975).

[12] On the referential/attributive distinction, see generally Keith Donnellan,

tion referentially is to pick out some one certain thing, even if the description used misdescribes the nature of that thing. I can use the description 'the star that appears in the morning' to refer to the planet Venus even though Venus is not a star. Such a referential use of descriptions is possible only if the things referred to have a nature that is *not* fixed by the properties used in the description. By contrast, I can use a description attributively to refer to something, but only as that thing has the property I attribute to it. If I use the description 'Jones's widow', for example, I could be referring to Mary, his present wife while he is alive; but to use the phrase attributively is to refer to whomever (if anyone) has the property married-to-Jones-when-he-dies. Misdescriptions become impossible for attributive uses—which is why there are no open questions for such uses—although the descriptions can fail to refer to anything.

The importance of this discussion in the present context lies in its implications for the equivalence thesis. If causally complex verbs like 'kill' are used referentially to refer to natural kinds, and if there are such natural kinds, then to verify the equivalence thesis we should seek the nature of the type of events that are killings as a matter of scientific fact. One has to gather up all the killings that are known to see what common nature they possess. Alternatively, if such verbs are used attributively to name nominal kinds, our investigation is into matters of linguistic convention: do the semantic conventions that in part guide the correct usage of 'killing' attribute the property causing-death-of-another, or do they not? Since I take killings not to be natural kinds, and since I take 'killing' to be used attributively in statutes as well as in ordinary speech, I take the investigation into the truth of the equivalence thesis to be a matter of linguistic investigation.

2. THE STRUCTURE OF COMPLEX DESCRIPTIONS OF ACTION

Both in ordinary speech and in the special part of the criminal law we are confronted by a bewildering variety of ways to describe actions. Alan White nicely captures this prolixity of possible description:

we can describe it [an act] in physical terms, for example moving an arm; or by its agent, for example, the act of a madman or a gentleman; or as being of a certain genre, whether moral (for example act of charity or injustice), or psychological (for example act of desperation or kindness), or legal (for example act of bigamy or perjury), or institutional (for example act of homage or loyalty). An act may be described in explanatory terms by reference to its purpose, reason, or cause, for example practicing the piano, obeying an order, an act of stupidity. Its description may contain a reference to its circumstances, for example its time (burglary, in earlier law, was stealing between certain hours) or manner (slouching, mumbling) or its conventions (cheating, castling) or the relation of the agent to the person affected (an act of piety is confined to one in a filial relation, rape is intercourse without consent); or to its consequences, for example homicide is an attack whose result is the death of the victim within a year and a day; or to some antecedent event, for example repetition, volleying; or to an intention, for example stalking, an act of fraud; or to some motive, for example an act of jealousy or greed; or to its object, for example treason is an act against the state, stealing is taking other people's property.[13]

Despite the (J. L.) Austinian scepticism that takes such prolixity of action description to belie any common nature to acting as such, many philosophers since Aristotle have attempted to isolate a common nature possessed by all of the various properties by virtue of which we categorize and describe human actions. As Nicholas Rescher indicates in his own attempt of this sort, 'actions appear to have only a rather small number of basic dimensions, representing the diverse avenues along which the fundamental aspects of action can be explored'.[14]

Rescher's own taxonomy of recurrent descriptive features of particular act-tokens is quite similar to Aristotle's, and if we combine the two, alter them both to apply to descriptions of act-types, and eliminate overlaps, we get:

(1) the identity of the agent: what class of persons can perform the action (e.g. only persons who are public officials can be guilty of taking a bribe);

[13] White, *Grounds of Liability*, 26–7.
[14] Nicholas Rescher, 'On the Characterization of Actions', in M. Brand (ed.), *The Nature of Human Action* (Glenview, Ill., 1970), 254.

(2) the identity of the victim: what class of persons must be affected (e.g. only persons who are police officers give rise to the offence of insulting a police officer);

(3) the manner, means, or instrumentality used: how was the action done (e.g. eavesdropping by means of any electronic device, excluding a radio);

(4) the end, motive, belief, or intention with which the action was done: what mental state caused or accompanied the act (e.g. an act of lying can only be done by one who knows the falsity of what is said);

(5) the circumstances in which the act was done: what must be true at the time the act was done (e.g. parking in an area posted 'no parking').

Unhelpfully Rescher adds a final category:

(6) act-type: what must an actor do to do an action of this type (e.g. to kill, hit, etc.)?

About this last item Rescher notes that 'this is the fundamental item in the specification of an action',[15] which is surely an understatement even as applied to descriptions of particular actions; such a category is of course entirely hopeless for taxonomizing recurrent features of descriptions of act-types, for all of the preceding five variables go into determining what type of act is being described.

If we eliminate this sixth category, we are left with the previous five. All of these, I now want to argue, should be collapsed into one category, that of the circumstances in which an action is done. The circumstances in which an act is done then includes any relevant facts about the identity of the agent or another, the manner, means, or instrumentality used, and the mental state with which the act was done. For all of these are not only conditions at the time the agent acts, but, more importantly, the agent acting does not cause such conditions to be present by his act. Rather, by acting in such circumstances, his act is an instance of the type of act that such circumstances define. If he moves his finger on to the 'record' button of a recording device, his finger will be an instance of the action, recording a communication intended by the participants to it to be confidential without

---

[15] Ibid. 249 n. 2.

the consent of all such participants by means of an electronic amplifying or recording device, except a radio,[16] if each of these circumstances is present at the time the finger moves.

The features that all of these circumstances share and that prima facie justify lumping them all together as circumstances are two: first, that they all exist when the basic act is done, and secondly, that none of them is caused to exist by the basic act. Such features are even true of Rescher's fourth category, what we might call the 'inner circumstances' of mental state. Imagine three sorts of inner-circumstance cases:

(1) A recorded a conversation the participants intended to be confidential
(2) B lied to the police officer.
(3) C fished for salmon out of season.

In (1), the mental state that must exist is a mental state of persons other than the actor. This is an easy case in which to treat such mental states as just another circumstance. In (2), the mental state needed for B to have lied is B's own; but again, if the state (of belief in falsity) existed at the time B acted, then B did unproblematically lie.

The third case is more troublesome. In giving the 'fishing' example, Alvin Goldman correctly notes that an arm movement that causes a fishing line to dangle in the water will constitute an act of fishing only if it is done with the intention that the fish be caught. Goldman thus concludes that 'an agent's desires, intentions, or motives often occur as circumstances'.[17]

The worry here is that fishing is not done simply by dangling one's line in the water where there are fish in the circumstance that one desires fish to be caught; in addition, such dangling is fishing only if one's desire for fish *causes* one to dangle the line. This circumstance does not seem passive in the way the others are; it is not an effect of the basic act, but it is one of the causes of the basic act.

Such considerations might lead one to distinguish a dimension of intentional complexity from that of circumstantial complexity, and, indeed, some philosophers have done so.[18] Yet for criminal-

---

[16] The *actus reus* of the California Privacy Act.
[17] Goldman. *A Theory of Human Action*, 27.
[18] See e.g. Bennett, *Events and Their Names*, 205–6.

law purposes we can ignore any differences here so as to treat the intention with which an action is done as just another circumstance that exists at the time the act is done. This is for two reasons. One we have already mentioned in the last chapter: for action descriptions like (2) and (3) above, we will penetrate the unity of such descriptions in ordinary usage and impose the criminal-law distinction of *actus reus/mens rea* upon them. Such 'circumstances' as B's belief in the falsity of his statements and C's goal of catching fish will thus drop out as elements of any kind in the *actus reus* of the respective offences, however much they are relevant to ordinary usages of these verbs. Secondly, for such circumstances as are involved in examples (2) and (3), we will not ask any *mens rea* questions about them. As I shall argue shortly, *the* policy guiding how we should taxonomize dimensions of complexity in action description for criminal-law purposes is the policy of allowing systematic and economical *mens rea* requirements to be stated in a criminal code. We do not impose *mens rea* requirements on elements that are themselves the actor's own mental states. We do not require that B have some purpose or belief about his belief that what he is saying is false; nor do we require the C have some purpose or belief about his purpose to catch fish.[19] We thus do not need to classify the mental states of B and C as circumstances, or as something else, because the only reason we would care what they are in this context is absent.

Primarily due to the enormous influence of Wittgenstein on contemporary philosophy of action, a dimension of complexity other than intentional complexity is often distinguished from circumstantial complexity.[20] This is called conventional complexity. The idea is that some types of action can only be performed if certain conventions (rules) are in place. An act of moving one's arm outside a car window, for example, will only constitute an action of signalling a left turn if there is a convention in place at the time whereby such movement is to be so regarded, and if

---

[19] The exception is for cases of wilful blindness, where under the common-law conception we ask, 'Did the defendant form a conscious purpose not to believe some relevant fact about his action?'

[20] See Goldman, *A Theory of Human Action*, 25–6; Thalberg, *Perception, Emotion and Action*, 89–92. Wittgensteinians about actions are so impressed with this kind of complexity that they often make it essential to action as such. See Melden, 'Action'; Peters, *The Concept of Motivation*; Peter Winch, *The Idea of a Social Science* (London, 1958).

the actor followed such convention as his reason for acting. A moving of one's arm upward will constitute the act of bidding at an auction only if there is a convention in place making such an act a way of bidding. In each instance the action-type is said to be conventionally complex because of the constitutive role played by certain conventions.

Yet as with the 'intentionally complex', there is nothing very special about conventions as circumstances.[21] Such conventions need only be a circumstance that exists at the time the act is done, as in the bidding example; or they must not only be in place but conformity with them must motivate the actor, as in the signalling example. In either case, such conventions and the actor's intention to follow them are simply circumstances in place at the time the actor acts, making his action the kind of action that it is. I conclude that we should collapse all such various conditions into the general category of circumstances, so that 'fishing', 'bidding', 'signalling', 'lying', 'recording confidential communications', etc. are all action descriptions that are circumstantially complex.

What we have thus far left out of our discussion of action complexities are the consequences of basic acts that get embedded in complex descriptions. A basic act of finger movement is not a killing just because that finger movement took place in the circumstances that the finger was next to a trigger and that a death occurred. No more than an act of recording a confidential communication takes place just because a finger movement occurs in the circumstances that a 'record' button is next to it, that a confidential conversation is going on, and that a recording of that conversation exists. Rather, my finger movements must *cause* the trigger to depress, which *causes* the death, in order for there to be a killing; my finger movement must *cause* the 'record' button to depress, and that must cause a recording to be produced by the machine in order for there to be a recording of a confidential communication. Such action descriptions possess a causally complex dimension as well a circumstantially complex one.

Once one sees this point, it is obvious that causal complexity infects many verbs of action. One cannot kill a person, hit a person, frighten a person, burn down a structure, break into and

---

[21] A point nicely brought out by Jonathan Bennett, in his *Events and Their Names*, 206–7.

enter a dwelling house of another, disfigure a person, steal from a person, etc. without having one's basic acts cause some consequence to come about. The consequence is not, of course, the killing, the hitting, the frightening, etc., for the basic acts involved *are* the killings, the hittings, the frightenings, etc. Rather, the consequence is some non-act state of affairs: the death of another, the contact with another's body, the state of another's fear, etc.

The difference between causal complexity and circumstantial complexity lies in the presence or absence of a causal relationship between some basic act and some event or state of affairs. Death of a human being is a type of event that could be either a circumstance or a consequence, depending on whether or not there must be a causal relationship between the basic act and that event. Compare two action descriptions: 'joking while another is dying', 'killing'. The first description treats death as a circumstance that must be present if one is to do an action of this type. With respect to death the description treats death as a consequence that must be caused if one is to do an action of this type. This description is thus causally complex.

I come, then, to the orthodox legal view of the dimensions of complexity of complex descriptions of action: there are only two such dimensions, one having to do with consequences, and the other, with circumstances. This was Bentham's view as well as the view of John Austin, John Salmond, and others. What I have yet to do is to justify slicing the pie this way, since we have seen along the way quite a few alternative slicings of the same pie.

My defence of the orthodox view is not metaphysical, nor can it be, for, as Alvin Goldman notes, 'the categorization of species of [complexity] is largely arbitrary'.[22] Remember that the types of action that criminal codes describe are only nominal kinds; the nature of each is constituted by the property(ies) conventionally made definitive of the type. About such mass of conventionally assigned properties, I have sought to abstract some dimensions that they have in common. Such 'dimensions' are accordingly no more natural than that of which they are abstractions. They are (abstractly described) properties that do recur amongst the properties that are conventionally definitive of various act-types.

---

[22] Goldman, *A Theory of Human Action*, 27.

So there is no natural categorization here. Rather, the justification for the orthodox view has to be based on the goals of the criminal law. Such justification has to answer the question of why it is useful to distinguish consequences from circumstances as different kinds of elements in complex descriptions of actions. The answer lies with the need to draw a distinction in moral culpability between circumstances and consequences. The moral distinction does not have to do with differential amounts of wrongdoing that one does, depending on whether one does a circumstantially complex action, a causally complex action, or an action that is complex in both dimensions. The amount of wrongdoing will indeed vary with the sort of circumstances and consequences that accompany, or are caused by, one's acts, but it will not vary with the dimensions of complexity as such.

The moral distinction has to do with culpability, not wrongdoing, and particularly with the primary measure of culpability, the *mens rea* with which an actor acts.[23] As is well known in both law and in morals, we grade culpability by the kind of mental state with which an action is done: doing the same act of shooting another is more culpably performed if done with the purpose that the victim die, than if done with some other purpose but believing with virtual certainty that he will die, and this latter shooting is itself more culpable than the same shooting would be if done by an actor who only believes that there is a substantial risk of death. This differential culpability is reflected in various gradings of offences by the mental states with which identical acts are done. Thus arises the problem: suppose a statute makes it criminal to assault another person with the intent to rape, and suppose another statute makes it criminal to assault a police officer with the intent to kill such police officer. Both statutes define inchoate crimes, since neither a rape nor a killing of a police officer need take place for the statutes to be violated. For that reason, such statutes are typically construed to require the most culpable mental state of purpose: only those whose assaults were motivated by rape or by death of a police officer are punishable under such statutes.

Bentham apparently thought that the complex descriptions

---

[23] This was Aristotle's motivation for taxonomizing modes of complexity in action description, for he does so in the context of discussing ignorance and intentionality. Aristotle, *Nicomachean Ethics*, 388–9.

'rape' and 'killing a police officer' presented a conceptual problem here, for each such description is circumstantially as well as causally complex: there must be a lack of consent for there to be a rape, and the victim of a homicide must have been a police officer at the time he was killed for there to be a killing of a police officer. And Bentham thought that there was no such thing as a purpose with respect to a circumstance:

> Now the circumstances of an act are no objects of intention. A man intends the act: and by his intention produces the act: but as to the circumstances, he does not intend them: he does not, inasmuch as they are circumstances of it, produce them. . . . Acts, with their consequences, are objects of the will [intention], as well as of the understanding [belief]: circumstances, as such, are objects of the understanding [belief] only. All he can do with these, as such, is to know or not to know them.[24]

If one takes this to be a psychological truism, then one has every reason to distinguish consequences from circumstances in a criminal code. One can then say, as the Model Penal Code indeed does say, that 'purpose' with respect to a consequence means purpose ('conscious object', motivation, etc.), but that 'purpose' with respect to a circumstance means only belief.[25]

This is not my justification for the distinction between circumstances and consequences, because I think Bentham's psychological premiss to be false. Bentham assumed that a circumstance, being already in existence when the actor acted, could not be caused by him to come into existence; only items we can produce by our acts can be the objects of our purposes ('intentions', 'will', motivations, etc.). Overlooked by Bentham is the possibility that a circumstance present when the actor acts will also continue to be present when the consequence ensues, and that the consequence itself becomes different from what it would be in the absence of the circumstance. Two examples:

> (1) A assaults a woman with the intent to have intercourse with her; she does not consent; it is important to him that she not consent, however, so that if she were to consent he would quit. His purpose in assaulting her is not just the desired consequence of intercourse, but equally the desired

[24] Bentham, *Introduction to the Principles of Morals and Legislation*, 88.
[25] Model Penal Code, §2.02(2)(*a*).

circumstance that it be unconsented-to intercourse. It surely is not psychologically impossible that some would-be rapists seek not just sex, but rape.

(2) B assaults C with the intent of killing C; C is a police officer, and B not only knows this but he is motivated by it. B is engaged in a cop-killing contest, and he scores only if he kills a cop. B thus desires not only the consequence of C's death, but that the death take place in the circumstance that C is a cop.

I see no basis to deny that A and B are motivated by desired circumstances as much as by desired consequences. The circumstance is what makes the consequence the object of their desire at all. Hence, I reject Bentham's psychological supposition.

Despite this rejection, Bentham was on the right track in his justification for the consequence/circumstance distinction. Contrast (1) and (2) above with (3) and (4) below:

(3) D assaults a woman with the intent to have intercourse with her; she does not consent. D does not care whether she consents or not, but he knows that she is not consenting. The circumstance that she is not consenting thus does not motivate D although he understands that the circumstance is present.

(4) E assaults F with the intent of killing F; F is a police officer, and E knows this. E is indifferent to the fact, since he is in the mood to kill anyone, cop or civilian status not mattering to him. The circumstance that F is a cop thus does not motivate E although he understands that the circumstance is present.

The correct justification for Bentham's distinction does not begin with a psychological premise, but with a moral one: A and D, and B and E, are close enough in the culpability of their assaults that they each should be liable for the appropriate form of aggravated assault. Given this equivalence of moral culpability, it becomes crucial to distinguish circumstances from consequences, so that one can say that 'intent' with respect to consequences means purpose (conscious object, motivation), but 'intent' with respect to circumstances is satisfied by either belief or purpose. Put simply: if the purpose/belief distinction in mental states differentiates culpability only with respect to consequences but not with respect to circumstances, then one had better be able to distinguish these two dimensions of complexity in action descriptions.

Whatever the need for this distinction (in order to mark distinctions in moral culpability properly), two conceptual objections have recently been urged against it. The first objection is that the distinction is unworkable because it is wholly indeterminate in its applications. The second objection is that the distinction is insufficient to do all of the work assigned to it, so that we need to mark out a third species of complexity in action description for criminal-law purposes.

Taking the first objection first: Consider, Smith and Hogan tell us, the crime of taking an unmarried girl out of the possession and without the consent of her parents.[26] How are we to characterize that part of the act-type that is the girl's being in the possession of her parent? Is it causally complex because one must cause a taking-of-a-girl-in-the-possession-of-her-parent? Or is it circumstantially complex because one must only cause a taking of a girl, in the circumstance that at the time of the taking she was in the possession of her parent? Indeed, why isn't the same query possible about the person taken being a girl, being under 16, or being without the consent of her parents? Does one cause an event with all of these properties—in which case they are all consequences—or does one cause an event of a taking without such properties—in which case they are only circumstances?

Even those sympathetic to the causally complex/circumstantially complex distinction have had a hard time with such queries. The American Law Institute's Model Penal Code is simply silent on such questions, giving no definition of what is a 'result' and what is a 'circumstance'. Glanville Williams concedes that for examples like the offence of assaulting of a police officer in the execution of his duties, the identity of the victim of the assault might equally well be regarded as a circumstance or as a result.[27] What Williams does not see is how perfectly general the problem

---

[26] J. C. Smith and B. Hogan, *Criminal Law*, 6th edn. (London, 1988), 39. See also Richard Buxton, 'The Working Paper on Inchoate Offenses: Incitement and Attempt', *Criminal Law Review* (1973), 656–73: 662–3.
[27] Glanville Williams, 'The Problem of Reckless Attempts', *Criminal Law Review* (1983), 265–75: 269.

is. The problem is to define how many properties of the overall act-type prohibited are to be regarded as parts of the consequence caused by a defendant's basic act and how many are to be regarded as parts of the situation or context in which such causing takes place. The causal relationship itself does not pick out how narrowly or broadly described is the event that is the consequence.

It might seem that the answer to this conceptual quandary lies in the temporal location of the element being considered. The suggestion is that consequences are events that occur *after* the actor has performed the relevant basic acts; elements of prohibited act-types are thus 'consequences' only if the relevant property comes into being in the future. Circumstance elements, by contrast, exist at the time the actor performs the relevant basic acts.

This temporal suggestion won't do. To begin with, it would classify the effects of those basic acts where there is instantaneous causation as circumstances, not consequences—despite the manifest metaphysical fact that they are consequences. Secondly, many of what for culpability-grading reasons we shall want to classify as results (as to which true purpose is required in specific-intent crimes) would by this temporal line be classified as circumstances. Thus, in the Smith and Hogan example, that it is a *female* that is taken, or even that it a *person* that is taken, are properties that are exemplified at the time the basic act takes place. Yet surely a purposive taking (as opposed to the less culpable knowing taking) should require that the actor be motivated not just to take something of another, but to take a female child of another. If the actor only knows that what he is taking is a girl, but isn't motivated by that fact, his act seemingly should be classified as a knowing taking and not the more culpable purposive taking.[28]

---

[28] In fact English law does not as such distinguish purposive versions of this crime from knowing versions. It might be forced to draw this purposive/knowledge distinction, however, should such a crime be the object of some would-be burglar's breaking and entering of a dwelling house of another or should the crime charged be attempted taking of a girl or aiding and abetting the taking of a girl; for, in these instances, specific intent (or purpose) would be required in many jurisdictions. In any case, the thought experiment is possible using this statute, even if English law as it happens never distinguishes purposive from knowing takers of girls.

What both of these worries reveals is that the presence of an element at the time the basic act is done can only be a necessary condition for that element being classified as a circumstance; it cannot be a sufficient condition because some of such co-present elements we must classify as consequences. We must thus supplement the temporal suggestion with some other criterion for the circumstance/consequence distinction. The way to begin is with a fact to which we already adverted in Chapter 5, namely, that action verbs are systematically ambiguous between their transitive and their intransitive senses. If something melts$_I$, there need be no human action causing this event, although there may be. As we shall see later in this chapter, causally complex action verbs are to be analysed as: if x melts$_T$ something, then x's basic act causes the melting$_I$ of something. To melt$_T$, in other words, is to cause a melting$_I$. Similarly, to sink$_T$ something is to cause a sinking$_I$; to open$_T$ something is to cause an opening$_I$ of something. Even where English does not provide an intransitive verb to match in form the transitive verb, it does supply us with another: to kill$_T$ something is not to cause a killing$_I$ but it is to cause a dying; to raise$_T$ something is not to cause a raising$_I$ but it is to cause a rising; to take$_T$ something is not to cause a taking$_I$, but it is to cause a moving$_I$ of that thing; etc.

We can now initially draw the consequence/circumstance line as follows: the result or consequence just is the type of event described by the intransitive verb that matches the transitive verb used in the definition of the offence. To kill a police officer while on duty is to cause a death (or dying) of a person, in the circumstances that the person killed be a police officer and that he be on duty. The consequence is a dying of a person; the circumstances are the status of the person as an officer and as an officer on duty. Analogously, to take an unmarried girl under the age of 16 out of the possession of her parents and without their consent is to cause a moving$_I$ of a girl, in the circumstances that the girl taken is under 16, is not consented to by her parents, and is while she is in possession of her parents. The consequence is the moving$_I$ of the girl; all else is circumstance.

It is important to see that the semantic match between transitive and intransitive action verbs in English only does part of the work in drawing the consequence/circumstance line as I have just drawn it in these examples. The semantic criterion tells us this much: *at least* the event-type described by the matching intransitive verb must be regarded as a consequence. Notice, however, that I included more than the event-type described by the matching intransitive verb in my classification of consequences. In the offence of killing a police officer while in performance of his official duties, the consequence element is not simply the dying of something; it is the dying *of a person*. Likewise, in the offence of taking a girl under the age of 16 out of the possession of her parents and without their consent, the consequence element is not simply the moving$_I$ of something; it is the moving$_I$ of a person, and a female person to boot. Some criterion in addition to the semantic criterion must be employed here in order to justify the enlarged characterization of the consequence element.

One suggestion about this non-semantic, supplementary criterion is that it is to be found in what is 'customarily regarded' a part of the accused's act.[29] Yet I doubt that there are any very precise customs here. And in any case, there is no reason to think that what customs there are answer to the purpose for which we are drawing the consequence/circumstance distinction.

The true supplementary criterion is a moral one. We need to ask of each act-type prohibited by a criminal code what, if anything (beyond the event-type described by the matching intransitive verb), must motivate an actor to render that actor most culpable in his doing of the overall act prohibited by the statute. It is this moral criterion that allows us to include 'person' in the event-type 'death of a person' as the consequence element of the first example above. Likewise, it is this moral criterion that allows us to include 'person' and 'female' in the event-type 'moving$_I$ of a female person' as the consequence element of the second example above. These properties—of the personhood or gender of the victim—have to motivate the actor to make him a fit candidate for those most serious punishments reserved for purposive (as opposed to merely knowing) cop-killers and girl-takers.

The second objection we can handle more quickly. It concedes that the consequence/circumstance distinction is a workable one, but urges that one other species of complexity must be recognized for culpability assessment purposes in the criminal law. The

---

[29] See Buxton, 'Circumstances, Consequences, and Attempted Rape', 31, where Buxton critically examines the suggestion.

American Law Institute's Model Penal Code seemingly adopts the Bentham–Austin two species of complexity,[30] but when it comes to using such species in order to mark culpability it adds a third species. In addition to results of basic acts, and circumstances in the presence of which basic acts are done, the Code recognizes that complex actions have a 'nature'.[31] Thus, 'purpose' means one thing with respect to the result *or* the *nature* of an action, and another with respect to the circumstances in which the action is done.

I take it that the thought behind this threefold taxonomy of complex action description is this: some complex action descriptions used in the criminal law are not causally complex, yet there is more to them than simply a basic act done in certain circumstances. Therefore, we need some label for this non-causal, non-circumstantial species of complexity. Hence, 'nature'. I take this same thought to lie behind the familiar view in English criminal-law theory that some crimes (called 'conduct-crimes') have no result elements but they do have a species of complexity beyond circumstantial complexity; therefore one talks about the act and its parts, as distinct from the circumstances in which it is done, where one means by 'act' more than a basic act of bodily movement.[32]

There is in fact no need for some catch-all, third species of complexity, whether it is called 'nature' of act or simply 'act'. Consider Glanville Williams's example of a supposed 'conduct-crime' that has no result elements: 'Abduction is a conduct-crime, not a result-crime; it is completed by the act of *taking* the girl . . . you do not have to wait and see if anything happens as a result of what the defendant does.'[33] Yet we do have to 'wait and see if anything happens as a result of what the defendant does'—so long as we don't beg the question by saying that what the defendant *does* is take a girl. What the defendant does is the *basic* act of moving$_T$ his limbs in various ways. Those basic acts have to cause a moving$_I$ of the girl before there can be a taking$_T$ by the

---

[30] Model Penal Code, §§1.13(2), 1.13(5), 1.13(9). See n. 5, *supra*, this chapter.

[31] Model Penal Code, §§2.02(2)(*a*)(i), 2.02(2)(*b*)(i).

[32] See the discussion and citations in Buxton, 'Circumstances, Consequences, and Attempted Rape', 28.

[33] Williams, 'The Problem of Reckless Attempts', 368. As another example, Williams considers penetration in rape to be the forbidden act or conduct, but not to contain a result element. Ibid. 366.

defendant. There *is* a result in such cases—the moving$_I$—so that we have no need for a 'nature', 'conduct', or 'act' species of complexity in them. The 'nature' of taking$_T$, in other words, is fully accounted for as a basic act causing a moving$_I$, so there is no need for this superfluous and confusing third species of complexity.

In the succeeding section I shall argue that all action verbs used in the criminal law are causally complex, as 'taking' is causally complex. There is thus never any need for some third species of complexity in order to mark properly the degrees of culpability.

The main policy guiding the drawing of the circumstance/consequence line is the policy of proportioning punishment to the differing culpability of purposive versus knowing wrongdoers. A subsidiary policy is to proportion punishment between purposive and merely reckless, negligent, or strictly liable wrongdoers. This subsidiary policy appears in the law of attempt and of accomplice liability.

With some dissent, both attempt and accomplice liability are crimes of specific intent, that is, they require purpose with respect to each material element.[34] Yet when the substantive crime being attempted or aided itself contains only *mens rea* requirements of recklessness, negligence, or strict liability, there arises the question of whether purpose is required as to all material elements after all. Take the crimes of attempted murder of a police officer and of aiding and abetting the murder of a police officer. Suppose the substantive crime of murdering a police office only requires recklessness, negligence, or no mental state at all with respect to the element that the victim was a police officer. Because this element is considered a circumstance, the Model Penal Code only requires that the attemptor or the accomplice have whatever mental state with respect to that circumstance is required for conviction of the substantive offence.[35] This might

---

[34] Model Penal Code, §§5.01(1), 2.06(3).

[35] Model Penal Code, §§5.01(1), 2.06(3). With respect to the latter provision, this is not expressly resolved. See American Law Institute, *Model Penal Code and Commentaries* (Philadelphia, 1985), Comment to §2.06 (Philadelphia, 1985), 311 n. 37. It is also easy to confuse this issue with respect to accomplice liability because of the *two mens rea* requirements for such an offence: (1) that the defendant have as his purpose to aid the offence; and (2) that at least with respect to result elements of the underlying offence, he have at least the *mens rea* required for conviction of that offence. The text discusses the first *mens rea* only: the defendant

be conscious awareness of a substantial risk that it is a police officer (recklessness), sufficient information that a reasonable person would believe it is a police officer (negligence), or no mental state at all, depending upon how the substantive offence is defined.[36]

When the consequence/circumstance distinction is drawn in these contexts, the semantic and moral criteria that I described earlier can be used because the same policy is being pursued. In both these cases of attempt and complicity liability, as well as in the earlier examples of other specific-intent crimes, one draws the distinction between consequences and circumstances in order to mark off what must be motivating an actor in order to deserve the most serious degree of punishment for the offence in question. That one is marking off such purposive culpability against recklessness, negligence, or strict liability, rather than against knowledge, does not require that the consequence/circumstance line be drawn differently here.

Unfortunately the consequence/circumstance distinction has been used to serve other purposes where this happy identity of criteria does not hold. Consider the gradations of culpability marked by knowledge (belief to a virtual certainty) versus recklessness (belief of a substantial risk). The question arises, how should we classify the phenomenon known as 'wilful blindness'? Wilful blindness occurs when the accused doesn't literally know some fact but he has some inkling that it might be true, and he deliberately refrains from finding out. Suppose that there are two grades of the crime of receiving stolen property: doing so knowing that it is stolen and doing so believing that there is a substantial risk that it is stolen. How should the wilfully blind fence be treated?

Notice that there are several complexities to the description 'receiving stolen property': first, that there be a receiving_I of something; secondly, that the thing received be property; thirdly, that the property has been stolen. Suppose two different wilfully

clearly has to be motivated by the result that he aid another; the question is whether he must also be motivated by the circumstances in which such aidings are being done, such as the circumstance that it is a police officer whose killing is being aided.

[36] On this issue in the US federal system, see *United States* v. *Feola*, 420 U.S. 672 (1975) (no *mens rea* required as to identity of federal officer for conviction of substantive offence of assaulting a federal officer).

blind fences: the first doesn't enquire about the stolen character of the property that he knows he is receiving, although he has some inkling that it is stolen; he doesn't enquire because he does not want to know. The second wilfully blind fence knows for sure that anything that he receives from Jones will be stolen property, but he is unsure about whether, as a consequence of his act of sending Jones money by mail, he will receive anything back (Jones being a thief and all). The second fellow refrains from thinking through the likelihood that Jones will send back something because he doesn't want to know for sure that he will.

If one's sense is that there is a difference between the culpability of these two wilfully blind fences, then one needs to mark that difference with a distinction between different species of complexity in action description. One might think, for example, that with a fact in existence at the time that he acted the first fence could verify its presence or absence, but that with a fact that will come into existence only after he has acted there is no equivalent checking procedure; being a fact true only in the future, the most the second fence can do is to try to *predict* the occurrence of that fact (say by some mental calculation about the motivations of repeat players and the like). In any case, one might conclude that the identical set of mental states—those called wilful blindness—result in different culpability depending upon whether those states are with respect to a present fact or a future fact.[37] One thus needs some distinction in species of complexity to mark this moral difference.

Unfortunately, some take the consequence/circumstance distinction either to be, or to be a good proxy for, the distinction between presently existing versus future facts. Yet, as we have seen, it is a poor proxy for this temporal distinction. Some of the properties exemplified when the actor acts are none the less properties ascribed to the consequence of the basic act, not its circumstances. In the offence of taking a girl, that it was a female is as true at the time the defendant begins to cause her moving$_I$ as

---

[37] It is this differential verifiability of present facts versus future facts that justifies the Model Penal Code's differentiation between these cases. §2.02(7)'s language of 'knowledge of the existence of a particular fact' is intended to capture only offences that require knowledge of a fact existing when the actor acted, not offences requiring knowledge of facts that come into existence after the actor acted. It is only the former sorts of case where wilful blindness is graded in culpability as knowing; in the latter case, it is graded as only recklessness.

when she has been moved; yet for the purpose of marking the differential culpability of purpose versus knowledge, this fact is properly allocated as a consequence.

Thus, if one is to use the consequence/circumstance distinction to do the work of marking how wilful blindness is graded in culpability—either as knowledge (with respect to circumstance elements) or recklessness (with respect to result elements)—this will be a different distinction from the semantic and moral distinction described earlier. It is preferable, therefore, to avoid this distinction entirely when the issue is how to grade wilful blindness, and to distinguish species of complexity in action description in this context solely by the temporal criterion.[38]

3. ARE ALL COMPLEX DESCRIPTIONS OF ACTIONS USED IN THE CRIMINAL LAW AT LEAST CAUSALLY COMPLEX?

The form of the equivalence thesis at which we have thus far arrived is the following: for any complex description of action V, to say that X V-ed is to say either that some basic act of X's caused some state S to obtain, or that some basic act of X's occurred in the circumstance that S obtains. Although this is a form of the equivalence thesis quite adequate to rebut J. L. Austin's scepticism about the prolixity of descriptions of actions, what I wish to examine in this section is whether the thesis may not be more simply stated.

Perhaps we can eliminate the disjunctive nature of this formulation of the thesis in favour of the following: X V-ed if and only if some basic act of X caused state S to obtain in circumstances C. For example, X killed a correctional officer while the latter was on duty and while X was an inmate at a correctional institution if and only if some basic act of X's done while X was an

---

[38] One also might reject the supposed difference in culpability between the two wilfully blind fences, thus eliminating the need for *any* distinction in species of complexity in action description in this context. As Antony Duff pertinently asks of this Model Penal Code grading distinction: 'Is there really a difference in culpability between, for instance, an agent who is wilfully blind to the (obvious) fact that what she is damaging is the property of another, and an agent who is wilfully blind to the (obvious) fact that her act will cause damage to what she knows to be property belonging to another?' Duff, private correspondence.

inmate caused the death of a correctional officer while the latter was on duty.

On this formulation C may be a null set. There need be no circumstances that are necessary to the correct application of some verbs (e.g. 'X killed Y') beyond the causation of certain results. But causation is not similarly dispensable here, for this form of the equivalence thesis asserts that all complex verbs are at least *causally* complex.

I shall in this section defend this causal version of the thesis. I shall do this only with respect to the complex verbs used in the criminal law. This restriction is partly because criminal law is our focus, but mainly because the causal version of the equivalence thesis is not plausible for all complex verbs of action.[39]

Even restricted to the verbs of action used in the criminal law the causal version of the equivalence thesis is commonly rejected. George Fletcher, for example, has often admonished his fellow teachers of substantive criminal law to end the traditional focus on homicide as the example of crime. Homicide prohibits the causally complex action of killing, which, Fletcher rightly notes, assumes causation of a 'harmful event that is conceptually independent of human action',[40] namely, the victim's death. Fletcher categorizes crimes like killing to be part of a 'pattern of liability', namely, the 'pattern of harmful consequences'. Fletcher's admonition is based on the worry that 'one pattern of liability—often the pattern of harmful consequences—is taken as paradigmatic of the whole criminal law'.[41]

This is a legitimate worry only if there are other 'patterns of liability' using action verbs that do *not* assume causation of a 'harmful event that is conceptually independent of human action'. Fletcher assumes that there are such patterns, what he calls the 'pattern of manifest criminality' and what he calls the 'pattern of subjective criminality'. Theft (which commonly pro-

---

[39] The existence of complex descriptions of action that are not causally complex can be seen by starting with a basic description and adding to it only a circumstance. For example, 'X raised his arm' is usually a basic description, while 'X raised his arm while on a boat' is a complex description adding only a circumstance. The complex action of boat-situated-arm-raising is not causally complex, only circumstantially so. (For further examples and analysis, see Thalberg, *Perception, Emotion and Action*, 96–7; Bennett, *Events and Their Names*, 219, 222, 231.) It is thus a contingent feature of the criminal law we have that it always is interested in *results* of basic acts.

[40] Fletcher, *Rethinking Criminal Law*, 388–9.   [41] Ibid. 390.

hibits the taking of property of another with intent to deprive) for Fletcher is an example of the first, and attempt (which commonly prohibits attempting to do an act already made criminal by some other statute) is an example of the second. Fletcher thus must assume that the act descriptions 'taking' and 'attempting' do not have the structure of descriptions like 'killing', namely, that of a causing of some event that is 'conceptually independent' of the act that caused it. It is the thesis of this section that this assumption is false, that all complex descriptions of action used in the criminal law require at least a causing of some state of affairs.

Fletcher's assumption is a very common one amongst criminal-law academics. The leading American case-book in criminal law, for example, contemplates that there can be crimes that are 'defined without regard to any result of the defendant's conduct (for example, attempt, conspiracy, burglary) [where] there is no need to face the issue of causation'.[42] Many English criminal-law theorists, as we have seen, also defend this thesis, urging that there are such crimes as 'conduct-crimes' when there are no result elements and thus no causation requirements.[43] Before I argue against this thesis, I first wish to clarify it.

First what does Fletcher mean by an event that is 'conceptually independent' of the act itself? It should be plain why Fletcher imposes this requirement. At least since David Hume, it has commonly been supposed that if two descriptions of two putatively separate events are conceptually connected in some way, then the two events cannot be causally related. Conceptual connectedness is a test for when causal relations are precluded.

We can approach what Fletcher might mean by 'conceptual connection' easiest by seeing what he cannot mean. Suppose I cause my arm to rise, and thus perform the action of raising my arm. Is the description of the event, 'my arm rising', conceptually independent of the description of the act, 'raising my arm?' It might be thought that a reason for answering this question negatively is that the event description 'arm-rising' is related to the

---

[42] Kadish and Schulhofer (eds.), *Criminal Law and Its Processes*, 587. The assumption also pervades Stephen Schulhofer, 'Harm and Punishment: A Critique of Emphasis on the Results of Conduct in the Criminal Law', *University of Pennsylvania Law Review*, 122 (1974), 1497–1607.

[43] See the discussion and citations in Buxton, 'Circumstances, Consequences, and Attempted Rape', 28.

act description 'arm-raising', because a common root is involved. Yet even when the same word is used, not just a common root, this is no reason to think that there is a conceptual connection here. The verbs of action, as we have seen, are systematically ambiguous between their transitive and their intransitive senses. 'He caused his arm to move' and 'He moved his arm' look quite similar, until we insert the subscript to mark the different senses of the word being employed. The first use of 'move' is intransitive, referring to the event of the arm moving$_I$; the second use is transitive, referring to the act of moving$_T$ the arm. There is no conceptual connection here between the moving$_I$ (or rising$_I$) and the moving$_T$ (or raising$_T$).

So in one sense the assumption of Fletcher and others is plainly false. If the volitional theory of action is true, all complex descriptions of actions require at least this much causation: a volition must cause a bodily movement$_I$. This weak causal thesis is not the thesis I wish here to defend, however, nor is its negation the assumption that Fletcher and most criminal-law academics have made. The weak thesis I already defended when I defended the volitional theory of action. By contrast, the thesis that rebuts Fletcher *et al.* is a much stronger thesis: all complex descriptions of actions used in the criminal law require that the volitionally caused bodily movement referred to by such descriptions itself cause some further event or state of affairs. The thesis is that all complex descriptions of actions share with 'killing' a built-in, *second* causal element: the bodily movement (that is caused by a volition) must itself cause some further, independent event to occur, like a death for 'killing'. It is this second, stronger causal thesis that is the equivalence thesis we are here examining, for it is only this stronger thesis that can make good the claim that all action descriptions are equivalent to the canonical description 'Basic act A caused state of affairs S'.

The strong causal thesis is highly plausible for most of the complex descriptions of action used in criminal law. 'Killing' is only the most prominent example, but equally obvious are: 'disfiguring' as used in the crime of mayhem (causing the disfigurement of the body of another);[44] 'maiming' as used in the crime of mayhem (causing the loss of some body part);[45] 'hitting' as used

[44] See Roland Perkins and R. Boyce, *Criminal Law*, 3rd edn. (Mineola, NY, 1982), 241.   [45] Ibid.

in the law of battery (causing contact with the body of another);[46] 'abuse' as used in child abuse statutes (causing physical injury to a child);[47] 'rape' as used in the crimes of forcible and statutory rape (causing sexual penetration of the female);[48] 'confining' or 'detaining' as used in the crime of false imprisonment (causing another to be confined or detained against his will);[49] 'abducting' as used in the crime of kidnapping (causing the transportation of another and causing the false imprisonment of that other);[50] 'burning' as used in the crime of arson (causing any part of a building to be destroyed or damaged by fire);[51] 'aiding' and 'abetting' as used in the theory of accomplice liability (causing another to be aided or helped in the perpetration of a crime);[52] etc.

This causal structure to complex descriptions exists also for verbs used in crimes that are part of what Fletcher calls the 'pattern of manifest criminality' and what English theorists call 'conduct-crimes'. Consider larceny and burglary. Larceny requires the *taking* of property of another. One performs the action of taking only when one *causes* the movement of such property, and, by causing such movement, one *causes* oneself (or another) to come into possession of such property.[53] Similarly, burglary requires the actions of breaking and entering the dwelling house of another. One 'breaks' within the meaning of the burglary statutes only when one causes an opening of some part of the dwelling (door, window, etc.);[54] one 'enters' only when one causes any

---

[46] Ibid. 152.
[47] See e.g. Annotated Code of Maryland, Art. 27, §35A(*b*)(7).
[48] See e.g. Model Penal Code, §213.0(2).
[49] Perkins and Boyce, *Criminal Law*, 224.　　[50] Ibid. 230.　　[51] Ibid. 278.
[52] See e.g. *State* v. *Tally*, 102 Ala. 25, 69, 15 So. 722, 739 (1894) (accomplice need not cause death to be accomplice to murder but he must cause aid to be rendered to the killer making the latter's killing easier).
[53] Perkins and Boyce, *Criminal Law*, 323, 302. For a close application of the requirement of 'asportation', see *State* v. *Patton*, 364 Mo. 1044, 271 S.W. 2d 560 (1954) (movement of concrete blocks by one other than the defendant held to be sufficient movement). Although the Model Penal Code (§223.2(1)) and the codes that follow its lead eliminate asportation as an element of theft, substituting 'exercise of unlawful control' by a defendant, this is only because the Code consciously merged attempted larceny with larceny. See American Law Institute, *Model Penal Code and Commentaries*, Comment to §223.2(2), 164. One then still must ask (of those new larcenies that would have been attempted larcenies before) whether the actor's basic act has caused a proximateness to movement and possession sufficient to be an attempt.
[54] Perkins and Boyce, *Criminal Law*, 248.

part of one's body to be inside the dwelling.[55] While theft and burglary also require that many circumstances be true at the time of these acts, this in no way diminishes the fact that these crimes require that certain states of affairs be caused by some basic act of bodily movement.

The inchoate crimes of conspiracy, attempt, and solicitation are most often mentioned as *the* counter-examples to the thesis here examined. Fletcher most obviously has such crimes in mind as his 'subjective pattern of liability'. Since attempt and attempt-like crimes are most obviously causative in their requirements, I shall begin with them.

Consider first the general category of attempt liability. As is well known, one is not guilty of an attempt (e.g. to murder) if one performed a basic act with the intent that it kill another. In addition, that act must cause a state of affairs to exist that is variously described as 'beyond mere preparation', 'in dangerous proximity to killing', a 'substantial step' towards killing, etc.[56] Attempt thus requires a causing of some further state of affairs beyond simply the moving of one's own body; that movement must itself cause enough other events to occur so as to satisfy these various descriptions of some further state of affairs.

One might think that the state of affairs (in some cases at least) will not be conceptually independent of the basic act that is its putative cause. Consider two variations on Squeaky Fromme's attempt to assassinate (then) President Ford. The first variation is close to the facts of the actual case: holding her pistol pointed at Ford she pressed her finger on the trigger with the intent of killing Ford; the trigger on the gun didn't move because it was jammed. Surely Fromme's movement took her beyond 'mere preparation' to the state called attempt, but not because the

---

[55] Perkins and Boyce, *Criminal Law*, 253.

[56] The variously described states that must exist before attempt liability attaches are discussed in Kadish and Schulhofer (eds.), *Criminal Law and Its Processes*, 633–55. For some of these tests of when an accused has gone far enough to cross over from 'mere preparation' into 'attempt', the line is not drawn in terms of actual proximity to the prohibited harms; rather, it is drawn in terms of how close to success was the actor if the facts were as he believed them to be. For example, Model Penal Code, §5.01(1)(c). Here the causal question is a rather recherché one: did the defendant's basic acts cause enough other events to occur in the actual world so that in the possible world created by his mistaken beliefs those events would have been close enough to success so as to be an attempt? Recherché, but a causal test none the less. (I owe the worry to Antony Duff.)

movement *caused* that state to exist—moving her finger caused nothing else (of relevance) to happen. Rather, the moving of her finger in these circumstances *constituted* the state of affairs referred to as an 'attempt', a 'substantial step', etc.—or so the argument goes.

In this first variation, the way round the argument is easy: Fromme did many basic acts that caused her to be in a position to pull the trigger of a gun pointed at President Ford. It is only because those acts *caused* various other events to occur that Fromme *attempted* to kill Ford. So make my hypothetical about Fromme harder (if much less realistic): she woke up finding that her hand had been wrapped round a loaded pistol that was pointed at Ford, decided for the first time at that moment to kill him, and moved her finder with that intent. As before, nothing else happened of relevance to Fromme's project of killing Ford, not even a contact of her finger on the trigger (she was stopped before her movement caused that contact to be made). Now the conclusion earlier advanced seems more plausible: the state of affairs that is referred to by 'dangerous proximity to killing Ford', or a 'substantial step towards it', or whatever, is constituted by the movement of her fingers in these circumstances.

Plausible, but false. If the only effect of Fromme's hypothesized finger movement$_T$ was that her fingers moved$_I$, *then* we should indeed conclude that the thesis here examined is false. For then the state of affairs caused is *part* of the basic action (volition-cause-movement$_I$) itself, so that Fromme's act (moving$_T$) could not *cause* Fromme's fingers moving$_I$. But Fromme's movement$_T$ must have caused a further state than that in order to be an attempt. The state is one of dangerous proximity to (or a substantial step towards) Ford's death. Ford, to give an old phrase a new meaning, must have been caused to have had a 'near-death experience' by Fromme's moving$_T$ before that moving$_T$ can be called an attempt. True, Ford's *state* of being near death is not an event, like a trigger moving; it is a relational state, but it is no less an *effect* for that. And as a state, it is not identical to the state of Fromme's finger moving^I, once we properly individuate states.[57]

---

[57] Consider an analogous example of Anthony Kenny's: 'when I learn French, I bring it about that I know French, i.e. that I have the capacity to speak French'. Anthony Kenny, *Action, Emotion, and Will* (London, 1963), 183. On

This same analysis applies to attempt-like crimes, such as simple assault, the aggravated assaults, and reckless endangerment of a person. In each case an actor must not only have caused a movement$_1$ of her body to occur, but that movement must itself cause a state of proximity (to contact, for assault, or to death, for reckless endangerment) to exist.

Is conspiracy any less causally complex in its *actus reus* requirements? The crime of conspiracy is often defined by statute as requiring that an 'overt act' be done in furtherance of the common plan, and such over-act requirement is then construed to require the causation of the state of affairs also required by attempt liability.[58] But at common law conspiracy has no overt-act requirement, so let us consider this more troublesome notion of conspiracy.

The *actus reus* of common-law conspiracy is variously defined as the forming of a combination, the entering of an agreement, or as an agreeing. These are not the same things, so let us consider each separately. Holmes opted for the first of these three views in his definition of conspiracy:

A conspiracy is constituted by an agreement, it is true, but it is the result of the agreement, rather than the agreement itself, just as a partnership, although constituted by a contract, is not the contract but is the result of it. . . . A conspiracy is a partnership in criminal purposes.[59]

On this view of conspiracy it is obvious that some state of affairs must be caused to exist by one's willed bodily movements (of speaking, nodding assent, or whatever): the combinational relationship with others. As Perkins and Boyce recognize, from Holmes's definition of the *actus reus* of conspiracy 'it follows that the verb "conspire," when used in the law, has reference to the formation of the combination'.[60]

The existence of the causal requirement is less obvious in the

Kenny's analysis, what I cause when I do the complex action of teaching myself French is a state, namely, a capacity to speak French correctly. The capacity to speak French is a state I am in even if I never utter a word of French. Such dispositions and abilities are like a proximity to speaking French. Such a state is 'conceptually independent' from all the basic acts I did in order to acquire such an ability; such acts may thus be the *causes* of this proximity state to exist.

[58] See Perkins and Boyce, *Criminal Law*, 685–7.
[59] *United states* v. *Kissel*, 218 U.S. 601, 608 (1910).
[60] Perkins and Boyce, *Criminal Law*, 683.

other two definitions of conspiracy. With the *actus reus* of 'entering an agreement', or 'agreeing', one might think that simply
nodding one's head, or saying 'Yeah' or 'Let's do it' are, in the
right circumstances, all there is to it. Yet even on a purely 'objective' theory of criminal contracts, this is false. Both agreeings and
agreements require either communication or at least attempted
communication. The vocal-cord movements that are involved in
making an oral offer, or making an oral acceptance, are not
enough for there to be communication. In addition, at a minimum such movements must cause sounds to occur. In oral communication, as J. L. Austin once noted, the '"production of
noises" is itself really a consequence of the minimum physical act
of moving one's vocal organs'.[61] For non-oral communications,
bodily movements must produce other effects in other mediums
(marks on paper, sounds on tape, etc.). More importantly, for
successful communication those sounds (or marks etc.) must
themselves be understood; that is, such sounds must *cause* audience belief of a certain sort. Even J. L. Austin, who generally
thought that acts like agreeing, warning, advising, etc. did not
involve the production of consequences (in the ordinary sense) of
some basic act of speaking, none the less conceded:

I cannot be said to have warned an audience unless it hears what I say
and takes what I say in a certain sense. An effect must be achieved on
the audience if the illocutionary act is to be carried out . . . the effect
amounts to bringing about the understanding of the meaning and the
force of the locution. So the performance of an illocutionary act involves
the securing of uptake.[62]

Both sounds and audience beliefs must be caused by some basic
acts of the would-be conspirator as much as for the would-be
warner, the would-be adviser, the would-be commander, etc.

To test this supposition of Austin's, suppose that I am con-
versing with an undercover police agent, who has no intention of
going along with my criminal scheme but who feigns acquies-
cence in order to gather evidence against me. (Under the Model
Penal Code's unilateral approach requiring only an agreeing, this

[61] J. L. Austin, *How to Do Things with Words*, 2nd edn. (Cambridge, Mass.,
1975), 114 n. 1.
[62] Ibid. 116–17.

might be a conspiracy, even though under the common law's bilateral approach requiring agreement, it could not be.[63] If I nod my head in response to an apparent offer of criminal agreement, does the audience of my nod (the undercover cop) have to understand its meaning in order for me to be *agreeing*? Does he have to hear my voice or see my nod? Does the 'audience' even have to be a person, or can I be nodding to a wall thinking that I am agreeing with a person? Granted, under the unilateral definition of conspiracy I can agree without causing there to be an agreement, but can I agree when I fail to cause audience understanding in any of these three ways?

The issue is an important one for the crime of solicitation as well as for conspiracy. The *actus reus* of criminal solicitation is to 'command, encourage or request' another person to engage in criminal conduct.[64] We face here the same question: can someone perform these illocutionary acts without securing (causing) understanding from the person solicited? I take this to be a doctrinal question, and doctrinally it is clear that the thing addressed by the putative solicitor or conspirator must be in fact a person. Both the Model Penal Code and the codes that follow it require that one guilty of solicitation command, encourage, or request 'another person';[65] such codes likewise require one guilty of conspiracy to agree 'with such other person or persons'.[66] So my last variation, where the speaker nods to a wall, can be ruled out even though he thinks he is communicating with another. But the Code treats differently my other two hypothetical variations.

J. L. Austin considered explicitly examples of illocutionary acts that constitute criminal solicitation, such as 'He urged (or advised, ordered, etc.) me to shoot her'.[67] Austin thought we must distinguish 'between the act of attempting . . . to perform a certain illocutionary act, and the act of successfully achieving or consummating or bringing off such an act . . . it is always possible, for example, to try to thank or inform somebody yet in different ways fail, because he doesn't listen, or takes it as ironical . . . and so on'.[68] Surely Austin's ear accurately captures ordinary, idiomatic usage of the illocutionary-act verbs that are used to define solicitation and conspiracy. Yet distinguishing

[63] Model Penal Code, §5.03(1).   [64] Model Penal Code, §5.02(1).
[65] Ibid.   [66] Model Penal Code, §5.03(1).
[67] Austin, *How to Do Things with Words*, 102.   [68] Ibid. 105–6.

attempted communications from successful communications does not answer the question of legal liability, and here the criminal law is clear: such attempted communications are both solicitations and conspiracies.[69]

This eliminates Austinian uptake as the effect the actor must cause in order to solicit or conspire. Yet it does so only by substituting another effect, namely, a proximity to uptake. Like other attempts, attempts to communicate require that the vocal-cord or other bodily movements of an actor cause a proximity to success (here, successful communication). Such actions thus retain a causal component, even though highly attenuated.[70]

It remains to consider Austin's own well-known reluctance to say that illocutionary acts like advising, encouraging, commanding, or agreeing include the causation of effects, 'at least in any ordinary sense' of cause and effect.[71] Austin's reluctance should be of importance to us, not only because of the enormous influence of his analysis to action theory,[72] but more specifically because his reluctance here grounds the belief of some criminal-law theorists that there must be 'conduct-crimes' or patterns of 'manifest criminality' that contain no causation requirements.

To see Austin's thought here, it is necessary to recall the bare-bones outline of his structuring of the actions we do when we speak. When we speak, Austin noted, we utter certain noises. That he called our phonetic act, carefully noting that such acts are already causative in that the movement of one's vocal organs must themselves cause production of sound.[73] Secondly, if we conformed to the syntactical rules of our language in making such noises, we can be said to have uttered certain words. This is our phatic act.[74] Thirdly, if we uttered these words in conformity with the semantic rules of our language, and did so with the right intentions, so as to give our words a sense and a reference, then

---

[69] Model Penal Code, §5.03(2); Saunders, 'Voluntary Acts and the Criminal Law', 471–2 n. 114.

[70] The survival of a causal requirement in the Model Penal Code's merger of attempted solicitation and solicitation parallels a like survival of a causal requirement whenever the Code does a like merger. See e.g. n. 53 *supra*, this chapter (merger of attempted larceny with larceny); Model Penal Code, §2.06(3)(*a*)(ii) (merger of attempt to aid and abet into aiding and abetting).

[71] Austin, *How to Do Things with Words*, 115.

[72] See e.g. Thomson, *Acts and Other Events*, 140–1, 223–4, who follows Austin in believing that all or most illocutionary act verbs are non-causal.

[73] Austin, *How to Do Things with Words*, 114 n. 7.   [74] Ibid. 95.

we can be said to have performed the action of saying something. This Austin termed our rhetic act.[75]

When we perform a phonetic, phatic, and rhetic act together, we perform, Austin concluded, the *locutionary* act of saying something. What kind of saying something we did depended on what *illocutionary* act we performed. In saying something, we could have been commanding, advising, requesting, promising, etc. These kinds of acts determine the illocutionary force of our utterance, whereas the locutionary act determines its meaning. Distinct from both of these is the *perlocutionary* acts we can perform by speaking. These are acts like persuading our audience, convincing them, scaring them, etc. These acts plainly require the causal production of effects, which is how Austin distinguished them from illocutionary acts.

Austin's taxonomic interests in maintaining the distinction between illocutionary acts and perlocutionary acts mislead him about the causal nature of the former. Austin thought that he had to say that 'there is clearly a difference between what we feel to be the real production of real effects [as in a perlocutionary act] and what we regard as mere conventional consequences . . . [as in an illocutionary act]'.[76] Austin correctly saw that an illocutionary act can be performed only in the circumstance that certain conventions (of appropriate utterance) were in place and utilized by the speaker: 'the illocutionary act is a conventional act: an act done as conforming to a convention'.[77] Yet Austin then assumed (like many philosophers of action who followed him in this) that the conventional complexity of illocutionary acts precluded them from also having a causal complexity. Compare, Austin at one point admonishes us, 'the distinction between kicking a wall and kicking a goal'.[78] Surely if we make the comparison Austin suggests, the difference is not that one action description is causally complex and the other is not; a leg-moving must cause an independent event in one case as much as the other. That one event must occur in the circumstance that a certain convention was in place and was followed in no way lessens the causal relation that must exist between that event and the bodily movement.[79]

---

[75] *How to Do Things with Words.*   [76] Ibid. 103.   [77] Ibid. 105.
[78] Ibid. 107.
[79] Austin sometimes seems to think that the difference a convention makes to

There *is* an important difference marked by Austin's enormously fruitful distinction between illocutionary and perlocutionary speech-acts. Since the advent of the Model Penal Code's unilateral approach to conspiracy (to take one legal example), there has been an important debate about whether the crime should require the perlocutionary act of securing agreement (the common law's position), or the illocutionary act of agreeing (the Model Penal Code's position). But this is not to argue about a distinction between causally complex versus non-causally complex action descriptions. Rather, the distinction to be drawn by both sides of the debate is between two kinds of causal effects we can achieve by speaking: having our audience understand what we said, and having the audience not only understand what we said but be convinced or persuaded by it.

I conclude that the *actus reus* requirements of the crimes of conspiracy, solicitation, and attempt are no more free of causal complexity than are the *actus reus* requirements of completed offences. The *actus reus* requirements of all crimes have hidden causation requirements built into them.

4. ARE ALL CAUSALLY COMPLEX DESCRIPTIONS OF ACTIONS EQUIVALENT TO DESCRIPTIONS OF BASIC ACTS CAUSING FURTHER STATES OF AFFAIRS?

In the section of this chapter just concluded we have established half of the causal version of the equivalence thesis. This we did in two steps. First, I argued at some length that all verbs of action used in the criminal law are causally complex. The second step we haven't taken, but it is an obvious one we should take now. It is to conclude that for any verb of action that is causally complex, a sentence containing such a verb entails another

causation lies in the fact that the person whose uptake is caused must *intentionally* employ the conventions in order to understand: 'the sense in which saying something produces effects on other persons, or *causes* things, is a fundamentally different sense of cause from that used in physical causation by pressure, etc. It has to operate through the conventions of language and is a matter of influence exerted by one person or another.' Ibid. 113 n. 1. The problem with this view is the *sui generis* nature of the causal relation posited to exist by Austin. Why should we think that human beliefs or behaviour are caused in some special sense of the word? To adopt such a dualistic view of causation makes us unnecessarily mysterious in how we interact with the natural world.

sentence of the form 'the actor's basic act caused some state of affairs to obtain'. That is, causally complex descriptions like 'X killed Y' entail causal expressions like 'X's basic act caused the death of Y'. Only an infelicitous selection of states or events caused by X's basic act could make one deny *that* entailment from causally complex verbs of action.[80]

These two steps together give us half of the equivalence thesis, namely, that for all verbs of action V used in our criminal codes, the sentence 'X V_T-ed' entails the sentence 'X's basic act caused a V_I-ing to occur'. Now we need to establish the converse entailment. We need to establish that expressions like 'X's basic act caused the death of Y' entail 'X killed Y'.

Most cases of causing-deaths-by-basic-acts that come to mind are clearly killings so that the second entailment holds. Where Jones moves his finger on the trigger, thereby causing the death of Smith by the bullet lodged in Smith's heart, there is no doubt of the propriety of redescribing Jones's action as 'killing Smith'. But there are a variety of related cases where we might think that Smith caused the death of Jones but that Smith did not kill Jones.

One such kind of case is where there is some spatial and temporal remoteness between the basic act and the death. For example, I pick up a stone in a field and finding it too heavy, drop it on the side of a hill on my way home. During the course of a year normal weather patterns cause sufficient erosion that the stone rolls down the hill on to a path, and a year after that Smith's horse trips on the stone, throwing Smith on to it, which kills him. One might be tempted to say that I caused Smith's death but there seems to be little temptation to say that I killed Smith.[81]

---

[80] Consider an example of Anthony Kenny's: 'washing the dishes is bringing it about that the dishes are clean'. Kenny, *Action, Emotion, and Will*, 177. One might well deny this on the ground that one can wash dishes so badly that one fails to get them clean. Such denial becomes less plausible if we formulate the result caused more accurately: washing$_T$ the dishes is causing them to be washed$_I$. Such (intransitive) washing$_I$ is not simply a having-been-washed$_T$-by-someone, any more than a door being open is an opening$_T$-of-a-door-by-someone. Such intransitive verbs name states independent of the acts that sometimes cause them just as 'death' names a state independent of killings. A washing$_I$ is the state of the dishes having water run over them, some pressure applied perhaps, all of which can be done naturally as well as by human actions of washing$_T$. See Ch. 5.

[81] The example is a variation of one found in Ginet, *On Action*, 5. See also Hornsby, *Actions*, 128.

A slightly different sort of case is presented where a death occurs as a result of some basic act, but that death comes about in a particularly freakish way. For example, I buy my wealthy uncle Smith an airplane ticket to a dangerous part of the world with the hope that something bad will happen to him so that I may accelerate my inheritance. He accepts my gift of a free vacation with gratitude, and, as it happens, his plane crashes on its return to New York because of a 'once-in-a-lifetime' wind-shear just off the runway at Kennedy Airport. I may have caused Smith's death, but did I kill him?[82]

Perhaps the most troublesome case for the thesis here considered is where another person's action is what I cause, and their action then causes the death of a third person. For example, I persuade Jones that Smith, having betrayed our organization, 'needs killing'; or I order Jones to kill Smith, Jones being my military subordinate; or I threaten Jones with death of his family unless he kills Smith; or I know that Jones intends to kill Smith if he can find him, and I tell Jones where Smith can be found; or I force Smith at gunpoint to step outside, knowing full well that Jones is waiting to kill him; or I tell Jones (truthfully or falsely) that Smith has been having an affair with Jones's wife, knowing full well Jones's explosive and uncontrollable temper and jealousy *vis-à-vis* his wife; and in all such cases, Jones shoots Smith dead. There may be some temptation in such cases to say that I caused Smith's death, but not that I killed him—Jones did that.[83]

There are of course two ways to deal with these cases so as to bring them into line with the equivalence thesis: either we can say that I did kill in each of these cases; or we can deny that I caused the various deaths. The first route I take to be foreclosed to us, both as a matter of ordinary English and as a matter of legal policy.

As a matter of ordinary English, this reluctance to redescribe a causing of a consequence C as a C-ing becomes more noticeable if we leave 'killing' as our example. As Sandy Kadish has observed, some actions are even more 'nonproxyable' than killings in the sense that they 'can be done by the actor only with his own body and never through the action of another'.[84] Some of Kadish's examples:

[82] See Kadish and Schulhofer (eds.), *Criminal Law and Its Processes*, 228.
[83] See Thomson, *Acts and Other Events*, 128.
[84] Kadish, *Blame and Punishment*, 172.

A sober defendant may cause an insensate and disorderly drunk to appear in a public place by physically depositing him there. But we could hardly say that the sober person has through the instrumentality of the drunk, himself committed the criminal action of being drunk and disorderly in a public place. A defendant may cause a married person to marry another by falsely leading the married person to believe his prior marriage was legally terminated. But the defendant could hardly be held liable for the crime of bigamy, since one does not marry simply by causing another person to marry . . . a visitor to a prison who abducts a prisoner could not be said to have committed the crime of escaping from prison by forcing a prisoner to do so.[85]

These ordinary-language observations themselves generate one legal-policy reason why judges ought not to construe 'killing' and other complex verbs of action so as to cover all these cases. The principle of legality (below discussed) should keep legal interpretations of 'killing' etc. close to ordinary interpretations of the words. Among other things, it gives citizens notice of what it is that is prohibited by the criminal law.

There is an additional policy reason to deter legal reinterpretations of 'killing' to include all of these cases as killings. Suppose there were a privilege in certain circumstances for me (but not for others) to do these killings. For example, in Texas in former years a husband could kill his wife without criminal liability if she provoked him by sexual infidelity. Or, to change acts so as to use a presently existing privilege, in some jurisdictions a husband still cannot be guilty of the crime of rape if the victim is his wife. Suppose H convinces R both that H's wife, W, would welcome R's sexual advances and that that welcome would exist even though W would appear to be resisting. Suppose further that R rapes W with the beliefs above described. If we were to say that H *raped* W, we would foreclose the possibility of holding H because of the marital privilege. So one might say that H did not rape W—R did that act, even though R did so with an innocent state of mind. That at least leaves open the possibility of holding H for *R's rape* on something like accomplice liability,[86] for it is

---

[85] Kadish, *Blame and Punishment* 172.

[86] H cannot be held as an actual accomplice because R is not himself guilty of rape because of his mistake as to W's consent. Still, one should hold H for causing a rape through an innocent agent, not for raping his wife, because that disallows the use by H of the marital privilege. This appears to be the theory of the Court of Appeal in *Regina* v. *Cogan* [1975-2-] All. Eng. Rep. 1059.

well established that the husband's privilege does not extend to helping another to rape his wife.

So we should put aside the first route to defending the equivalence thesis in the face of these examples. The second route is more promising. In none of these cases should we say that I (or my basic acts) *caused* the deaths. This route demands, of course, that we employ some notion of 'cause' that is more restrictive than the counterfactual test lawyers regularly identify as 'causation-in-fact': 'but for my acts, would death have occurred?' By that minimalist notion of causation, I did cause the death in each of the above examples.

Some linguists and philosophers despair of there being any more restrictive notion of causation available to rule out these apparent counter-examples to the equivalence thesis. Zeno Vendler, for example, sees the problem but refuses 'to trail off into the foggy regions in which causists and lawyers ply their trade'.[87] Jerry Fodor, who apparently has visited such regions', brings back the following report for his fellow philosophers and linguists:

'X caused Y to die' doesn't entail 'X killed Y.' Consider the case where X causes Y to die by getting someone else to kill him. It is usual to reply to this sort of objection by invoking a special relation of 'immediate causation' such that, by fiat, 'X immediately caused Y to die' does entail 'X killed Y.' It is this relation of immediate causation that is said to figure in the definition of verbs like 'kill.' It is a mystery (apparently one which is to remain permanently unexplicated) what, precisely, this relation is. (In the most obvious sense of 'immediately cause' what immediately causes one's death isn't, usually, what kills one. If it were, we should all die of heart failure.)[88]

Fodor sceptically concludes that there is only one equivalence to be found here: 'of all the species of X causing Y to die, there is one and only one which is necessary and sufficient for making "X killed Y" true: viz., X's causing Y to die by killing Y'.[89] In such an equivalence, of course, all the work of analysing 'X killed Y' is done by the uselessly circular criterion 'killing Y'.

[87] Zeno Vendler, 'Agency and Causation', *Midwest Studies in Philosophy*, 9 (1984), 371–84: 380.
[88] Jerry Fodor, *The Language of Thought* (Cambridge, Mass., 1975), 130–1 n. 23.
[89] Ibid.

Yet the law may not be the murky region that Vendler, Fodor, Chomsky, and other linguists seem to think.[90] Take the first two classes of cases, those where the death is spatially or temporally remote or where it is any event freakishly brought about. Where it is too remote or too freakish, in law I did not (proximately) cause the death of Smith. Since I didn't kill him either, the two expressions remain equivalent. To be sure, there is great vagueness about 'too remote' or 'too freakish' limitations of proximateness. Legal definitions, like that of the Model Penal Code, do not much reduce that vagueness: an act causes a result only if that result is 'not too remote or accidental in its occurrence to have a just bearing on the actor's liability'.[91] Yet the vagueness of 'causing' here may be a virtue, not a vice. After all, 'killed' is vague in its application too: the less remote or less freakish the death is from the act, the more likely that we should call the act a killing. If 'causing death' and 'killing' are indeed used equivalently, then one should expect a vagueness in the former if there is such vagueness in the latter.[92]

It might be thought that there is not a matching vagueness in the two classes of expressions because complex descriptions of actions are more restrictive (less freakish) in the means that can be employed in order to do such actions. Donald Davidson (whose objection we shall later examine in more detail) objects to equivalence on this ground. He urges that 'The doctor's basic act caused the patient to have no appendix' does not imply the action description 'The doctor removed the patient's appendix'— for the doctor could cause the patient to have no appendix 'by running the patient down with his Lincoln Continental', in which case we would not 'say that the doctor removed the patient's

<hr/>

[90] Some linguists have long sought to analyse causally complex action verbs (what they call 'causatives') in terms of causing certain states of affairs. Noam Chomsky's scepticism about the possibility of such an analysis has prompted a large literature. See Noam Chomsky, *Studies on Semantics in Generative Grammar* (The Hague, 1972), 72; J. L. Morgan, 'On Arguing about Semantics', *Papers in Linguistics*, 1 (1969), 69–70; G. Lakoff, 'On Generative Semantics', in D. Steinberg and L. Jakobovits (eds.), *Disciplinary Reader in Philosophy, Linguistics, and Psychology* (Cambridge, 1971); J. D. McCawley, 'Prelexical Syntax', in P. A. M. Seuren (ed.), *Semantical Syntax* (London, 1974).

[91] Model Penal Code, §§2.03(2)(*b*), 2.03(3)(*b*).

[92] A point also made by Wilson, *The Intentionality of Human Action*, 64, in his defence of this entailment.

appendix'.[93] Yet, we have to specify the effect caused with more precision than does Davidson here. We should compare 'The doctor removed$_T$ the patient's appendix' with 'The doctor's basic act caused the removal$_I$ of the patient's appendix'. The description 'removed$_I$ appendix' is not the same as the description 'having no appendix' in just the way that Davidson infers: the first description is more means-restrictive than is the second. So we should test the equivalence thesis with the first effect, not the second.

If we modify the example in this way, it is less plausible that a causing of a removal$_I$ would not be a removing$_T$. Suppose the doctor runs the patient down with his Lincoln Continental, hitting the patient in just the way that causes his appendix to fly out of his body. Has the doctor caused a removing$_I$ of the appendix? Has he removed$_T$ the appendix? Suppose we concede *arguendo* that the answer to the second question is in the negative, that a removing$_T$ requires something like a surgical cutting as the means of getting the appendix out. Then I take such a means restriction to be built into removing$_I$ too. Only something like a cutting (by something or someone) can be a removing$_I$. So the doctor did not cause a removing$_I$ of the appendix either. Alternatively, if the doctor's old Lincoln has sharp edges that cut out the appendix with the deftness of a surgeon's scalpels, then the doctor did cause a removing$_I$ of the appendix. Of course, he also in such a case removed$_T$ the appendix, and the equivalence remains.

The third kind of example is different, for we do not sense the same uncertainty about 'killing' when we have other people doing the killing. If Jones persuades another to kill Smith, and that other person does it, Jones did not kill Smith, his agent did. Yet the crispness of this conclusion is matched by a like crispness about causation: Jones's encouragement may have caused another to kill Smith, but it did not (proximately) cause Smith's death. The other's action was an intervening, voluntary act 'breaking the causal chain'.[94] In law, in morals, and in everyday ascriptions of causation, we recognize the difference between enabling conditions and full-fledged causes. Jones's telling another where Smith may be found, convincing the other to do

---

[93] Davidson, *Essays on Actions and Events*, 110–11.
[94] By far the most sophisticated treatment of the law's notion of an intervening cause remains Hart and Honoré, *Causation in the Law*.

it, and the like, are only enabling conditions. The other's action was the cause of Smith's death just as surely as it was the killing of Smith.

Suppose, by way of contrast, that Smith died in the following way: Jones pushed another person from the top of an embankment on to Smith, who was standing below, and the impact killed Smith. Here there is no intervening voluntary act by the other, only the use of his body as the instrument for causing death by Jones. So here Jones clearly did cause Smith's death. With equal clarity, however, Jones also *killed* Smith.

There are of course many intermediate cases of lesser clarity. What if the intervening actor acted in killing Smith, say by shooting him, but: that actor had a disability (insanity, uncontrollable temper, pressure of a threat, ignorance about Smith being still alive at the time he is shot, intoxication, youth) making his choice to kill Smith less deliberate or intentional. Where Jones utilizes the intermediary's known disability, he looks somewhat like the Jones who uses the intermediary's body as an instrument for the *causing* of the death of Smith; still, the intermediary in all such cases does act and (in some) even choose to kill. So there will be doubtful cases of intervention, rendering doubtful our judgement whether Jones (proximately) caused Smith's death. Yet are we not equally doubtful about whether Jones killed Smith? Don't we indicate such doubt with expressions of the form 'Jones as good as killed Smith himself?'[95] Again, if the vagueness of 'X caused Y's death' is matched by a vagueness of 'X killed Y' that is an argument in favour of the equivalence, not against it.

It is important not to confuse the proximate-causation criterion with our overall judgements about moral responsibility or about legal liability. When we say that Jones did not proximately cause Smith's death when Jones persuaded another to kill Smith, we have not said that Jones is not responsible, morally or legally, for Smith's death. He is—but not because he killed Smith. Jones persuaded another to kill, or aided the other to find the victim, so he is an accomplice in that killing. Accomplice liability exists, as Sanford Kadish has shown in detail, precisely to make people like Jones liable for deaths that they do not cause.[96]

[95] Wilson, *The Intentionality of Human Action*, 63.
[96] Kadish, *Blame and Punishment*, ch. 8.

The same is true for those of our doubtful cases (where the act that intervenes is less than fully deliberate or intentional) where we may conclude that Jones did not cause Smith's death. We may well think Jones to be morally responsible for Smith's death, liable to punishment as a murderer, anyway. In law this gets reflected either in accomplice liability (where the disability of the intervening actor does not excuse him) or in liability for causing an agent to cause death (where the disability does excuse the intervening agent, meaning that Jones cannot be held as the latter's accomplice).[97] In either case, we may hold people liable for deaths despite the fact that they neither caused them nor killed.[98]

All of this holds good for verbs other than 'kill'. Take one of Kadish's examples, the verb 'marry'.[99] We are quite clear that Kadish's supposed instigator of another's bigamy is not himself a bigamist, for he did not *marry* someone while still being married to someone else. Yet isn't the causal conclusion equally crisp: the instigator did not cause a state of marriage to exist between *himself* and another while *he* (the instigator) was still married to someone else. The instigator may have caused such a secondary marriage to exist for someone else, and that itself may be a crime; but he did not *cause* himself a secondary marriage any more than he *married*.

---

[97] See e.g. Model Penal Code, §2.06(2)(*a*), which makes one liable for the conduct of others even when one is not an accomplice of those others. If one 'causes an innocent or irresponsible person to engage in such conduct', then one is as guilty as if one did it oneself.

[98] These different bases of legal liability complicate the *actus reus* requirement. One way to be guilty of homicide is by *killing*. This is a complex description of action whose nature we have been unpacking. Alternatively, however, one may be liable for homicide: (1) by soliciting or aiding another to kill, who does then kill; (2) by agreeing with another that a third should be killed, who is then killed by the other; (3) by causing an innocent or irresponsible agent to kill, who does then kill. In addition, some states eschew use of complex action verbs like 'kill', and prohibit: (4) causing the death of another human being (for example, Model Penal Code, §210.1(1)). The equivalence thesis we have been examining certainly holds for these alternative bases of liability. With the fourth possibility the equivalence is obvious, but with the first three it is equally so if we keep clear just what is the prohibited state of affairs that an actor must not cause. In (1), the effect is either understanding (Austinian 'uptake') of the person solicited, or aid being given to the killer; in (2), it is either the understanding required for agreeing or the conviction required for agreement; in (3), it is the killing by another agent. From what we have earlier said, these bases of liability are equivalently prohibited as 'No basic act that causes S'—so long as we remember that state (or event) S need not be the death of the victim.

[99] Kadish, *Blame and Punishment*, 172.

Try 'escaping'. Kadish is surely right to conclude that the visitor to a prison who abducts a prisoner cannot himself be guilty of *escaping* from prison. Yet it is equally clear that the visitor did not cause an escape from prison either. The prisoner did not escape, because he was abducted; the visitor didn't escape because he was not a prisoner, and only prisoners can escape. So there was no escaping by anyone that the visitor caused. His abduction may be criminal, but it is not the crime of escape.

The key to both of the examples (and to many others like them using different action verbs) may be seen in the third of Kadish's examples, when the sober person causes the drunk person to be in a public place by carrying him there. The crime involved requires there to be a state: being drunk in a public place. The sober person is not in that place, and so cannot be guilty of that crime. Notice that reaching this conclusion has nothing to do with the supposed peculiarities of action (versus causal) descriptions, because the statute requires that an accused be in a *state* of public drunkenness. One cannot violate this statute by acting in such a way so as to cause another to be in the prohibited state—because no matter how much one causes the other to be in the prohibited state, one did not cause *oneself* to be in it.

Now notice that the 'marrying' and the 'escaping' prohibitions are no different. The states of affairs for these verbs are respectively the state of one's being married and the state of one's being outside the prison walls. One doesn't *marry* unless one reaches the state of being married; one doesn't *escape* unless one reaches the outside of the prison. The trick is to see that these verbs require that the 'one' who is doing the marrying (escaping) be the same as the 'one' who is in the state of being married (being outside the prison). So when we frame the causal question in order to test the equivalence thesis, we should make sure we keep clear what state it is the actor must cause. We should ask: did X cause X to be married? Did X cause X to be outside the prison? We should not ask: did X cause *anyone* to be married? or, did X cause *anyone* to be outside the prison? We should not ask these latter questions because these are not the states of affairs the action descriptions require. To test the equivalence thesis with these causal questions is like testing it for 'suicide' by comparing 'X committed suicide' with 'X caused the death of someone'. These last two of course are not equivalent descriptions, but then, that should be no surprise when the effect in the causal statement is this badly misdescribed.

Sometimes the objection here considered is framed slightly differently from the way I have hitherto framed it. Suppose we do not focus on comparisons of immediacy of causation between 'X killed Y' and 'X's basic act caused the death of Y'. Rather, perhaps the second expression does not entail the first because of an implicit means restriction present in the first but not the second. 'Kill' is not a verb that very plausibly has such means restrictions built into it, so let us again shift examples.

Donald Davidson, as we have seen, tests the entailment with the verb 'remove'.[100] Does 'The doctor's basic acts caused the removal$_I$ of the patient's appendix' entail 'The doctor removed$_T$ the patient's appendix?'[101] Davidson thinks not: 'The doctor may bring it about that the patient has no appendix by turning the patient over to another doctor who performs the operation; or by running the patient down with his Lincoln Continental. In neither case would we say the doctor removed the patient's appendix.'[102] Davidson's first variation (with the intervening doctor) we have already implicitly considered when we discussed intervening causes. It is the second—the one we explicitly discussed before—that raises a distinct objection. The objection is that the verb 'remove$_T$' has a means restriction built into its meaning that 'causing a removal$_I$' does not. Removals, Davidson apparently thinks, must be something like surgical removals if they are removals at all; cuttings out of appendixes by sharp edged Lincolns do not qualify.

Jonathan Bennett, who is otherwise quite sympathetic to the equivalence thesis,[103] none the less thinks that some causally

---

[100] Davidson, *Essays on Actions and Events*, 110–11.
[101] I have reworded Davidson's example somewhat to eliminate competing features making the entailment tested less plausible.   [102] Ibid.
[103] Bennett, *Events and Their Names*, 221–9. Bennett analyses causally complex descriptions of action in a way very similar to that presented above. According to Bennett, 'X killed Y' is equivalent to: (1) 'X (or X's movement$_T$) brought about Y's death' (ibid. 224); and (2) X's movement is pictured as being, in the circumstances, the sole input into Y's death (ibid. 224–6); and (3) no wholly intervening agency intervenes between X's movement$_T$ and Y's death (ibid. 226–8). Bennett's restrictions (2) and (3) are his analogues to my proximate-cause limitations on when acts cause results, and to the same end: causally complex verbs can be given a causal analysis so long as causation is properly analysed.

complex action verbs have individual names to them that are quite 'unmanageable'.[104] One of his examples: 'Suppose I contribute to a tree's falling by digging the earth away from its roots and burning them through; do I *fell* it? I am inclined to say no, but that seems to be partly because in communities with which I am familiar that is not a standard way of getting trees to fall.'[105] As a third example, consider a class of causally complex verbs to which Judy Thomson has drawn attention, verbs like 'kick', 'kiss', 'swallow', 'bite', 'slap', and 'punch'.[106] Such verbs seem to have quite explicit means restrictions built into them. A kicking seems to require foot movement, for example, as a kissing requires lip movement, a swallowing, throat movement, a biting, teeth movement, etc. The suggestion then is that 'X did some basic act that caused contact with Y's shin' does not entail 'X kicked Y's shin', for X might have caused such contact without moving his foot (i.e. without *kicking*). For example, 'If I start up a machine that a minute later thrusts my booted foot into your midriff, it need not ever be the case that I have kicked you.'[107]

My own view is that none of these examples do the work claimed for them. 'Remove', 'fell', and 'kick' do generate audience expectations about normal ways of removing appendixes, felling trees, or kicking others. These expectations get heightened and particularized when we are told that it was a doctor who did the removing, a lumberjack who did the tree-felling, or a little boy who did the kicking. Such expectations give us a kind of picture or stereotype of what such actions should look like. Yet I see no reason to honour such stereotypical expectations with *semantic* dignity. Why should these individual nuances be part of the meaning of these verbs? Surely these are at best pragmatic features of appropriate utterance on certain occasions. If one doctor tells another, 'Remove the patient's appendix', he does not mean, 'Use your sharp-edged Lincoln'; if a foreman tells a lumberjack to fell a tree, he does not mean, 'Dig out its roots and burn them until the tree falls'; if one boy tells another, 'Let's kick someone', he does not envision setting up a kicking machine. Yet the situations in which such utterances take place

---

[104] *Events and Their Names,* 230.   [105] Ibid. 229.
[106] Thomson, *Acts and Other Events,* 220–2. See also Bennett, *Events and Their Names,* 222–4.
[107] Bennett, *Events and Their Names,* 222.

rule out these unusual removings, fellings, and kickings only because they are too unsafe, too time-consuming, or otherwise too impracticable. To my ear, each is none the less a literal removing, felling, or kicking.

The 'kicking' class of verbs may seem to evade this response somewhat, for it may seem that 'kicking', 'kissing', etc. involve specific body motions in the way that 'removing' and 'felling' do not.[108] Yet this is an illusion brought about by a failure to attend carefully to *what* must be caused in order for there to be a kicking, a kissing, etc. The actor's specific body parts are involved in such actions, but not as necessary means. Rather, some relation to such body parts is required as the state of affairs caused. The state of affairs involved in kissing is not just contact with the lips (or elsewhere) of another; it is contact between the lips of the actor and the lips or other body part of another. The state of affairs involved in kicking in the shins is not just contact with the shins of another; it is contact between the foot of the actor and the shins of another; etc. There is no requirement that the actor move his lips in order to kiss (it may well be a bad kiss if he doesn't move his lips, but a kiss none the less); the only requirement is that the actor move his body in some way so as to cause lip contact. There is similarly no requirement that an actor move his foot (as a basic act) in order to kick; granted, he will have to put the leg in motion by *some* basic act of his, but how he does this is irrelevant to his *kicking*.[109]

I conclude that there are no means restrictions built into the meaning of causally complex verbs of action that make them more restrictive in their application than the corresponding causal idioms. If there are no tighter causal requirements built into such verbs either, as I have also argued in this section, then

---

[108] Thus, Thomson, *Acts and Other Events*, 220–1: 'Notice that you kick a thing only if you move something (you have to move your foot); you punch a thing only if you move something (your fist); you kiss something only if you move something (your lips); and so on. . . . It seems very plausible that every event that is a kicking of something is a moving of a foot.'

[109] There may be other restrictions one wants to add to the kind of state of affairs caused in order to capture the meaning of verbs like 'kick'. Soft contacts on another's shins, for example, may not be the kind of contact required for a kick (as Thomson suggests, ibid. 222); but such refinements of the consequence that must be caused in order for there to be a kick cannot affect the causal analysis of 'kick'.

we should conclude that 'X's basic act caused a V-ing$_I$' entails 'X V-ed$_T$' as much as vice versa, for all causally complex verbs of action V. This means that the two forms of expression are equivalent, despite the objections we have considered.

9

THE NORMATIVE BASIS FOR THE *ACTUS REUS* REQUIREMENT

Having clarified the *actus reus* requirement and the kind of complex action descriptions that it requires, it remains to justify it. As with the act requirement, so with the *actus reus* requirement we need to distinguish two different aspects of the requirement that need justifying. The first is the limiting, or negative, aspect of the *actus reus* requirement: no one may be punished unless their act was an instance of the type of action prohibited by some statute in place at the time the act was done. The second is the motivating, or positive, aspect of the requirement: one should be punished if (along with certain other requirements) one has done an act that was an instance of the type of action prohibited by some statute in place at the time the act was done. To justify the first aspect is to show what good is served by *not* punishing those who do *not* satisfy the requirement; to justify the second aspect is to show what good is served by punishing those who do satisfy the requirement. It is helpful to separate these two aspects of the requirement, not because some good cannot justify both, but because (as with the act requirement) the good(s) that in fact justify only the negative aspect are commonly taken to justify the positive as well.

The usual justification offered for the *actus reus* requirement is in terms of the 'principle of legality'. The principle of legality is in reality a mixture of four values that jointly justify a wide variety of criminal-law doctrines.[1] The values are those of fairness, liberty, democracy, and equality, which are unpacked as follows. It is *unfair* to surprise citizens with liability to criminal sanctions when they reasonably relied on their actions not being criminal

---

[1] I discuss the principle of legality in Moore, 'The Limits of Legislation'. See also John Jeffries, 'Legality, Vagueness, and the Construction of Penal Statutes', *Virginia Law Review*, 71 (1985), 189–245.

at the time they were done. It impedes *liberty* if citizens cannot know the content of the criminal law well enough to take into account the possibilities of penal liability in planning their actions. The value of *democratic decision-making* requires that elected legislatures decide what is and what is not criminal, and (not electorally responsive) courts would frustrate that value if they were to take it upon themselves to make conduct criminal without statutory authorization. *Equality* dictates that those who are in all morally relevant respects alike be treated alike, and this requires that neither legislatures nor judges single individuals out for arbitrarily different treatment.[2]

In Anglo-American law the doctrinal expressions of these values are numerous, and many of such doctrines are over-determined by more than one of the four values just described. The main doctrines are nine in number:

1. The prohibition against 'common-law crimes', or 'crimes by analogy', holds that courts are without power to create new crimes, either from whole cloth or by analogy to crimes already prohibited by statute.[3] The primary value furthered by this doctrine is democracy, because the justification for restricting criminal law-making to legislatures is largely due to the more democratic selection of legislatures over judges. However, fairness and liberty are also served by courts refusing to use (retroactively applicable) adjudication as the occasion on which to announce new crimes.

2. The doctrine that *ex post facto* laws are unconstitutional[4] plainly serves the values of fairness and liberty. Retroactive criminal laws would both unfairly disappoint reliance on an activity not being criminal when it is done, and chill liberty by the fear that such surprise might be forthcoming.

3. The constitutional (due process) prohibition of retroactive judicial enlargement of criminal liability,[5] like the *ex post facto* limitation on legislation, plainly serves the values of fairness and liberty. For the purposes of these values, it does not matter

---

[2] For a more detailed explication of these legality values, see Moore, 'A Natural Law Theory of Interpretation', 313–18.

[3] See e.g. *Keeler* v. *Superior Court*, 2 Cal. 3d 619, 470 P. 2d 617 (1970).

[4] US Constitution, Art. I, §§9 and 10.

[5] *Bouie* v. *City of Columbia*, 378 U.S. 347 (1964).

whether the surprise comes from law-making by a court or by a legislature.

4. The void-for-vagueness doctrine requires legislatures to frame acts with sufficient clarity that they can be understood by those to whom they are directed.[6] Such failure of understanding can occur not only because of vagueness in some statutory predicate, but also because of: ambiguity in the predicate; inconsistency between this statutory provision and some other; inability of most citizens to be able to verify factually whether the behaviour they contemplate will or will not have the characteristics the statute prohibits.[7] Any of these inabilities to understand makes a criminal statute 'void-for-vagueness' under the due process clause. The main values served here are fairness and liberty, but the equality value is also served by not allowing vague statutes to grant such discretion to enforcement officials that discriminatory application is made possible.[8]

5. The prohibition against bills of attainder[9] also serves the equality value by requiring of criminal legislation that it be general in the sense that it apply to some *class* of persons. Such generality is the first step towards equal treatment, because such generality precludes the numerical distinctness of persons being used to single individuals out for criminal punishment. Because bills of attainder are also retroactive in effect, the prohibition also serves the values of fairness and liberty.

6. The common-law maxim that criminal statutes must be strictly construed[10] serves the values of fairness and liberty by disallowing prosecution in the vague penumbra of some statute's *actus reus* prohibition. Such a maxim gives citizens the benefit of the doubt in the interpretation of criminal statutes, so that any reasonable mistakes of interpretation due to the vagueness of the statutory language do not result in criminal liability.

[6] See e.g. *Papachristou* v. *City of Jacksonville*, 405 U.S. 156 (1972).
[7] As was argued to be the case about drunk-driving prohibitions that define drunkenness by percentage of alcohol in the blood, a legal line some argued to be unverifiable by ordinary citizens. See M. Moore, 'Drunk-Driving Law: Precise is Vague?', *Los Angeles Times*, 8 June 1983, pt. 2, p. 5.
[8] As Douglas emphasizes in his opinion in *Papachristou* v. *City of Jacksonville*, 405 U.S. 156 (1972).
[9] US Constitution, Art. i, §9.
[10] See e.g. *Keeler* v. *Superior Court*, 2 Cal. 3rd 619, 470 P. 2d 617 (1970), for the hardiness of this maxim even in the face of a statutory repeal of it. (On the latter, see California Penal Code, §4.)

7. Mistake of law by an accused is allowed as a defence when that defendant's mistake was induced by his reliance on (mis)-advice about the law received from some governmental official.[11] Here the value is mostly fairness, because the defendant is unfairly whip-sawed between one governmental agency (which gave the advice) and another (the court trying him).

8. Some state constitutions require that criminal statutes be in English,[12] again with the plain rationale that citizens must be able to find out what is punishable in advance. The values served are again fairness and liberty.

9. Mistake of law is also a defence to criminal prosecution in situations where the law making an act criminal was not publicly promulgated.[13] The doctrine also has as its rationale the values of fairness and liberty. What isn't publicly available cannot be known, so that when applied such a statute unfairly surprises, and the potential for such surprise chills the liberty of those not knowing whether their act is secretly criminal.

The four values, as they converge to jointly justify these nine doctrines, are what is misleadingly referred to as 'the' principle of legality. Such principle prohibits punishment except for actions prohibited by some public, prospective, general, clear, consistent, verifiable, strictly construed, legislatively created law. The four values that collectively make up the principle of legality give good reason to impose the *actus reus* requirement as a prerequisite to punishment. Indeed, without the requirement that no act be punishable unless it is an instance of a type of act prohibited by law, it would be hard to make sense of the nine doctrinal expressions of the principle of legality. That is, without the requirement of fit between act and action description (which is what the *actus reus* requirement requires), what point would there be to requiring criminal laws to be legislative in origin, prospective, public, clear, etc.?[14]

Considered as a limitation on punishment, the *actus reus*

---

[11] Model Penal Code, §2.04(3)(*b*).

[12] See e.g. former California Constitution, Article 4, §24 (repealed 1966). See e.g. *In re Lockett*, 179 Cal. 581 (1919). (Latin descriptions of oral sexual practices not enforceable because not in a language those who might engage in such practices can understand.)   [13] Model Penal Code, §2.04(3)(*a*).

[14] A point Lon Fuller understood when he added congruency of official interpretation to his conditions of legality. Lon Fuller, *The Morality of Law*, 2nd edn. Stanford, Calif., 1969), ch. 2.

requirement thus can be justified by the values served by legality. For the mixed theorist about punishment, that may be justification enough, but a retributivist about punishment should want more. Granted, it is unfair to punish those who have not been afforded sufficient opportunity to find out what is prohibited; but why is it good to punish those who have been afforded such opportunity? For a retributivist the ultimate good served here must be the attainment of retributive justice, i.e. seeing that those who deserve punishment get it.[15] Two conditions are necessary before retributive justice is achieved by punishment: the act for which punishment is inflicted must have been morally wrong to do, and the actor must have been morally culpable in the doing of it. Does satisfying the *actus reus* requirement ensure the satisfaction of either of these conditions of retributive justice?

My answer is a qualified yes. Assuming a large overlap between the content of the criminal law's prohibitions and the content of morality's prohibitions, then the likelihood that those punished are both culpable and have done wrong is increased by the existence of the *actus reus* requirement. With regard to culpability, the function of the *actus reus* requirement is a heuristic one in that citizens can learn morality from the criminal law. Assuming the overlap above stipulated, and assuming that some kind of knowledge of right and wrong is a condition of moral culpability, citizens punished in a regime where legality (including the *actus reus* requirement) prevails are more likely to be culpable than in a regime where it does not.

With regard to wrongdoing, the *actus reus* requirement serves two functions. First, it also serves a heuristic function here, but this time for judges rather than for citizens: assuming the overlap above stipulated, judges can best punish moral wrongs by punishing legally prohibited acts. For with a large overlap between moral wrongs and legally prohibited act-types, punishing at least those complex actions prohibited by a criminal code gives that much of moral wrongdoing its (retributively) just deserts.[16]

---

[15] Moore, 'The Moral Worth of Retribution'; id., 'Closet Retributivism'.

[16] It may seem that this doesn't justify the *causally* complex structure I urged in Ch. 8 to be essentially present in all act-types prohibited by Anglo-American criminal codes. Yet in so far as moral norms themselves prohibit only causally complex act-types—and not just bodily movements done in certain circumstances—then punishing legally prohibited act-types *is* a good heuristic for punishing moral wrongs and thus achieving retributive justice.

Secondly, some criminal statutes solve co-ordination problems that everyone is morally obligated to solve, such as the rules of the road. Such statutes create moral obligations where none existed prior to the statute. To violate such statutes is accordingly itself a moral wrong, without regard to the moral quality of the act prior to the passage of the statute. For such statutes, punishing those who satisfy the *actus reus* requirement directly serves the function of achieving retributive justice.[17]

Of course, if there were no great overlap between what morality prohibits and what Anglo-American criminal law prohibits, the two heuristic functions noted above would not be served. But then, absent such an overlap, there would be little retributive justice achieved by punishment anyway (except for co-ordinative crimes). In situations when the whole institution of the criminal law lacks justification, it should be no surprise that the *actus reus* requirement would lack justification too.

[17] On the role of criminal sanctions in solving co-ordination problems, compare John Finnis, *Natural Law and Natural Rights* (Oxford, 1980), with Leslie Green, 'Law, Co-ordination, and the Common Good', *Oxford Journal of Legal Studies*, 3 (1983), 299–324.

10

THE METAPHYSICS OF COMPLEX ACTIONS I: THE DEPENDENCE OF COMPLEX ACTIONS ON BASIC ACTS

The metaphysical question about complex actions is the question of what, essentially, they are. The answer I shall here defend is a straightforward one: all such actions are (identical to) bodily movements$_T$. There are, that is, no complex actions that are not basic acts.

Before defending this answer, it may be well to recall the importance of the question. There are two legal concerns that justify criminal-law theoreticians' paying attention to this question. The first has to do with the act requirement. Whether there is a univocal act requirement that underlies the diverse actions prohibited both by morality and by our criminal codes depends on this metaphysical question. For if mental acts or omissions, for example, could constitute the actions of killing, maiming, etc., or if a killing may be an event distinct from the event of finger movement that on that occasion causes the death, then there is no univocal act requirement. Rather, there would be killings by omission as well as killings by commission, killings by mental action as well as killings by physical action, etc. Alternatively, even for killings done by the commission of bodily movements$_T$, the killing would be distinct from the basic act that caused death. Since this would also be true for actions of raping, maiming, kidnapping, etc., none of these types of complex actions could share *this* essential nature (of being bodily movements$_T$). Since such diverse kinds of complex actions don't plausibly share any other essential nature, there would be no univocal act requirement to either the criminal law or the morality that underlies it.

The answer that we give to this metaphysical question thus determines the answer we can give to the moral and legal questions about there being an act requirement. The answer I shall defend I earlier called the exclusivity thesis, which I showed to be part of the Bentham–Austin theory of action. As I stated earlier, in Chapter 5, to defend that answer is to defend two propositions: (1) that no non-bodily movement$_T$ (such as an omission or a mental act) can be a complex action like killing; and (2) that for those complex actions (say of killings) that plainly involve bodily movement$_T$, that movement$_T$ (and not the death it causes, nor the circumstance in which it takes place) *is* the complex action of killing.

Given the answer that I shall defend, we thus must discuss the metaphysics of complex actions in two parts, which I shall do in this and the succeeding chapter. Before doing so, however, it remains to point out the second importance of the metaphysics of complex actions to criminal-law theoreticians. This has to do with the *actus reus* requirement of the criminal law, not the act requirement.

The equivalence thesis at which we laboriously arrived in the preceding chapters asserts that for all complex action descriptions V, to say that x V$_T$-ed is equivalent to saying that some willed bodily movement by x caused a V$_I$-ing to occur. This unified formula for complex action descriptions goes by the board if there is some other way in which x can V$_T$ except through some willed bodily movement. At a minimum, the formula would have to be altered by substituting some disjunction of initiating causes for willed bodily movements. For example: 'x V$_T$-ed if and only if x's willed or involuntary bodily movements, or his thoughts, or his omissions, caused a V$_I$-ing to occur'. More likely, the obliteration of the equivalence thesis would be more complete than this, for it could not plausibly be maintained that any involuntary bodily movement that causes a death, or any omission that fails to prevent a death, can be a killing. One may well be forced all the way back to J. L. Austin's kind of scepticism about actions, namely, that each action description is unique and whole; each description must thus be analysed by itself; it is not complex because it does not have any dimensions of complexity; and *a fortiori* action descriptions do not have *shared* dimensions of complexity, causal, circumstantial, or otherwise.

In addition to obliterating the unity to complex action description described by the equivalence thesis, finding that complex actions like killing were not (token–token) identical to willed bodily movements would affect certain procedural aspects of the *actus reus* requirement that I have hitherto ignored. Criminal prohibitions typically have certain procedural niceties attached to their valid application. One set of niceties makes relevant *where* an action takes place: if Jones kills Smith in Nevada, California may not prosecute him for it; alternatively, if Jones kills Smith outside the walls of a correctional institution, Jones may not be held for 'killing a correctional officer within a correctional institution'. For jurisdictional, choice-of-law, venue, and interpretive purposes, it matters *where* the killing of Smith took place. Another set of niceties has to do with *when* an action takes place: if Jones kills Smith in California, but did so (1) when the California murder statute had been voided for unconstitutionality; or (2) before any murder statute was passed; or (3) so long ago that the statute of limitations has run; or (4) when Smith was no longer a state correctional officer; or (5) when Jones reasonably believed Smith was already dead, then Jones cannot be punished for intentionally killing a state correctional officer.

Under the current law of jurisdiction etc. this hodgepodge of issues is made to depend on the correct metaphysical answer as to the time and place of complex actions like killings. The correct metaphysical answer as to when and where a killing takes place in turn depends on whether killings are or are not identical to the bodily movements that cause the relevant deaths. Suppose my proverbial Smith was a California correctional officer working at a California correctional institution located in California on the border with Nevada. As Smith leaves work for his home in Nevada, he is shot by Jones from within the prison. Smith staggers home, goes into a coma, and eventually dies. He was on the verge of retirement anyway, and at some point between Jones's finger movement on the trigger and Smith's death Smith was automatically retired from California's correctional service; in addition, during this same period the California statute criminalizing the killing of a state correctional officer within a state correctional facility was repealed. If the killing of Smith by Jones is identical to the moving of the trigger finger by Jones, then there are no barriers to prosecuting Jones in California under the California law making it a crime to kill a state correctional officer within a state correctional facility. If the killing of Smith by Jones is not identical to this movement event, then the killing presumably has greater spatio–temporal boundaries (perhaps even including the death of Smith), and these may well give rise to jurisdictional, choice-of-law, venue, legality, and other technical objections to prosecution. It thus matters to such issues (as they are currently framed) whether complex actions like killings are identical to bodily movements or not.

In this chapter I turn to the first of the two metaphysical issues, whether all complex actions can be done only through the doing of a willed bodily movement. If one were to list several hundred complex action-types, and then call to mind all of the instances of those types that one had ever experienced or could imagine, one would start to have a database adequate to answer our present question. Life being short and this book already being too long, I shall not adopt this procedure. Instead, let us pick one complex action-type, killing, and ask whether an act of that type can be done in any way except by the doing of a basic act of bodily movement$_T$.

Since we are dealing with the thesis that every action (here, of killing) is at its core a willed bodily movement, our question is usefully divided into: (1) cases where there are bodily movements causing death, but there is no willing (i.e. cases of involuntary bodily movements); (2) cases where there is no bodily movement but there is at least something like a willing and there is at least something like a causal relationship between that (almost) willing and death (mental acts of a certain kind, and intentional omissions); and (3) cases where there is neither a willing nor a bodily movement (statuses, unintentional omissions, mental states and mental events that are not acts). I shall treat each of these three cases in turn.

1. NON-WILLED ('INVOLUNTARY') MOVEMENTS AS COMPLEX ACTIONS?

I once divided involuntary movements into three classes:[1] (1) those where our bodies move but our muscles do not cause such movements, (2) those where our muscles move our bodies but there is no intelligent pattern to such movements, and (3) those

---

[1] Moore, 'Responsibility and the Unconscious', 1572.

where our muscles move our bodies but there is an intelligent pattern to such movements. Examples of each are: (1) Another throws my body down a well shaft, I land on top of you, thereby causing your death. (2) Another rigs my foot to the trigger of a gun, and causes my foot to move and thereby to discharge the gun, causing death, either by: (*a*) tapping my knee, inducing a reflex movement; or (*b*) inducing an epileptic seizure in me; or (*c*) inducing a hypoglycemic episode in me by injection of too much insulin. (3) I walk in my sleep, and deliver two deadly and accurate blows to my daughter's head with an axe, thereby causing her death.

It is the examples of the third kind that many theorists wish to categorize as *actions* of killing despite the seeming lack of any willing or volition. It is therefore worth examining this third category with some care. There are five sorts of conditions that are the leading examples of this third sort of involuntary bodily movement.

Perhaps best known is the kind of somnambulistic behaviour just described. A person while asleep seemingly executes very complicated routines having a recognizable, waking-life pattern to them. In two of the leading somnambulistic cases, the pattern is one of a waking-life killing. In *Fain* v. *Commonwealth*,[2] the defendant pulled his gun, aimed it at the victim, and shot him dead. In *Cogdon*,[3] the defendant dreamt that her daughter was being attacked by North Koreans and that she must defend her; in reality, the defendant delivered two forceful and accurate blows to the head of her daughter thereby causing the latter's death.

A second class of cases occurs when the defendant is unconscious from a blow to the head or a shot to the abdomen, and despite being unconscious manages to shoot the victim in an intelligent manner just as one would do if one were consciously trying to kill the other. In the highly publicized trial of Huey Newton, the Black Panther,[4] my former colleague at Berkeley, Bernard Diamond, testified that Huey Newton could have been

---

[2] 78 Ky. 183, 39 Am. Rep. 213 (1879).

[3] *Cogdon* was an unreported Australian case described in the *Daily Telegraph*, 20 Dec. 1950. It is discussed in Norval Morris, 'Somnambulistic Homicide: Ghosts, Spiders, and North Koreans', *Res Judicatae*, 5 (1951), 29–33.

[4] *People* v. *Newton*, 8 Cal. App. 3d 359, 87 Cal. Rptr. 394 (1970).

rendered unconscious by the shock administered to his system, the shock coming from Newton being shot in his thoracic cavity; this, Diamond testified, was consistent with Huey Newton none the less delivering five accurately placed shots into an Oakland police officer, thereby killing him.

A third class of cases has to do with hypnosis. While hypnotized, the defendant might kill, as in the once popular film *The Manchurian Candidate*, where the subject is directed to kill while under hypnosis. Alternatively, the defendant might kill while no longer in a hypnotic trance, but in response to a post-hypnotic suggestion. There is considerable dispute about the degree to which hypnotic or post-hypnotic suggestion can cause people to do that which they are not already inclined to do, but the thought experiment relevant for action theory is best conducted by conceding a great deal of causal power to such suggestions.

Contrast both hypnotic and post-hypnotic suggestions, on the one hand, with ordinary suggestions, on the other. An example of a completely non-coercive suggestion is where we know another to be excessively proud of some feature of his life and, as a result, excessively prone to recounting the same tale about it when it comes to mind. We often can manipulate the tale-telling simply by the most indirect of suggestions, e.g. mentioning a key-word that triggers the association. Despite the causation of the tale-telling by our suggestion, we still think of that telling as an action by our suggestible speaker.

A slightly more coercive example is provided by the well-known Milgram experiments.[5] In getting the subjects to deliver what they must have thought were lethal doses of electricity, Milgram and his associates used the coercion of their authority to get the subjects to do the 'killings'. Even here, it seems uncontroversial to think that these actions of pushing the button (supposedly causing death-dealing jolts of electricity to go into another) are just that, *actions*, even though caused by suggestion; had the supposed victims died of electrical jolts, it would surely be correct to say that the subjects who administered the electric jolts *killed* them.

Hypnotic and post-hypnotic suggestions are supposed to be more powerful than these kinds of more ordinary suggestion.

[5] Stanley Milgram, *Obedience to Authority* (New York, 1974).

Indeed, although there are few actual decided cases, the Model Penal Code and scholarly commentary both take the position that bodily movements caused by hypnosis are not acts and, thus, that when a death ensues, there still is no killing.[6]

The fourth class of cases has to do with 'brainwashing', or coercively induced beliefs, that themselves bring about the death of another. The most discussed modern case is the Patty Hearst trial, although the treatment of American prisoners during the Korean War also gave rise to much discussion.[7] Patty Hearst, an heiress to the Hearst fortune, was subjected to assaults and misinformation techniques for two weeks in the closet of the 'Simbionese Liberation Army'. As a result of the experience, she emerged from the closet as 'Tania', an SLA revolutionary who helped rob banks in order to finance the operations of the SLA. Some have wanted to argue that because Patty Heart's bodily movements were in response to desires and beliefs (in the goodness of the SLA) themselves coercively introduced into her belief–desire system, those movements were not her actions, and thus she did not perform the complex action of robbing a bank.

The fifth set of examples comes from the literature of dynamic psychiatry. Freud wrote that there were three main types of behaviour that evidenced the existence of the unconscious: dreams, parapraxes (slips of the tongue, miswriting, misreading, forgettings, awkward movements), and neurotic symptoms. Consider an example of each.

Suppose it were treasonable to contemplate ('compass') the killing of the king. Suppose that such a crime punishes only the (mental) action of calling to mind images of the king's death so that it does not punish the passive thoughts that just come to one unbidden. Recall the Roman citizen foolish enough to admit that he dreamt of the death of the Emperor.[8] Did the citizen perform the (mental) action of contemplating the death of the king?

In discussing those kinds of parapraxes that are apparently accidental or spastic movements, Freud gave an example from

---

[6] Model Penal Code, §2.01(2)(c). For the hesitation of the drafters on this, see American Law Institute, *Model Penal Code and Commentaries*, Comment to §2.01, 221. See generally Katz, *Bad Acts and Guilty Minds*, 128–35; Williams, *Criminal Law,*, 768–9; Note, 'Hypnotism and the Law', *Vanderbilt Law Review*, 14 (1961), 1509–24.

[7] See Katz, *Bad Acts and Guilty Minds*, 63–4, 79–80.

[8] Scholz, *Schlaf und Traum*, trans. as *Sleep and Dream*, 62.

his own life.[9] Freud disliked the ink-well given him by his sister. He accordingly wished to replace it with a new one but was restrained from doing so by considerations of hurt feelings. Freud found that one day he accidentally brushed his arm across his desk in a deftly executed destruction of the ink-well. Because the movement fulfilled his desire as well as a consciously willed movement would have, Freud concluded that he had performed the action of 'executing' (his description) the undesired ink-well.

One of Freud's better-known conversion hysterics was the patient Freud gave the pseudonym 'Dora'.[10] Dora's neurotic symptom was the sore throat she experienced. Freud's explanation of the symptom was in terms of Dora's repressed sexual desire for a certain gentleman, who had made a pass at her earlier in her life. Dora's hysterically sore throat was caused by her imagined oral copulation of the gentleman in question, bolstered by the secondary gain from the symptom (of getting the attention of an invalid). Was Dora's making her throat sore an action, if we credit all of Freud's explanation?

The question to be asked of all five of these kinds of examples is whether they are the complex actions of killing, of robbing, of contemplating, of destroying property, or of sore-throat-making? The question is not directly one of either moral responsibility or legal liability. Although the answer about the action question is relevant to the responsibility–liability question, these are not the same questions; after all, one might well conclude that some of these five sorts of movements are actions and yet excuse the actors from both moral responsibility and legal liability.

The most popular route to affirming that these are actions has been based on observations of ordinary language. Consider the first of these five sorts of examples, bodily movements occurring during sleep. Herbert Hart once recognized that:

The meaning of the words used in defining the offence . . . [would include a basic act requirement] *if* it were the case that, whenever we have an active verb like 'drives', this implies, as part of its meaning, the existence of the minimum form of conscious muscular control, upon which the general doctrine [of willed bodily movements] insists. As a

---

[9] Sigmund Freud, *The Psychopathology of Everyday Life, Standard Edition of the Complete Psychological Works of Sigmund Freud* (London, 1960), vi. 167.
[10] Sigmund Freud, *Dora: An Analysis of a Case of Hysteria* (New York, 1963).

matter of ordinary English this is however not the case. The phrase 'sleep-walking' is alone sufficient to remind us that if the outward movements appear to be co-ordinated as they are in normal action, the fact that the subject is unconscious from whatever cause does not prevent us using an active verb to describe the case . . .[11]

Hart seems to think that the sleep-walker indeed performs the action of *walking*, the unconscious driver indeed *drives*, the hypoglycemic assaulter indeed *hits* his victims, the hypnotized subject told to raise his arm, and who does so, does indeed *raise* his arm. Hart's reason for thinking that involuntary bodily movements such as these are actions is that it is idiomatic in ordinary English to describe them with the ordinary verbs of action. Ordinary usage certainly seems to be with Hart here. In support of Hart, consider Shakespeare's description of the queen's somnambulistic behaviour in *Macbeth*:

Lo you! here she comes. This her very guise; and, upon my life, fast asleep. . . .
What is it that she does now? Look, how she rubs her hands.
It is an accustom'd action with her, to seem thus washing her hands.[12]

Bernard Williams has recently joined Herbert Hart in thinking that such ordinary-language examples show that it is 'appropriate beyond dispute' to describe somnambulistic and other like behaviour in 'the language, not just of action, but of purposive action'.[13]

Not only does the data of ordinary English usage not show this 'beyond dispute', I do not even think it counts as an argument tending to show this at all. Consider three points about such usage. First, ordinary usage is misleading here because it glosses over a crucial ambiguity in our use of action verbs. I may with perfect linguistic propriety say: 'The tree *shed* its leaves'; or, 'The water *sought* its own level'; or, 'The sulphuric acid *dissolved* the zinc'. Are these natural events all actions just because ordinary verbs of action can be used to describe them? Or are not these verbs used differently in the foregoing sentences than they

[11] Hart, *Punishment and Responsibility*, 109. For a more cautious enthusiasm for this view, see Husak, *Philosophy of the Criminal Law*, 90.

[12] *Macbeth*, v. i.

[13] Bernard Williams, 'Voluntary Acts and Responsible Agents', *Oxford Journal of Legal Studies*, 10 (1990), 1–10, 1.

are when we say: 'He *shed* his clothes in a hurry'; or, 'She *sought* her husband in the crowd'; or, 'He *dissolved* the zinc in a sulphuric-acid solution?' Our use of the verbs of human action is rich in metaphorical extension to the 'actions' of inanimate objects. Yet surely these metaphorical actions need not be accounted for in any theory of *human* action.

Nor, I shall now argue, need the metaphorical actions of sleepwalking, unconscious driving, etc. One can see that these are only metaphorically actions by seeing the second feature of ordinary usage relevant here, namely, the often unnoticed ambiguity in the nouns and pronouns by which we refer to persons.[14] I may say: 'I hit the ball', and 'I am six feet tall', yet the thing referred to in these two sentences is different. The first 'I' refers to me as personal agent, the second, only to my body as inanimate object. I, the person, and I, my body, are not the same things: if I say, 'I love her', I mean I, the person, has a certain emotion towards another; make it clear that the 'I' refers to my body, and you get a considerably different meaning.

Thirdly: because we use the same words to refer to our bodies and to our personal selves, and because we use action verbs like 'killed' to refer to natural processes that can in no way be human actions as well as to refer to human actions, it is quite idiomatic to use the verbs of action metaphorically with our inanimate bodies as their subjects. 'Jones hit her while he was asleep' may well use 'hit' the same as in 'One billiard ball hit the other'. 'Jones' need not refer to Jones the person but to the body of Jones, as is referred to in 'Jones is six feet tall'. Similarly, while 'Jones killed Smith' idiomatically may describe the situation where Jones (the person) willed some movement of his body that in turn caused the death of Smith, it may with equal idiomatic propriety be used to describe the body of Jones falling on Smith and killing him. Indeed, this last example shows that the idiomatic propriety of saying 'Jones killed Smith' is not limited to the situations where Jones's body goes through some complicated, seemingly intelligent routine whereby Smith dies; any

---

[14] On the dual applicability of person-predicates and physical-predicates to human beings, see Peter Strawson, *Individuals* (London, 1959). I have earlier used this ambiguity in who *we* are to excise the metaphorical usage of action verbs from further consideration by those interested in *human* action. Moore, 'Responsibility and the Unconscious', 1637–8.

dumb, brute causal contribution to Smith's death made by Jones's body will be enough to license 'Jones killed Smith'.

The upshot is that ordinary usage provides no reason to think that involuntary bodily movements are actions. Is any better reason provided by the fact that the movements in question seem so intelligently designed, so responsive to shifts in the environment? These features certainly make such movements look like actions. Such features certainly make us infer the existence of feedback loops by virtue of which bodily motion is adjusted in response to perceptual cues. Yet for those examples where we are not able to infer volitional causation of such movements, then such movements would be only illusory actions. They would be the fool's gold of actions: they would look just like actions, but they would not share the essential nature of actions, so they would not be actions. The only way to overturn this point is by coming up with some alternative theory about the essence of actions, and showing that these involuntary bodily movements have that essence too.

There is of course no dearth of such alternative theories. Indeed, I mentioned and put aside a number of them in Chapter 6. Among them, that acts are those bodily movements that are willed, where willing is itself an action, not a non-actional state (as are volitions); also, that acts are those bodily movements that are directly caused by states of belief and desire, without the mediation of any third kinds of states like intentions or volitions; etc. I gave reasons in Chapter 6 to prefer the volitional theory to these competitors, but I needn't rely on such reasons here. Rather, the point is here simpler: there is no comfort to be found in such theories for those who wish to hold sleep-walking etc. to be human actions.

Take the belief–desire causal theory, according to which a bodily movement is an action if and only if it is caused by a belief–desire set whose contents form a valid practical syllogism. To be plausible in getting round deviant causal-chain counter-examples, such a theory will have to say that the causal route between the belief–desire states and the movements was of the right (i.e. 'actional') kind.[15] Even if the 'right causal route' means something other than 'through the actor's executory volitions',

---

[15] See e.g. Goldman, *A Theory of Human Action*, 61–2.

such a theory still will not allow sleep-walking, hypnotic movements, parapraxes, etc. to qualify as actions any more easily than will the volitional theory. Both theories will ask a scientific question about such behaviours: the volitional theory will ask if the movements are caused in the right way by volitional states; the belief–desire theory will ask if the movements are caused in the right way by some belief–desire set.[16]

Take Freud's 'execution' of the ink-well from his sister. Freud tells us that he *desired* its destruction because he *desired* to get a new one. Even if we credit that and we suppose it likely that, in addition, he *believed* that if he moved his hand in a certain way, the ink-well would fall to the floor, be destroyed, and allow him to buy a new one—even if we grant him that there is some causal connection between the desire and beliefs, on the one hand, and the awkwardly accurate movement, on the other; still the theory must allow for the possibility that this movement might not have been an action because, for example, Freud got so excited as he thought of the new ink-well he could purchase if this one were broken that his arm trembled in the way that it did. Beliefs and desires can cause bodily movements that are not actions—my desire to beat Bobby Fischer in chess, or my belief that there is a prowler outside the window, can cause my heart to pound—but that does not make my heart's pounding my action. So the belief–desire causal theory will have to ask and answer the further question—of execution through the right causal route—that is also asked and answered by the volitional theory.

In any case, the questions whether sleep-walking is a kind of walking (as an action), or whether post-hypnotic movement with death as its consequence is a kind of killing (as an action), are only answered by one's best theory of action. Ordinary-language observations cannot overthrow such a theory, no matter how ordinary and idiomatic may be the employment of our verbs of action to describe such phenomena.

By the theory of action defended in the preceding chapters, some of the five examples given undoubtedly are examples of actions. If that is true, it is only true by application of the theory of action. Such examples being examples of action cannot thus

---

[16] For a very nice application of the belief–desire theory of action to these cases, see Schopp, *Automatism, Insanity, and the Psychology of Criminal Responsibility*, ch. 5.

overthrow the theory, since it is use of the theory that shows them to be actions.

Although not necessary to defending the theory of action, it may none the less be useful here to apply it to these five examples. The first three examples—somnambulistic behaviour, behaviour while unconscious, and behaviour during or after hypnotic trance—each raise a common issue. This is the issue of how altered or divided consciousness affects the existence of volitions in such cases. For in each of these three examples, the subject's body engages in complex routines requiring perception and readjustment in order to achieve certain goals; in each, it might well be the case that the 'readiness potential' (of negative shift in the supplementary motor area of the brain) has a pattern generally indicative of willed bodily movements.[17] Therefore, one might infer the existence of states of volition executing more general beliefs, desires, and intentions of the agent. Yet in each sort of case, the execution is not done by the *conscious* will, for the conscious will is 'elsewhere': asleep, unconscious, or hypnotized.

There are two, complementary reasons to think that Anglo-American criminal law has made the right assessment of these cases as not being cases of action. One is built on a sense of who the self is that wills actions. Consciousness seems essential as part of our self-boundaries, so that if *we* (our conscious selves) are asleep or are otherwise not active, then *we* don't will anything—even if volition-like states execute certain of our background states of desire, belief, and general intention.[18] It may well be that subpersonal agencies within us are achieving quite complex functions in these three kinds of case. Yet without consciousness (including that accessibility to consciousness that Freud called being *pre-conscious*)[19] these subpersonal agencies do not represent *our* willing or *our* action.

---

[17] 'Readiness potential' is the brain indicator of voluntary movement studied by physiologists. See the discussion on p. 163, *supra*.

[18] Moore, 'Responsibility and the Unconscious'. See also id., *Law and Psychiatry*, 337–43.

[19] The parenthetical qualification is important, for one does not want to exclude acts of skill carried on when the actor's 'mind was on something else'—like manœuvring a car down a mountain road while thinking through philosophy puzzles. For those acts, the agent is morally responsible. I take it that the Model Penal Code is attempting to capture this same point in its inclusion of the 'effort or determination, conscious *or habitual*', as action for which the accused is prima-facie liable. Model Penal Code, §2.01(2)(*d*).

The second reason complements the first, because the second reason tends toward the conclusion that the states that mediate between our emotions, beliefs, desires, and general intentions, on the one hand, and our bodily movements during sleep, unconsciousness, or hypnosis, on the other, are not (or are not fully) volitions. Recall that on a functionalist theory of mind volitions are states whose essence is given by their functional roles. As we examined in Chapter 6, one of those roles is the executory role of causing just those movements to take place that would achieve the objects of one's desire and general intentions if the world is as one believes it to be. This function *is* served by whatever state immediately causes the kinds of movement we are here considering, which is why such movements look so much like actions. Yet the other essential function definitive of volitions is not served by such states. This is the resolving function. In the face of ordinary conflict, Buridan conflict, or akratic conflict—the kinds of conflict discussed in Chapter 6—volitions are part of the hierarchy of bare intentions that resolve, all things considered, what to do now. To serve such a resolving function, volitions must be responsive to all (or at least a fair sample) of what one desires, believes, and intends. And this is what being asleep, being unconscious, or being hypnotized prevents. These states seem to break the unity of consciousness that allows volitions to be formed that are responsive to all of one's desires, beliefs, and intentions, and not just responsive to a small subset. These states of altered or disassociated consciousness seem to 'seal off' a certain subset of one's mental states for execution in bodily movement, without there being the kind of resolution volitions are supposed to represent.[20]

Both of these reasons are not only metaphysical reasons for concluding that there are no actions done by sleep-walkers, unconscious shooters, and hypnotized subjects. Because of the relevance to moral responsibility of the metaphysical question of when there is an action (that we explored in Chapter 3), these are also moral reasons for not holding these 'actors' morally responsible. After all, moral agency tags along obediently behind the metaphysical boundaries of self, so that if it's not *me* that 'walks in my sleep', it is not my moral agency to whom that movement gets attributed. Similarly, if that state that immediately causes somnambulistic or hypnotic movements does not resolve conflicts between my mental states because it is not responsive to most of them, then its execution of some of my mental states in movement does not speak for me—such a state is not a *choice* worthy of moral praise or condemnation, absent such resolving function.

The upshot is that on the volitional theory it is doubtful whether somnambulistic behaviour, behaviour while unconscious, and behaviour during hypnosis are actions. We at present do not know enough, either about volitions in general or about the modes of initiation of these three kinds of behaviour (particularly in hypnosis), to resolve the issue definitively one way or the other. In drafting criminal codes, however, we should place our bets in the direction chosen by the American Law Institute's Model Penal Code: bet that such movements are not volitionally caused, and therefore are not actions. The alternative, adopted by a few courts, is to bet the other way, making lack of full consciousness while acting an affirmative defence, not a basis for negating the prosecution's showing of action.[21]

About the fourth example, 'brainwashing', I think the evidence is the other way. If 'brainwashing' consists essentially of belief–desire implantation—so that, for example, Patty Hearst before her experiences in the closet thought the SLA to be a silly organization, but the 'Tania' that emerged from the closet believed in the justness of the SLA cause—that does not touch the fact that these beliefs and desires were executed by Ms Hearst's volitionally caused bodily movements. Even if she were not thought to be responsible for her new beliefs and desires, that is irrelevant to the action question: did she execute some belief–desire set by a volitionally caused bodily movement? Seemingly she did, and she therefore performed the complex action of robbing the bank.

The only way to undercut this conclusion is by likening cases of coercive indoctrination to hypnosis, sleep, and unconsciousness. This would be to say that brainwashed subjects are in an altered state of consciousness. Patty Hearst, the argument would go, was no more present when 'she' robbed the bank than Huey

---

[21] See *State* v. *Caddell*, 287 N.C. 266, 215 S.E. 2d 348 (1975).

Newton was when 'he' shot Officer Frey, for both had no conscious awareness of what they were doing; further, the state that
executed her newly implanted pro-SLA desires and intentions did
not resolve the conflict of those new states with her still-existing,
larger set of desires and intentions (which were definitely not pro-
SLA), because her altered state of consciousness prevented resolution of this conflict. On these two suppositions, one might well
deny that her movements were volitionally caused.

The evidence is against this view of brainwashing. However
much one might think that there is a discontinuity in character
between 'Patty' and 'Tania', there is little evidence of that sharp
break in consciousness that there is in cases of sleep, unconsciousness, and hypnosis. Thus, even if one bet on the latter turning out to be cases of non-action, Patty's behaviour would
remain her (consciously willed) action.

This conclusion is of course compatible with affording coercively indoctrinated subjects like Ms Hearst an affirmative
defence. Such a defence would be like that urged on behalf of
newly arrived foreigners with quite different cultural backgrounds
from ours, and like the defence of infancy: in all cases we might
think that the actor must be given a fair chance to learn that her
proposed behaviour is both immoral and prohibited. In a case
like Patty's, the opportunity such a defence would afford would
not be to learn of the moral and legal norms, for those she knew;
rather, it would be the opportunity to integrate the newly
implanted beliefs into her character, or to reject them in light of
other of her beliefs.[22] Whether such an opportunity should be
allowed by way of affirmative defence, of course, does not impact
on the conclusion relevant to action theory, namely, that subjects
like Patty Hearst do perform actions when they respond to their
conditioning.

I have elsewhere dealt at length with the fifth and last set of
examples here, those of the Freudian kind.[23] I can therefore be
brief. The essential Freudian insight of relevance to action theory
is what I have called the insight of extended memory. Freud's
striking claim was that we experience various mental states and
mental events even though those experienced states and events

---

[22] For suggestions in this direction, see Dennett, *Brainstorms*, 248–53.

[23] Moore, 'Responsibility and the Unconscious'; Moore, *Law and Psychiatry*,
chs. 9–11.

are not easily accessible to consciousness, and, further, that by free association, transference, or other memory-jogging techniques, we can be made to remember such states and events. One might for example, experience the emotion of anger (as shown, say, by shouting at someone, 'I am not angry'), and only later come to remember that one was angry at the earlier time. Such remembering is not to be confused with inferential conclusions we might reach from observing our own behaviour; we, like an analyst, can infer that we must have been angry, but that is not the same as remembering (via re-experiencing it) that we were angry.

Freud made the same claim about the mental state involved in acting. About dreaming, for example, Freud adopted the common-sense view that we can dream and not be aware on awakening that we have dreamt, but that further associations can recapture seemingly forgotten dream contents. More importantly, what we sometimes remember about a dream is not only its content, but also that we created the content, as we might create a play for our own amusement. We remember having a supervenient attitude towards the dream while we were dreaming it, so that we remember redoing the dream in certain ways to make it better.

To take a personal example: While writing a paper on Freudian dream theory,[24] I discounted Freud's claim about supervenient attitudes in dreams, arguing that they rarely if ever occurred. The night I wrote that passage in the paper I produced a textbook example of a supervenient dream, an anxiety dream in which I remembered redreaming the anxiety-provoking situation so as to heighten the anxiety. Have very many of those experiences, and you join Freud in having the plastic impression of a second will operating, without consciousness.

Such extended memory of having acted increases the stock of our phenomenal evidence for when we truly act. One might well think of supervenient dreaming as a mental action persons perform while asleep. Again, that need not, *contra* Freud,[25] make

---

[24] Moore, 'The Nature of Psychoanalytic Explanation', *Psychoanalysis and Contemporary Thought*, 3 (1980), 459–543, rev. and repr. in L. Laudan (ed.), *Mind and Medicine* (Los Angeles, 1983).

[25] Freud, 'Moral Responsibility for the Content of One's Dreams', in *Collected Papers*, ed. James Strachey (New York, 1959), 154–7.

one morally responsible for the content of dreams (as, for example, a playwright is responsible for the content of a vulgar or boring play); for the lack of consciousness as one is dreaming bars responsibility, even if dreaming is sometimes an action. Still, Freud's striking claim of extended memory sometimes makes plausible that we act unconsciously.

Despite this, most of the Freudian examples remain non-actions. Most parapraxes are like Freud's supposed 'action' of executing his ink-well: although there is some evidence that the movements are caused by the subject's beliefs or desires, there is no evidence—certainly none of the phenomenal kind like that for supervenient dreams—that these beliefs or desires are (consciously or unconsciously) executed by volitional movement. Rather, Freud simply posits there to be unconscious actions here in order to increase the breadth of his theory. Such theoretical positing is particularly glaring in Freud's conclusion that neurotic symptoms like Dora's sore throat are actions by her. Making one's throat sore is not within anyone's basic act repertoire. We can of course make our throats sore as complex actions—we can pour hot liquid in them, clear them constantly, etc. But no biofeedback laboratory has yet shown that we can simply will our throats to be sore. It is thus especially implausible to view such bodily changes as acts.

However one comes out on these five sets of examples, it bears the emphasis of repetition to say that there will be no impact on the truth of the volitional theory of action. For we have been using the volitional theory to pick our way through such examples, showing some to be actions and some not. Even if all such examples turned out to be actions because of unconscious or otherwise hidden volitions, that would not in the least refute the volitional theory.

2. WILLINGS WITHOUT BODILY MOVEMENTS AS COMPLEX ACTIONS?

Now we turn to the class of supposed counter-examples where there appears to be no bodily movement, but there is some mental state that might be thought of as a willing. There are two major classes of such supposed counter-examples: where mental acts effect real-world changes, but not through the willing of bodily movements; and where intentionally omitting to prevent some harm like death is equated with causing death and thus with the complex action of killing.

2.1. Mental Acts as Complex Actions

About mental events, there are two concessions to be made at the outset: first, that some mental events are actively called forth (such as calling up an image) and not merely passively suffered; secondly, some mental doings of this sort are *complex*, such as remembering a name *by* calling to mind certain images. So there are complex actions of a certain kind that involve no volitionally caused bodily movements. Yet the complex action descriptions in terms of which criminal laws are framed are not of this kind. Criminal law prohibits complex actions like killing, scaring, hitting, abusing, etc.—actions where an effect takes place in the world outside the actor's mind. So the relevant question here is not: can *any* complex action be performed without the performance of a volitionally caused bodily movement? Rather, the question is: can any of the complex actions *prohibited by Anglo-American criminal law* be performed except by volitionally caused movement? Can we, for example, *kill* or *frighten* another by any kind of mental action alone?

Before things get complicated, the answer to this latter question is clearly no. By and large, such complex actions as killing cannot be done by anyone without that person using his body as an instrument in some way. Even inducing others to kill requires that one move some part of one's body in such a way as to communicate to others that they are to kill. Usually there can be none of the effects of the kind required by the complex actions prohibited by the criminal law without bodily movement. If by thinking poorly of someone we could thereby kill him, defame him, scare him, etc., this wouldn't be true; but we have no such magical powers. We can only change the world through those volitionally caused bodily movements called basic acts.

Now let us make things more complicated by returning to the sorts of hypotheticals that we considered in Chapter 5. Suppose that D is a witch-doctor of reputedly magical powers; V believes that D can kill a person just by wishing him dead. On a given occasion D is staring at V trying to recall who V is. D tries to call up a mental image of V in a group of people, and succeeds, and that (mental act) allows D to recall that V is the son of his most hated enemy; that recollection prompts a wish in D that V die (along with a feeling of regret that it is not in D's power to kill V by simply wishing him dead). V, correctly reading the behavioural manifestations of D's wish, believes that he, V, will indeed die. This thought so scares him that he indeed does die of heart palpitations. Hasn't D performed the action of *killing* V without performing any basic action (volitionally caused movement)? True, D did not *intentionally* kill V, but that is not our question here. Also true, D's body did have to move₁ in order for V to believe that D wished him dead, setting in motion the sequence of events leading to V's death; but the facial expressions and body language by virtue of which V understood D's wishes were not willed movements of D's. If D were trying to kill V by this means, such movements would have been willed movements; but D is not an intentional killer. He may well seem to be a *killer*, however, without any volitionally caused bodily movement on his part.

Leo Katz has suggested to me a related sort of hypothetical, which I shall call the example of the reluctant executioner. A state executioner who has read thus far into this book wishes to do his job yet not perform the action of killing. He accordingly connects the guillotine that is used in his line of work to a device that reads the electrical conductivity of his skin. He learns how to use the device in two stages: in stage one, he learns that by fantasizing about something very pleasant, he can raise the conductivity of his skin; in stage two, he learns to dispense with the fantasy and simply to will his skin conductivity to increase. At either stage, the increase in skin conductivity causes the guillotine to drop, killing the condemned prisoner. In each case has not the executioner killed, even though he did not move his body in order to do so?

The second stage of this second example (of the squeamish executioner) is easy. If the executioner can simply will a change in the conductivity of his skin, and that change in conductivity can have real-world effects, then we should construe bodily movements to include such changes of conductivity. For he also surely kills when he wills such a change of bodily states in order to release the guillotine, which is in fact released and beheads its victim. He has used his body as the means to effect a change in the world which he desires to bring about, and it shouldn't matter whether the means involve a literal motion of a whole limb or only the microscopic motions within some limbs or other organ. All such willed changes of bodily states, when used as a means to change the world, are actions.

The first hypothetical, where D is the witch-doctor of reputedly magical powers, is also easy in light of our conclusions in Chapter 5: there is no action of killing V by D. True, V died as a result of D's bodily movements$_I$; true, those movements were themselves the effect of D's wishing V dead, which was itself the effect of D's mental act of calling V's image to mind. Yet D did not will the bodily movements that killed V, nor did D intend to kill V and select those movements as his chosen means; D's mental act of calling the image to mind was thus not a choosing to have effects in the world outside his head, needed for there to be actions of moving$_T$. If such movements$_I$ are not basic actions, then they cannot be instances of more complex actions like killing when such movements happen themselves to cause a death.

On the other hand, the squeamish executioner in the first stage of the second hypothetical does seem to perform the action of killing. That is because the squeamish executioner does intend to cause the death of the condemned prisoner, and he does choose to use his bodily movements$_I$ (the increased electrical conductivity of his skin) as his chosen means. The only thing missing is that he does not will the increase in electrical conductivity in his skin; he wills a fantasy as the most basic means by which he causes his electrical conductivity to go up (which is in turn his means of killing).

One might urge that the only object of any willing or volition (in the first stage of the second hypothetical) is the fantasy, and that the change in electrical conductivity is an *intended* effect of such willing (just as death is an intended effect of such movement$_I$). Such would be one interpretation of the intention/ volition distinction based on their differing Intentional objects, discussed in Chapter 6.

Two things require consideration in order to assess this alternative interpretation of the intention/volition line. One is whether volitions can ever be complex in their objects (as in the hypothetical), or whether they must have as their object 'the simplest thing we know how to bring about' even when that thing is a purely mental state. Think of an individual who has two injuries: his right leg has been amputated, and the nerve to his left leg has been diseased to the point of paralysis. He has the phantom-limb experience for his right leg, although not for his left, given the deadened nerve. Taking their inspiration from the split-brain experiments, doctors connect the nerve leading to his (former) right leg to the healthy part of the nerve going into the left leg. Now when he wills to move his right leg, the left leg moves; when he wills to move his left leg, nothing happens. Once he has mastered the only means available to him of moving his left leg, how should we construe the object of his volitions? We could say he *wills* only to move his right leg, even if he *intends* by doing so to move his left leg; much as we properly say that when we raise our left arm by lifting it with our right arm, we *will* the moving of our right arm and *intend* by doing so to cause a rising of our left arm. Alternatively, and I think preferably, we could say that he wills the movement of his left leg, even if he has learned a new way to bring such a thing about, namely, by means of willing the movement of his (no longer existent) right leg.

If this is right so far, how then should we construe his mental state when we substitute the mental act of calling a fantasy to mind for the willing of the non-existent right leg? Are volitions a plausible ingredient for such mental acts as they are for physical acts? In Chapter 3 we noticed that we might will some omissions in the sense that one might be tempted, for example, to rescue another and only by the mental act of distracting one's own thoughts on to other topics could one bring oneself to refrain from rescuing. Such mental actions have some of the phenomenal feel of volitions, and this phenomenology constitutes some evidence that there are volitions present in such cases of omission. Such mental acts also play some of the executory role of volitions, being the last state needed to execute our more general and prima-facie beliefs, desires, and intentions into a particular and all-things-considered choice. Lacking, however, in the case of both chosen omissions and mental acts is the further specification of the functional role of volitions, namely, that they initiate bodily movement. Also missing, presumably, is any of the physiological data now associated with volitions, including prominently blood flow patterns and slow negative shift ('readiness potential') in the supplementary motor area of the brain.[26]

On balance, those considerations militate against thinking that volitions are present either for intentional omissions or for mental acts like calling a name to mind. Only when that mental act itself is done in execution of an intention to cause a movement₁ of one's body, and thereby to effect changes in the world, is there a volition. In such a case the object of such a volition is not only the causing of a fantasized image, but also the bodily movement₁ such image is called forth to cause. There is in such cases, in this complex way, a willed bodily movement, and thus a complex action of killing when that movement itself causes death. There remains no complex action of killing where there is no willing of the bodily movements₁ that cause death, as in the first hypothetical of the witch-doctor.

2.2. Omissions as Complex Actions

Perhaps the most commonly supposed counter-example to the thesis that all complex actions require willed bodily movements stems from cases of omission. The argument is that in certain circumstances we *cause* death by intentionally refraining from preventing a death, and if we cause a death by our intentional refraining, we *kill*. George Fletcher's examples: 'a parent might starve a child to death by refusing to feed it. A water company might poison the public by systematically omitting to purify the water. These "motionless" activities are appropriately described by an affirmative verb of killing.'[27] *Ergo*, the desired conclusion: some complex actions can be done without the presence of a volitionally caused bodily movement because they can be done by an omission.

There has been a considerable debate about whether omissions

---

[26] To my knowledge, none of the now very extensive studies of the supplementary motor area of the brain have found any of the indices of willed action in purely mental acts. See generally Goldberg, 'Supplementary Motor Area Structure and Function', and Libet, 'Unconscious Cerebral Initiative and the Role of Conscious Will in Voluntary Action', and the extensive peer commentary on both review papers. Libet has found the readiness potential characteristic of volitions when subjects intend to move, then veto the idea—but that 'mental act' is of course movement-directed even though it does not cause movement.

[27] Fletcher, *Rethinking Criminal Law*, 601.

can be causes.[28] How one answers this question obviously depends on what one means by 'cause'. In the familiar necessity (or *sine qua non*) test of cause-in-fact, omissions can be causes. 'But for the failure of someone to throw a life preserver to another who is drowning, would the latter have died?' is a seemingly sensible question that sometimes has an affirmative answer.

Those who oppose the idea that omissions can be causes (on the necessary-condition conception of causation) often voice the objection that omissions are never necessary to the production of some result because other conditions are already sufficient to produce that event. Francis Bohlen captured this deliverance of common sense as follows: 'by failing to interfere in the plaintiff's affairs, the defendant has left him as he was before; no better off, it is true, but still in no worse position; he has failed to benefit him, but he has not caused him any new injury nor created any new injurious situation'.[29] Yet on a necessity conception of causation, those who omit to interfere (say, to rescue the plaintiff) have caused his worsened position. For the other conditions that also caused the plaintiff's harm were *not* sufficient to produce it; also necessary was that the defendant omitted to prevent the harm from occurring (which, by hypothesis, he could have done).

A better objection to omissions being necessary conditions for the occurrence of some harm is a kind of *reductio ad absurdum*: if the *absence* of an action can be a necessary condition to the happening of some harm, and thus a cause of it, then there are an indefinitely large number of causes for anything. Eric Mack's example: 'Brown's failure to assassinate Carter, Carter's not killing himself, Carter's failure to beat his wife in public, the public's not believing that Carter beats his wife, the non-explosion of

[28] On this topic, see Mack, 'Bad Samaritanism and the Causation of Harm'; Epstein, 'A Theory of Strict Liability'; Eric Mack, 'Causing and Failing to Prevent', *Southwestern Journal of Philosophy*, 7 (1976), 83–9; Jonathan Harris, 'Bad Samaritans Cause Harm', *Philosophical Quarterly*, 32 (1982), 60–9; id., 'The Marxist Conception of Violence', *Philosophy and Public Affairs*, 3 (1974), 192–220; Weinryb, 'Omissions and Responsibility', *Philosophical Quarterly*, 30 (1980), 1–18; James Rachels, 'Killing and Starving to Death', *Philosophy*, 54 (1979), 159–71; P. Smith, 'Recklessness, Omission, and Responsibility: Some Reflections on the Moral Significance of Causation', *Southern Journal of Philosophy*, 27 (1989), 569–83: 571–2; Husak, *Philosophy of the Criminal Law*, 160–5; Joel Feinberg, *Harm to Others* (Oxford, 1984), 171–86. See generally B. Steinbock ed., *Killing and Letting Die* (Englewood Cliffs, NJ, 1980).
[29] Francis Bohlen, 'The Moral Duty to Aid Others as the Basis of Tort Liability', *University of Pennsylvania Law Review*, 56 (1908), 217–44, 316–38: 219.

the solar system in 1937, the non-collapse of the galaxy in 1936, and so on' are all causes of Jimmy Carter's election as President in 1976 on the view here considered, for all of these omissions were necessary conditions of Carter's electoral success.[30] This absurd conclusion leads Mack and others to recoil from the premiss that apparently generates it: that omissions can be causes.

The problem with the 'promiscuity objection', as I shall call it, is that it misconceives its target. Even if one restricts the things that can be causes to events (thus excluding the absence of events, i.e. omissions), one gets absurd causal statements from the necessary-condition conception of causation. It was a cause of President Carter's election: that a certain act of intercourse happened when it did between Carter's parents; that the economy in the South with regard to peanuts had the upswing that it did at a certain time; that young Jimmy Carter listened to the religious teachers that he did; and so on. The causal statements are not much less numerous, and thus not much less absurd, than Mack's; so, *if* they are all absurd, the fault lies with the necessity conception of causation, not with its application to omissions.

Thus, if the promiscuity objection is well-founded, then it is the necessary-condition analysis of causation that must go. In any event, we have other reasons for thinking the necessary-condition analysis of causation to be inadequate. Chief among these is the well-known over-determination objection.[31] Imagine two scenarios in each of which the plaintiff's house is destroyed by fire. In both scenarios, there are two fires, one set by the defendant, and the other of natural origins; each is headed for the plaintiff's house and each would be sufficient to burn down the plaintiff's house in the absence of the other. As it happens, in the first scenario the two fires join, and the fire that results destroys the plaintiff's house; in the second scenario, the two fires do not join, one fire burning down the plaintiff's house just before the other fire reaches it.

By our primitive, unanalysed notions of causation, we have very clear intuitions about these cases: in the first scenario, the

---

[30] Mack, 'Bad Samaritanism and the Causation of Harm', 244.

[31] See the discussion of this problem in Richard Wright, 'Causation in Tort Law', *California Law Review*, 73 (1985), 1735–1828. It was the over-determination problem that led the American Law Institute to abandon the necessary-condition analysis of causation for the 'substantial factor' conception in the *Restatement of the Law of Torts*.

defendant's act of setting the fire caused the destruction of the plaintiff's house; in the second scenario, the defendant's act of setting the fire either clearly did cause the loss of the house (if that was the fire that reached the house first), or it clearly did not cause the loss of the house (if it was the second fire to reach the spot where the house once was). The necessary-condition analysis of causation seems unable to match these intuitively crisp conclusions, for in over-determination cases of this kind, the defendant's act (and his fire) were not necessary for the loss of the plaintiff's house by fire; but for the defendant's fire, the plaintiff still would have lost his house by fire.

A third problem with the necessary-condition analysis of causation stems from the well-known indeterminacy of counterfactuals[32] to which I adverted briefly in Chapter 2 where discussing how omissions should be conceptualized. Put in the form of a *reductio*,[33] the problem can be raised thus:

(1) x is a cause of y if and only if x is a necessary condition of y (the necessary-condition analysis of 'cause');

(2) x is a necessary condition of y if and only if, if x did not occur, y would not occur (the counterfactual analysis of necessity);

(3) if x did not occur, y would not occur if and only if, 'if x did not occur, y would not occur' has a truth value (a weak version of Tarski's disquotational device allowing semantic ascent);[34]

(4) 'if x did not occur, y would not occur' has a truth value if and only if some choice is made with respect to all features of the situation surrounding x, distinguishing those that are varied (along with the absence of x) from those that are

---

[32] See Goodman, *Fact, Fiction, and Forecast*, 9–17; Roderick Chisholm, 'Law Statements and Counterfactual Inference', *Analysis*, 15 (1955), 97–105; Nicholas Rescher, 'Belief-Contravening Suppositions', *Philosophical Review*, 60 (1961), 176–96; Robert Stalnaker, 'A Theory of Conditionals', in N. Rescher (ed.), *Studies in Logical Theory* (Oxford, 1968), for discussions of the problem of indeterminacy for counterfactual conditionals. The last three essays are also collected in E. Sosa (ed.), *Causation and Conditionals* (Oxford, 1975), which contains a discussion of the problem in its introduction, 12–14.

[33] The *reductio* form of the argument is from Moore, 'Thomson's Preliminaries about Causation and Rights', 506.

[34] For a good, brief discussion of using Tarski-like truth sentences simply as 'disquotational' devices allowing semantic ascent, see Michael Devitt, *Realism and Truth* (Cambridge, Mass., 1984), 28–31.

held fixed as part of the background (the element of choice in imagining the possible world in which the counterfactual is to be tested);

(5) if a choice is made with respect to all features of the situation, distinguishing those that are varied (along with the absence of x) from those that are held fixed as part of the background, then there is a convention or expectation guiding the choice of possible worlds in which the causal statement is tested.

Therefore, from (1), (2), (3), (4), and (5):

(6) x is the cause of y only if there is a convention or expectation that guides the choice of possible worlds in which the causal statement is tested.

The conclusion, (6), is an unwanted conclusion to anyone with any metaphysically realist commitments about causation. For it asserts that whether there is a causal relation between two events depends on the existence of certain conventions. Whether x causes y, in other words, does not depend on there being some convention-independent relation in the world but at least partly on there being conventions.

If that conclusion strikes one as absurd, as it does me, then one of the premises that generates it must be abandoned. Since (2) and (3) are surely true, the only way to rescue (1), the necessary-condition analysis of causation, is to abandon (4) or (5). Yet (4) and (5) have the plausibility that they do because of the indeterminacy of counterfactuals. We can see their plausibility in the following way. When we say, contrary to fact, that 'But for x, y would not have occurred', there is a lot that we have left unsaid. Two items we have not mentioned are, first, what are we to imagine happened in place of x? Consider the example where a man named Smith walks up the steps to the courthouse and *causes* the collapse of one of the steps. The counterfactual analysis of that causal statement would be 'If Smith had not set foot on the steps, the rotten beams under the steps would not have collapsed'. In such statement, what replaces 'Smith setting foot on the steps' to test the truth of this counterfactual? If we replace it with 'Smith using a different stairway', the counterfactual may well be true; if we replace it with 'Smith bounding up the steps at just the time he actually walked up the steps', it may seem to be false. Secondly, even once we choose, say, 'Smith using a different stairway' as the replacement, how many other things in the world are we to imagine having also changed along with Smith's change of entrance? For example, what had to change for Smith to have used a different entrance? Perhaps Jones, who is Smith's chauffeur and who is heavier than Smith, dropped Smith off at the safe entrance only to run Smith's forgotten briefcase up the rotten one. If this is what we vary, then the counterfactual may seem to be false; other events, neutral to weight equal to or greater than Smith's being on the stairs at the relevant time, would leave the counterfactual true.

On both of these bases one might find plausible the idea that the truth of a counterfactual depends on what we hold fixed and on what we vary—thus, (4). Further, one might think that what we do fix or vary is a matter guided solely by convention and not by any limitations imposed by the nature of the causal relation. One convention we likely follow is: 'Hold fixed normal features of the situation'. Following such a convention, we would thus hold fixed Smith's weight and thus not ask any counterfactual question where what replaces the real Smith stepping on the stairs is a lighter or a heavier Smith doing so. John Mackie, for example, finds that common-sense views of causality employ an 'all-or-nothing' convention in considering what is to replace matters susceptible of continuous variation such as weight. In discussing whether a hammer-blow is necessary for the crushing of a chestnut, Mackie holds that:

we regard the hammer-blow as a unit, and simply do not consider parts or subdivisions of it or quantitative alterations to it. The alternatives considered are that I strike the chestnut in the way described and that I do not. In constructing possible worlds, in considering what might or would have happened, we either plug in the hammer-blow as a whole or leave it out as a whole.[35]

Likewise, the variations on walking (jumping, bounding, treading, etc.) would also be ruled out on similar grounds. Hence, the plausibility of (5).

I have elsewhere attempted to question this objection to the necessary-condition analysis of causation, as well to question the

[35] John Mackie, *The Cement of the Universe* (Oxford, 1980), 44.

preceding objection, by employing a highly fine-grained view of event individuation between possible worlds.[36] The defendant's fire in the preceding example is a necessary condition of the destruction of the plaintiff's house if either that fire joins the second fire, or that fire reaches the plaintiff's house first; because the destruction of the plaintiff's house that would have taken place in the possible world where there was no defendant's fire (but only the other fire) would have been a different destruction. Likewise, Smith's walking up the stairs can, without resort to convention, be said to be a necessary condition for the collapse of the stairs; for in all other possible worlds (including those where Smith bounds up the stairs, or his chauffeur goes up the stairs instead of Smith), there would not be *this* kind of collapsing of the stairs.

It is unclear to me whether one can make the individuation of events across possible worlds fine-grained enough to do this work without smuggling in relational properties (like a Smith-caused-collapsing) that trivializes the effort. But even if this fine-grained rescue worked, notice how many causes there are for anything. Think of how many things are necessary for there to be a given collapse of the stairs if what we mean by such collapse is the event-token with all of its attributes. Even if this defence holds the indeterminacy of counterfactuals and over-determination objections at bay, it greatly increases the power of the promiscuity objection to the necessary-condition analysis of causation.

A fourth major problem with the necessary-condition analysis of causation is the blindness of that analysis to one of causation's basic features, namely, its asymmetry with respect to time.[37] The events that consisted in the starting of the two fires that join together surely caused the loss of the plaintiff's house, but the destruction of the plaintiff's house did not cause the starting of the two fires. Yet the necessary-condition analysis would suggest otherwise. The starting of the two fires, together with conditions like wind patterns, oxygen in the air, the presence of combustibles on the route to the plaintiff's house, etc. were jointly *sufficient* for the occurrence of the destruction of the house. Logically, therefore, the occurrence of the destruction of the house was *necessary* for the set of events and states consisting

---

[36] Moore, 'Thomson's Preliminaries about Causation and Rights', 510–12.

[37] See generally Taylor, *Action and Purpose*, chs. 1–3.

of the starting of the two fires, the presence of oxygen, the wind patterns, etc. On the necessary-condition analysis of causation, therefore, the destruction of the house *caused* this set of earlier existing states and events to occur. True, one could avoid this unwelcome conclusion by stipulating that causes are necessary conditions of effects only when the two are simultaneous, or when the latter precedes the former in time. Yet such a stipulation is purely *ad hoc*. Moreover, such a stipulation is made to fit some pre-existing idea of causation, an idea which different from the necessary-condition idea, which does not remotely suggest such a temporal stipulation.

A fifth argument often advanced against the necessary-condition analysis of causation is that if necessary conditions were causes, then omissions could be causes. *That* argument, of course, we cannot urge here, but it does highlight the dependence of the idea that omissions can be causes on a very problematic notion of causation. If we reject the necessary-condition conception of causation because of the first four considerations here advanced, and supplant it with some other conception, omissions drop out as candidates to be causes of anything.

The problem lies in coming up with some such alternative notion of cause that does not soundly beg the question about omissions. Consider notions of mechanistic causation by way of example. It has often been observed that our ideas of mechanistic causation—cause as a making some effect happen—stem from our own first-person experience as actors, making events in the world happen via our bodies.[38] This has led some[39] to conclude that human actions are our paradigms of the causal relation in which I have elsewhere called the strong sense:[40] not only are actions like hitting exemplars from which we learn of the existence of the causal relation (the weak sense of 'paradigm'); such actions also give the meaning of 'cause' in the manner of the old 'paradigm case argument'. On this view, human actions are the only clear cases of causation, all causal relations between natural events being such only by analogy to these strong paradigms.

Like all uses of the paradigm case argument, the argument chain here is embarrassingly short: omissions are not causes

---

[38] Hart and Honoré, *Causation in the Law*, 28–32.
[39] Taylor, *Action and Purpose*; Epstein, 'A Theory of Strict Liability'.
[40] Moore, 'The Semantics of Judging', 285.

because only actions are paradigms of causation, and omissions are not relevantly like actions. I too share the intuition that omissions cannot be killings or other actions, but such a conception of causation adds nothing to that intuition since such a conception is built on such an intuition to start with.

In default of having ready to hand a better-fleshed-out conception of the causal relation, perhaps we can bypass the thorny question of causation in the following way. Suppose we focus, not on the meaning of 'cause' in expressions of the form 'x causes y', but, rather, on the domain of the variables x and y. Perhaps one could defend the thesis that omissions cannot be causes by showing that only events can be causes, and omissions are not events.

The intuitive appeals here are to the ideas that: (1) what causes cause are always changes in the world, and (2) only change can beget change. *Onsets* or *cessations* of states can be causes, on this view, but states themselves can no more be causes than can objects.[41] Just as 'The bullet caused the death' is elliptical for 'Some event(s) involving the bullet caused the death', so 'His courageous character saved the city from destruction' is elliptical for 'Some courageous act(s) of his saved the city from destruction'. If states cannot be causes because they do not involve change, neither can omissions be causes, for they involve no change either.

Unfortunately, the intermediate conclusion of this line of analysis is implausible. Why cannot states be causes? Wasn't the presence of oxygen a cause of the fire and of the deaths in the Apollo capsule years ago (the capsule was supposed to be oxygen-free), even though it also required the event of a spark occurring to cause these events? Why must we say that it wasn't the state (of oxygen being present) that caused the deaths but, rather, the event of the oxygen being introduced? What if the oxygen was always present, even if it should have been vacuumed out? Further, the plausibility of the states-cannot-be-causes thesis depends on the plausibility of the states-cannot-be-effects thesis; and why cannot states be effects? Why, in other words, must effects always be changes (events)? Are there not causes for a ship floating quietly at a certain level—causes in terms of an

---

[41] Judy Thomson explores this in her 'Causality and Rights'.

equilibrium between the force of gravity and the buoyant force due to the displacement of a certain amount of water?

It may be that one could work out a metaphysics of causation whereby only events could be causes or effects, but the task is daunting; for one would have to show how our considerable use of 'cause' with states is to be paraphrased into more regimented talk. That is also a task that cannot be done here, if it can be done at all.

Despite the blindness of this second alley too, perhaps the comparison between omissions and states can throw some light. Compare the *state* of oxygen being present in the Apollo capsule with the *omission* of certain workers to pump it out. Are these one and the same thing—the omission just being the presence of the unvacuumed oxygen—so that if the state is a cause of the explosion, so is the omission? Such an identification does seem to be the root of the temptation to think of omissions as causes. Perhaps we can remove this temptation to judge that omissions can be causes, even if one cannot decisively defeat the latter judgement with a definitive analysis of causation.

The identification of omissions with the states such omissions leave unchanged tempts one to the judgement that omissions can be causes, because it transforms omissions into things that exist. The state of oxygen being present is at least a thing that exists in the world, and thus is something eligible to be a cause. If omissions are this sort of thing, they too gain eligibility as causes. Yet omissions are not states. Indeed, as I argued in Chapter 2, they are not any things. An absent elephant is not a special, ghostly sort of elephant with ghostly powers; nor is the omission of a certain elephant to trample the grass to be identified as the state that the healthy, untrampled grass is in.

Let us see if we can make this more precise. As we discussed in Chapter 4, the state of oxygen being present in the Apollo capsule is the possession of a static property (presence of oxygen) by an object (the Apollo capsule) at a time (just before the explosion). As we also saw in Chapter 4, the action of pumping the oxygen out of the Apollo capsule is the possession of a dynamic property (pumping of oxygen out) by an object (the Apollo capsule) at a time (whenever the pumping takes place). Since dynamic properties are simply changes in static properties, an act of pumping the oxygen out would involve a change from the possession of the static property of oxygen being present to the static property of no oxygen being present. Since omissions are absent actions, one might then think that the omission to pump the oxygen out just is the absence of any change in the static property of oxygen being present, and further, that the absence of any change in a static property just is the state in which that static property is possessed by the relevant object at the relevant time.

The problem with this otherwise plausible identification of omissions with the states those omissions fail to change is that omissions to do things, like pump out oxygen, are not simply the absence of any change in the static property of oxygen being present. Rather, such omissions are omissions to do *one* sort of act that would change the presence of oxygen; there are many other ways that the oxygen could be lessened or removed from the Apollo capsule—e.g. by starting a fire. Omissions are absent actions of a certain type; they are not non-existent changes of state (which is just the state itself), even when that state is just the state that would have been changed had the action not been omitted.

Return to my favourite sort of elephant, the absent one. The state of grass around a water-hole being tall and straight may well be identical to there being an absence in any change of that state (the 'absence of a change' operating like double negation). Yet such an absence of *any* change of such a state is hardly the same as the absence of elephants trampling the grass down; such an omission is only an absence of *one* way in which there could be a change of such state.

I thus conclude that while states like oxygen being present may well cause events like explosions, omissions to change such states are not the same as the states themselves, so no temptation to treat omissions as causes can legitimately arise on this basis. If I am correct that much of the temptation to treat omissions as causes comes from this identification of non-existent actions (omissions) with plainly existent things (states), then showing this identification to be false may be reason enough to reject the temptation to which it gives rise.

To be sure, some of the temptation to treat omissions as causes stems also from the necessity conception of causation. But we have seen reasons to reject this conception, even if we are not in a position to see clearly the alternative conception of causation. It is probably also true that some of the attraction of both the necessary-condition conception of cause and the idea that omissions can be causes stems from a moral judgement: that people can be as morally responsible for harms they fail to prevent is for harms they cause. We have seen in Chapter 3 reasons for rejecting much of this moral claim as well. But however much reason there might be to hold people morally responsible for harms they fail to prevent, such reason is no justification for gerrymandering one's metaphysics in order to make one's moral conclusions look more palatable. Omitting to save is not causing death, and no view of moral responsibility gives us any reason to pretend otherwise.[42]

3. NON-WILLED NON-BODILY MOVEMENTS AS COMPLEX ACTIONS

The easiest class of cases is that that involves neither the movement of some subject's body nor anything like a mental state of volition by that subject. I refer to what the criminal law terms the *status* of a defendant. About status, we can be much more brief than about chosen omissions, mental acts, or involuntary bodily movements, for the question 'Can one kill, not by doing anything, but by being something?' barely makes sense. How could one perform the action of killing simply by being of a certain status? One can assuredly be a killer, and being in that status does entail that one has killed, but that isn't to answer the

---

[42] There is a *reductio* on the judgement that omissions cannot be causes that goes like this: if A's omission to rescue V when he could easily do so is not a cause of V's death, then B's *act* of holding A (who is otherwise about to rescue V) is also not a cause of V's death—for B did no more than cause A to omit to rescue V, and if A's omission did not cause V's death then the cause of that omission (B's act) also did not cause V's death. The assumption is that it is absurd to think that B did not kill V. Perhaps some very mechanistic notion of causation is committed to the conclusion that B did not cause V's death. But it also would be committed to denying that B caused V's death where B changed any feature of the world on which V's survival depends; e.g. B steals all of V's food or water; B takes down the trapeze on to which V had expected to jump; B drains the water of V's diving pool; etc. Whatever conception of causation one uses to replace the necessary-condition conception, it should not deny that there is a causal connection in this class of cases, including the case where B prevents V's rescuer from rescuing.

relevant question. How can one kill by being a killer or by being anything else?

It is true that the possession of properties by one person can be a cause of another person's death. X's being addicted to drugs so grieves his mother that she dies; Y's being so frighteningly ugly causes another to be in such fear that he dies; Z's wealth causes another such envy that he dies; etc. We might in each case say, 'X killed his mother', 'Y killed the fearful stranger', and 'Z killed the envious neighbour', but surely the 'killings' are wholly metaphorical. The names of persons like 'X', 'Y', and 'Z' are elliptical for the properties possessed by such persons, so that it is the states themselves that are the causes of their respective deaths. With this substitution, the 'killings' are in reality only the causing of deaths by an inanimate thing, a state. States 'kill' like bullets 'kill', which is to say, not at all like persons kill. There are no complex actions where a state of the actor can do the action-constituting work of a willed bodily movement by the actor.

11

THE METAPHYSICS OF COMPLEX ACTIONS II: THE IDENTITY OF COMPLEX ACTIONS WITH BASIC ACTS

1. THE METAPHYSICAL ISSUE OF IDENTIFYING EVERY COMPLEX ACT-TOKEN WITH SOME BASIC ACT-TOKEN

We come now to the second of our two questions, the question of identity: assuming (from the just-concluded discussion) that every complex action requires the doing of a basic act (bodily movement$_T$), is that complex action identical to that basic act? There are two well-armed philosophical camps on this question. Those who hold a 'coarse-grained' theory, such as Donald Davidson, Michael Bratman, Alan White, Lawrence Lombard, Jennifer Hornsby, and Elizabeth Anscombe, answer affirmatively; those who hold a 'fine-grained' theory, such as Alvin Goldman, Jaegwon Kim, Lawrence Davis, Monroe Beardsley, Carl Ginet, Irving Thalberg, and Judith Thomson, answer negatively, holding killings to be distinct from the finger-movings by virtue of which such killings are done.[1]

---

[1] The coarser folks are: Davidson, *Essays on Actions and Events*; Michael Bratman, 'Individuation and Action', *Philosophical Studies*, 33 (1978), 367–75; White, *Grounds of Liability*; Lombard, *Events*; Hornsby, *Actions*; Jonathan Bennett, 'Shooting, Killing, Dying', *Canadian Journal of Philosophy*, 2 (1973), 315–23; C. Arnold, 'Conceptions of Actions', *Sydney Law Review*, 8 (1977), 86–106; and Anscombe, *Intention*. Fine-grained theorists prominently include: Kim, 'On the Psycho-Physical Identity Theory'; id., 'Events and Their Descriptions: Some Considerations', in N. Rescher (ed.), *Essays in Honor of Carl G. Hempel* (Dordrecht, 1969); Monroe Beardsley, 'Actions and Events: The Problem of Individuation', *American Philosophical Quarterly*, 12 (1975), 263–76; Judith Thomson, 'The Time of a Killing', *Journal of Philosophy*, 68 (1971), 115–32; id., *Acts and Other Events*; Lawrence Davis, 'Individuation of Actions', *Journal of Philosophy*, 67 (1970), 520–30; id., *Theory of Action*; Thalberg, *Perception, Emotion and Action*; Ginet, *On Action*. There are considerable differences between fine-grained theorists. Goldman and Kim are usually thought of as extremely fine-grained theorists, since they would hold 'he raised the glass' and 'he raised the glass slowly' to refer

It may seem that we have already answered this question when we took a position on the ultimate nature of events, but this is only partly true. Where one came out on the large choices about the nature of events only loosely influences the identity conditions of events, because too many details of those choices were left open. For example, a very fine-grained theory of event individuation is compatible with thinking of events as properties, propositions, states of affairs, or facts about events. The fact (or proposition) that there was a killing is different from the fact (or proposition) that there was a shooting; the state of affairs (or property) of being a killing is different from the state of affairs (or property) of being a shooting.[2] Such a fine-grained theory is also compatible with the view that events are particulars of a certain sort, namely, instances of dynamic properties. Since a particular killing is an instance of the property of being a killing, and a particular shooting is an instance of the property of being a shooting, and since killing and shooting are different properties, then an instance of the one may not be identical with an instance of the other.[3]

Alternatively, the coarse-grained theory is also compatible with the view that events are instances of properties. Although being a killing and being a shooting are distinct properties, there is nothing that necessitates that an instance of one on a certain occasion is not also an instance of another on that same occasion. Indeed, if the conjunction of properties is itself a property—as seems quite plausible[4]—then there is the property of being a killing and a shooting, and *this* property could have a single instance that is the action in question.[5]

to distinct actions. 'Moderate' fine-grained theorists like Davis, Thomson, Thalberg, and Beardsley differ between themselves in how finely to individuate actions and in why such limits are imposed. I assume in what follows that the moderate theorists are like Thalberg and Thomson, who hold that actions like killings have parts and that death is one such part of a killing. This is the 'componential' or 'moderately fine-grained' view of actions.

[2] On the individuation of facts, see Bennett, *Events and Their Names*, 9–12. Bennett persuasively argues that the Kim–Goldman view is best seen as a theory treating events as facts, even though that is not what they say they are saying. Ibid. 73–87. See also Lombard, *Events*, 61.

[3] This is what Kim and Goldman take themselves to be saying.

[4] See Armstrong, *A Theory of Universals*, ii. 30–6.

[5] Of course, on the coarse-grained view, the property of which a discrete action is an instance is a much more complex conjunctive property. In the example

There is thus no direct inference we may draw about the identity conditions of events from what we have already said about their nature. We will thus need some independent argument for the coarse-grained or the fine-grained theories.

It may also seem that we have already resolved the issue of event identity when we defended the equivalence thesis earlier. For, as Alvin Goldman once observed, it may seem that if the sentence 'x killed y' is equivalent to the sentence 'x moved his finger (and that had y's death as a consequence)', then the event referred to by 'the killing of y by x' must be identical to the event referred to by 'the moving by x of his finger'. This, on the supposition that if two sentences are equivalent, then the nominals derived from them cannot refer to different events.[6]

We cannot dispose of the identity issue this easily, however, for the premiss that drives the above argument is surely false. That premiss assumes that there are unique nominals to be derived from the two sentences describing complex and basic actions. Yet from these two sentences, one can as easily derive the nominals 'the killing of y by x' and 'the death of y (that, incidentally, was caused by the moving of x's finger)'.[7] And surely the events referred to by these two phrases are not identical. The upshot is that the equivalence thesis I earlier defended provides no basis for defence of the exclusivity thesis here in issue.

So the orthodox theory of action of Austin and Bentham has to argue for its exclusivity thesis, which is to say that it has to argue for the coarse-grained view of action individuation. Jones's killing Smith is identical to Jones's wounding Smith, is identical to Jones's shooting Smith, is identical to Jones pulling the trigger, is identical to Jones's basic act of moving his finger. Jones did but one action here, not five distinct actions. Thus far I have only given a kind of common-sense argument for the coarse-grained view—the common intuition being that we do only perform *one* act in scenarios like that of Jones killing Smith. We seem, on the fine-grained view, to be much busier than common sense would allow.

---

given, it might include: is a trigger-pulling, is a finger-moving, is a bullet-firing, etc. See Bennett, *Events and Their Names*, 94; Lombard, *Events*, 54–5.

[6] Goldman, 'The Individuation of Actions', 766.

[7] See Bennett, *Events and Their Names*, 192. For other reasons, see also Ginet, *On Action*, 60.

Fine-grained theorists have none the less rejected common sense for a variety of reasons.[8] Perhaps chief among these are what I shall call the temporal and spatial arguments. The temporal argument goes like this: Jones cannot kill Smith until Smith dies; yet Jones completes the basic act of moving his finger well before Smith's death. Therefore, since identical particulars must share all of their properties, the event that was Jones's killing of Smith is not identical to the event that was Jones moving his finger. The spatial argument is quite similar. The argument assumes that the place of Smith's killing must at least include the place at which he dies, yet that is not the location of the moving of Jones's finger. Therefore, since the killing and the moving do not share this spatial property, they cannot be identical.[9]

The focal point for discussion of both of these arguments lies in their assumptions about the time and place of a killing. Fine-grained theorists assume that no killing occurs until the victim dies. This locates a killing temporally either: (1) as existing at the time of the victim's death, because that is when the property 'been killed' is exemplified (the extreme fine-grained view); or (2) as existing during a period that begins with the defendant's willed bodily movement (t_1), that continues during that movement's intermediate effects upon the victim's body (t_2), and that ends with the death of the victim (t_3), these all being parts of the killing (the moderate fine-grained view). It also locates a killing spatially as either: (1) being wherever the victim is when he dies since that is where the property 'been killed' is exemplified; or (2) beginning wherever the defendant is located (s_1), continuing wherever the moving of the defendant's body has its first impact on the victim's body (s_2), and ending wherever the victim dies (s_3), these all being parts of the killing. A coarse-grained theorist, by contrast, locates a killing temporally as occurring whenever the killer executed those basic acts that caused the victim's death

[8] For survey discussions of the quite considerable number of arguments, see Beardsley, 'Actions and Events'; Davis, *Theory of Action*, 28–38; Bennett, *Events and Their Names*, 188–202.

[9] Like most arguments on this issue, both the temporal and the spatial arguments presuppose what is commonly called Leibniz's Law: for any two particulars x and y, x is identical to y if and only if x and y have all the same properties $(x)(y)[(x = y) \supset (F_x \equiv F_y)]$. Both arguments also assume that expressions like 'A's shooting of B' and 'A's killing of B' are used in extensional contexts where they refer directly to event-particulars. Neither of these assumptions should be questioned.

(t_1). A single movement of a trigger finger may constitute such a killing, but so may repeated blows given by the killer over some period of time. Analogously, the place at which a killing takes place is wherever the defendant was when he did those basic acts that caused the victim's death (s_1).

These implications about the spatial and temporal location of actions are often used to argue for or against the theories of identity and individuation that imply them. Often the arguments given by fine-grained theorists proceed from some observations about ordinary usage. About the temporal dimension of the problem, Judy Thomson, for example, finds three closely related implications of the coarse-grained view of the matter odd.[10] (1) The tense problem: It is odd to say at t_2 that the defendant has killed the victim, because the past tense of the verb indicates that the killing is over and done with. This would be an odd thing to say at t_2 when the victim is not yet dead. This would be like saying, 'The defendant caused (past tense) the death of the victim', while the victim is still alive. It is even odd, Thomson adds, to use the present or present-perfect tenses of 'killing' at t_2 to describe what the defendant is doing. It is only what Thomson derisively dubs 'Hollywood' English that allows the (obviously alive) victim to say, 'You kill me', or even, 'You are killing me'. (2) The date problem: It would be odd to say at t_4, sometime after the victim dies, that the defendant killed the victim when he moved his finger on the trigger (t_1). (3) The temporal-order problem: When the victim's death occurs several hours after the defendant initiated the sequence of events that caused it, one can easily say, 'V's death occurred several hours after defendant's shooting'. But, Thomson argues, we cannot say, 'V's death occurred several hours after V's killing by defendant'. Yet if the killing was the shooting (was the trigger finger movement), we should be able to so relate it to the victim's death.

About the spatial dimension of the problem, Monroe Beardsley makes an analogous observation.[11] He imagines a case mentioned by John Dewey,[12] wherein the defendant in New York kills his victim in California by means of sending poisoned candy from the one place to the other. According to Dewey and Beard-

---

[10] Thomson, 'The Time of a Killing', 116–19.
[11] Beardsley, 'Actions and Events', 265.
[12] John Dewey, *Experience and Nature* (New York, 1958), 198.

sley, 'the locus of the act now extends all the way from New York to California'. The reason for including California as one of the places where the killing occurred is the same as that noted by Thomson: otherwise, if the killing only occurred in New York, that seems to suggest that there can be killings before there are deaths. More reasonable, thought both Beardsley and Dewey, was the conceptualization reached by the California Supreme Court in this actual case: *part of* the killing occurred in each state.[13] The preparation of the poisoned candy, and the death, were each part of the killing, so that its locus extended across the country.

The main response to these observations of oddity has been the *tu quoque* response: the fine-grained theorist's temporal and spatial location of actions commits them to saying some rather peculiar things too. Donald Davidson was the first to notice this. In imagining a Hamlet-like poisoning in which the queen kills her sleeping husband by pouring poison into his ear, Davidson asks: 'Is it not absurd to suppose that, after the queen had moved her hand in such a way as to cause the king's death, any work remains for her to do or complete? She has done her work; it only remains for the poison to do its.'[14] Yet the fine-grained theorist seems committed to saying that the queen is killing her husband while she is sitting quietly watching him—or even while she is doing something else such as playing croquet—or even after she herself had *died* (so long as her poison is still working on the king)! It surely is odd to be committed to the view that we can continue to *do* anything after we are dead, but the fine-grained view of actions so commits one.

These juxtaposed oddities put us at an impasse about the time and place of a complex action. The impasse is to be broken only if we can asymmetrically explain away one or other set of the seeming oddities. Suppose we first assume that complex actions like killings are identical (token–token) to basic acts like finger movements. If we assume this metaphysics, that locate killings wherever and whenever the identical bodily movements take place. Can we then explain away the oddities of speech to which Thomson and Beardsley have drawn our attention? I think that we can. The trick is to see that events can acquire certain

[13] *People* v. *Botkin*, 132 Cal. 231, 64 Pac. 286 (1901).
[14] Davidson, *Essays on Actions and Events*, 57–8.

properties over time that they did not have before, and still remain the same events (numerically) as they were before.[15] After all, objects are commonly so treated. University Fellow Jones may become Fellow *and* Professor Jones at some point in time, but he is still the same person; he has simply acquired a new property. Similarly, a finger movement can become a killing at some later time (namely, when the victim dies), and remain the same event even with that later acquisition of the property 'cause of death'. This is surely very common with events other than human actions, so that, for example, the flooding of a town becomes, at some later time, the most discussed event of the decade.

That a finger movement later becomes a killing in no way dates or places the killing at that later time or place. That would be like saying that the most discussed event of the decade—i.e. the flood—took place, in part, wherever and whenever those discussions took place. Yet such discussions are only consequences of the flood, just as death of a victim is only a consequence of the killing of that victim.

The distinction—between deaths being *parts* of killings, and causing deaths being *properties* of killings acquired afterwards—takes away any metaphysical oddity to the coarse-grained view of action identity. This does not quite dispel the three linguistic (or usage) oddities that Thomson references. Yet this metaphysical distinction does pave the way to explain away Thomson's usage oddities as just that, usages that are *pragmatically* odd. The *semantics* of complex action expressions can then be drawn in line with the coarse-grained metaphysics of the Austin–Bentham view.

Consider what Thomson calls the tense problem. The oddness of describing the defendant's shooting at t_2 as a past tense 'having killed' or a present-perfect tense 'is killing' is due to the suggestion that the victim has died by the time of utterance, when he has not. Such suggestion arises because a speaker at t_2 cannot know for certain that the action of shooting that is completed

[15] See Bennett, *Events and Their Names*, 197–8; Bennett, 'Shooting, Killing, Drying'; Grimm, 'Eventual Change and Action Identity', *American Philosophical Quarterly*, 14 (1977), 221–9; John Vollrath, 'When Actions Are Causes', *Philosophical Studies*, 27 (1975), 329–39; Alan White, 'Shooting, Killing, and Fatally Wounding', *Proceedings of the Aristotelian Society*, 80 (1980), 1–15.

has the property of causing death that will make it a killing, for the victim hasn't yet died. It is an odd way to pick out any particular to use a property of it that the speaker knows it does not yet possess at the time he is speaking. It is like picking out George Bush by the description 'President Bush' before Bush was elected. Bush at that time was not yet President, although one could predict that he almost certainly would be. Analogously, it would be odd right after a monumental flood to pick out such an event with a property that it does not yet possess, even if it will possess it in the near future—as in describing it as 'the most discussed event of the decade' before any discussion has yet taken place and before the decade is over.

Because it is odd to pick out either objects or events with descriptions of properties that they do not yet (at the time of utterance) possess, a listener may assume that use of 'killed' in any tense is done only when the action has the property 'caused death'. The oddness of using 'killed' or 'killing' when no death has yet taken place is only odd because it violates this pragmatic convention or expectation about how one picks out particulars with descriptions. That pragmatic surprise hardly affects the semantics of 'killing' or the metaphysics of killings. Sometimes an utterance will eliminate any surprise on the part of the audience, pragmatic or semantic. Consider this recent bit of English prose: 'The South killed Lucy Bondurant Chastain Venable on the day she was born. It just took her until now to die.'[16] Although the author is playing with our normal pragmatic expectations, this explicit use of the past tense of the verb 'to kill' does not jar our sense of linguistic propriety.

Thomson's second problem, the dating problem, in fact presents no oddness of usage, pragmatic or semantic. I can say at t_4, after the victim has died at t_3, 'The defendant killed him at t_1, when he moved his finger on the trigger', with no oddness whatsoever. By the equivalence thesis defended in Chapter 8, to say that the defendant killed the victim at t_1, is to say that some basic act of the defendant, such as his moving his finger on the trigger, at t_1 caused the victim's death; t_1 is when the defendant did the act that caused the death, and there is nothing odd about saying that at t_4.

---

[16] Anne Rivers Siddons, *Peachtree Road* (New York, 1988), 3.

The *only* oddness of saying any of these things arises because of a potential ambiguity of a question like 'When did the defendant cause victim's death?'[17] One reading of such a question focuses on the victim, so that the question really is 'When did the causal process that the defendant initiated, and which terminated with the victim's death, end?' The answer to *that* question is, of course, t_3, because the process of causing death does not end until there is a death. One might thus confuse this reading of the dating of causing death (in this sense) with the dating of killing. Such confusion can hardly affect the semantics of 'killed'. 'Killed' refers to an act, not to a process.[18] The act of killing, as opposed to the process of dying, ends when the defendant's finger movement at t_1 ends, even if that need not of course be when *what* he causes (dying) ends.

Thomson's temporal-order problem is perhaps the most jarring bit of usage for which we have to account. It is decidedly odd to say 'A killed B and then B died' or 'A's killing of B caused B to die several hours later'. Yet the oddity here too is only a pragmatic feature of this usage, not affecting the semantics of the verb 'to kill'. It is like saying that the most discussed event of the decade (referring to a large flood) caused more discussions to take place about it over the next ten years than about any other event. The oddity is that the information contained in the predicate is already conveyed by the phrase chosen to pick out the subject. The oddity is the same as is to be found in trivial identity statements like 'The Morning Star is the Morning Star'— true, but uninformative given the way in which Venus is originally picked out by the first usage of the phrase 'the Morning Star'. It is less odd, because more informative, to say 'Venus is the Morning Star'; but the truth values of the two sentences are the same. Analogously, it is less odd to say 'The movement of Jones's finger caused Smith's death' than it is to say 'The killing of Smith by Jones caused Smith's death', but the truth values of the two sentences are the same.

The supposed temporal and spatial oddities of locating Jones's killing of Smith at t_1 and s_1 thus have nothing to do with the

---

[17] On this, see particularly Vollrath, 'When Actions Are Causes'.

[18] George Wilson finds there to be a *process* sense of 'killed', which must be distinguished from its sense as referring to an action. Wilson, *The Intentionality of Human Action*, 72.

semantics of phrases like 'the killing of Smith'. The *reference* of such phrases is to the basic act that causes Smith's death. The only oddities are *pragmatic*, such as repeating in the predicate information already conveyed by the terms used to pick out the subject. Such pragmatic oddities should not in any way deflect one from the true metaphysics of the situation: complex actions take place where and when the basic acts with which they are affiliated take place.[19]

Can the fine-grained metaphysician about actions explain away the oddities of speech that follow upon his metaphysics of action? Fine-grained theorists have certainly attempted this task. One such attempt is based on Monroe Beardsley's distinction between actions and activities. With regard to the dead queen continuing to kill her husband via a poison working after her death, Beardsley concedes that 'of course activities cease with death, but not necessarily actions: for the event identical to the action has to run its course, and some events, once started, carry through of their own impulse, so that initiating the event is all that is required to act'.[20] This seems to say that actions have a life of their own, independent of that of the actors whose actions they are; we can kill, though we be dead.[21] As Jonathan Bennett says of this swallowing of the *reductio* by Beardsley: 'when put this plainly, it is incredible'.[22]

---

[19] This general line is also the response to the other main argument often advanced against the coarse-grained view, namely, the argument that it is false to say that the defendant's killing of the victim caused the defendant's gun to fire. (See Goldman, *A Theory of Human Action*, 2; Thomson, *Acts and Other Events*, 49.) The response is that it is only pragmatically odd to refer to an event e_1 (that caused another event e_2 to occur) with a description of e_1 that (pragmatically) presupposes that that same effect (e_2) has occurred. It is odd in the same way as it is odd to say: 'The most discussed event of the decade caused people to start talking about it'.   [20] Beardsley, 'Actions and Events', 270.

[21] The US Supreme Court once said something like this in *Pinkerton* v. *United States*, 328 U.S. 640 (1946). In *Pinkerton* two brothers were found guilty of the crime of conspiring to run bootleg liquor. Under the doctrine announced in *Pinkerton*, a conspirator is guilty not only of the crime of conspiracy, and not only of all substantive crimes he either committed or aided others to commit; the conspirator is also guilty of all substantive crimes committed by his co-conspirators. In the case of Daniel Pinkerton, that meant that he was guilty of all the bootlegging crimes done by his brother Walter even though Daniel was locked up in the penitentiary when Walter did these offences. Yet the *Pinkerton* doctrine is rightly regarded as a doctrine of vicarious liability because it makes a conspirator liable for crimes he did not do or aid others in doing. It is not a (metaphysical) doctrine about actions extending to spatially and temporally remote effects.

[22] Bennett, *Events and Their Names*, 196.

Judy Thomson also tries to defuse the 'no further effort' objection by an analogy to actions that are plainly performed over
time because they require a sequence of bodily movements to
complete them. Thomson's example is one wherein she performs
the action of cleaning her floor by: pouring Stuff on it (t_1); waiting for the Stuff to dry (which drying action picks up the dirt)
(t_2); and then vacuuming up the Stuff residue (t_3). Surely she has
not *cleaned the floor* until t_3, so that at t_2 while she is waiting she
is still cleaning her floor.[23] Thomson then analogizes this scenario
to another, wherein her neighbour cleans her floor with Superstuff. Superstuff works just like Stuff, except that the Superstuff
residue evaporates and thus needs no vacuuming. Surely, Thomson concludes, her neighbour has not yet cleaned her floor when
she pours on Superstuff at t_1 any more than when she poured on
Stuff; her action of cleaning is completed only when her cleaning
agent has finished its work.[24]

This is surely a very bad analogy. Consider a like one by the
Supreme Court of Indiana.[25] Harry was convicted of voluntary
manslaughter of John. He had struck John in one county, where-
upon John died three days later in another county. Harry was
tried and convicted in the county where death occurred, despite a
state constitutional provision requiring 'trial in the county in
which the offence was committed'.[26] In upholding the conviction,
the Indiana Supreme Court analogized the effects of the blow
working in John's body to a succession of blows on John's body
by Harry. Because the later blows would surely be as much a
part of the killing as were the earlier ones, so the effects within
John's body that immediately caused his death were part of the
killing:

The crime here charged was not completed by the blow, but the blow
and its effects continued to operate, like a succession of blows, until it
resulted in death. In other words, the blow and its effects continued to
operate, beginning in Hendricks county and extending into Marion
County, where it finally accomplished the complete crime by causing
death.[27]

---

[23] Thomson, *Acts and Other Events*, 50–5.   [24] Ibid. 56–7.
[25] *Peats* v. *State*, 213 Ind. 560, 12 N.E. 2d 270 (1938).
[26] Constitution of Indiana, Art. I, §13.
[27] *Peats* v. *State*, 12 N.E. 2d at 273.

What makes such analogies hopeless is their initial reliance on further basic acts by the actor at t3 (which of course triggers out intuitions that the whole action is not yet completed) only to have just that feature subtracted in the actual case we must judge. A cleaning or a killing is of course not completed until the last vacuuming causing cleanliness or the last blow causing death is done, but how does that speak to the date of completion of those cleanings or killings where no further vacuuming or hitting needs doing? This is like arguing that x is F because y is F— when x lacks just that feature possessed by y that makes y so clearly F.

The third response by some fine-grained theorists—the 'componential' or moderate ones—is to soften the separateness they posit between basic and complex actions. A killing of V by D on this view consists of parts: a shooting of V by D, and the death of V. The shooting, in turn, consists of parts: a moving of D's trigger finger, and a bullet striking V. The killing, the shooting, and the moving are distinct actions only 'in the uncontroversial and unexciting sense that an ear is distinct from its lobe'.[28] The idea is that actions like moving one's finger can 'take on broader dimensions and more parts' over time,[29] so that a moving at one time can become a killing at a later time. All of this is intended to blunt the oddity of saying that someone's action may go on after they are dead or otherwise engaged. When such further 'parts' become part of the whole, the actor 'accomplishes more— not that he does something extra, such as smiling in addition to crooking his finger'.[30]

It is not clear how this kind of attempt to demotivate the question, one act or many, could blunt the 'no further effort' objection. For on the view here considered, an actor's complex actions concededly 'have more extensive spatio-temporal dimensions than his basic performance which is their fountain head'.[31] There is thus no (whole) killing until the part that is the victim's death has occurred. Therefore, a new event comes into being—a killing event—when the victim dies, and we are left with our original absurdity: on the fine-grained view, someone's action of killing

---

[28] Thalberg, *Perception, Emotion and Action*, 111–12.
[29] Ibid. 110–11.
[30] Ibid. 106–7. See also Ginet, *On Action*, 61, for a similar view.
[31] Thalberg, *Perception, Emotion and Action*, 107.

must be said to be going on even if he is immobile, doing something else, dead, or unconscious.[32]

Moderately fine-grained theorists such as Thalberg have it
almost right here. What they have to give up is the idea that the
deaths of the victims are *parts* of the killings that kill them;
rather, such deaths—or more exactly, such causings-of-deaths—
are properties that actions of killing acquire at a time after such
actions have themselves ceased. It is giving up the notion of
death as *part* of the killings that allows one to avoid the absurdity of thinking that killings can go on long after the killer is
dead. Yet to give up the part/whole idea is of course to give up
the moderate fine-grained metaphysics of events.

I conclude that the location implications of either the extreme
or the moderate fine-grained metaphysics of complex actions
leads to absurdities that cannot be explained away, whereas the
location implications of the coarse-grained metaphysics of complex actions lead only to apparent absurdities that can be
explained away in a satisfactory manner. There are many pragmatically odd things that people can say that are none the less
perfectly true. 'Jones killed Smith, and then Smith died' is one of
them.

Aside from the location arguments that we have just discussed,
there is an extensive literature addressing many other arguments
pro and con on the exclusivity thesis.[33] We cannot hope to deal
adequately with all such arguments, presupposing as they do
issues that themselves require extensive treatment, such as the
extensionality of causation. I shall thus rest content with my limited defence of the exclusivity thesis in terms of its getting the
location of actions correct. The intuition that locates actions at
their source is, I think, the root intuition behind the coarse-
grained metaphysics of action: actions are things people *do*, and
the movements of their bodies are the only doings persons
directly control, so that when they cease moving they cease
acting.

---

[32] See Davis, *Theory of Action*, 36; Hornsby, *Actions*, 29–32.
[33] For summaries, see the references in n. 8 to this chapter.

2. THE LEGAL LOCATABILITY OF ACTIONS

2.1. The Spatial Locations of Actions in Law

It would be nice if the legal doctrines of Anglo-American criminal law unambiguously reflected the coarse-grained metaphysics of complex actions together with its implications for the spatio-temporal location of actions. In fact those doctrines are pretty much across the board in the apparent metaphysics that they presuppose. Take the problem of the spatial location of action first. The main issues requiring courts to locate actions spatially are those of jurisdiction, venue, choice of law, and interpretation of spatially described circumstances of various *actus reus*. To revert to my example near the beginning of Chapter 10, of the murdered prison guard: these are, respectively, the issues of whether California has the power to try the murderer, whether California (or some county within it) is the proper place for trial, whether California law will be applied if the trial does take place there or elsewhere,[34] and whether the killing took place *in* the prison (as the hypothetical statute requires). Following what is called the 'territorial principle',[35] legal doctrines typically require that only the state where the killing took place has power (jurisdiction) to try the killer, only the state (or country) within which the killing took place is the proper place for trial, and only the law in force wherever the killing took place is the law that can be applied to such a killing. Although 'the place where the wrongdoer was at the moment of perpetration might well have been selected as the situs of the crime'[36] under the territorial principle, the common law did not develop in such a way as to adopt this coarse-grained metaphysics. Rather, the common law developed a view about the locus of actions that is difficult to make any metaphysical

---

[34] In criminal cases jurisdiction of a court to adjudicate ('jurisdiction') and jurisdiction of a legislature to legislate ('choice of law') usually go hand in hand, so that the question of whether California's criminal law can apply to a given action answers and is answered by the question of whether a California court can exercise jurisdiction to try a given defendant. This is not as often true internationally, and is not true at all in the creation and application of tort law.

[35] See generally Roland Perkins, 'The Territorial Principle in Criminal Law', *Hastings Law Journal*, 22 (1971), 1155–72; Walter Wheeler Cook, 'The Logical and Legal Bases of the Conflict of Laws', *Yale Law Journal*, 33 (1924), 457–88: 463.

[36] Perkins, 'The Territorial Principle in Criminal Law', 1159.

sense of: 'the situs of criminal homicide was held to be where the fatal force impinged upon the body of the victim'.[37] This is not where the wrongdoer moved his body (s_1), nor where the death occurred (s_3); rather, this is where the body of the victim was hit, by blows, bullets, or whatever (s_2).

Notice what this requires one to say about the nature of a complex action like killing. One has to think that the essential nature of a killing lies neither in its source (in a basic action), nor in its result (death), nor in a continuous causal process connecting the two extremes. Rather, the victim is killed when and where the dying process within his body begins. Such a metaphysics is not only counter-intuitive on its face—which it is, since it honours an arbitrary middle point as the essence of a killing; such a metaphysics also leads to *both* sets of absurdities mentioned earlier. That is, if a killing takes place when and where the hitting of the victim's body take place, then one must say such things as: (1) the defendant killed the victim (at t_2, when the blow fell), and then the victim died; *and* (2) the defendant was killing the victim (at t_2, when the poison was taken by the victim) even though the defendant had previously taken the poison himself and died.

The typical American statutory alternative to the common law makes as little metaphysical sense as the common law that it replaced. As Perkins notes: 'The most common enactment for this purpose has been the so-called "in whole or in part statute." The wording is in some such form as that punishability under the laws of the state shall include "all persons who commit, in whole or in part, any crime within this state." '[38] This statutory alternative adopts entirely the moderate fine-grained (or 'componential') metaphysics of actions, and is subject to the same objections as those we have seen.

Court interpretations of the various *actus reus* prohibitions that happen to include spatially described circumstances also make very little metaphysical sense. Consider the English decision of *Jemmison* v. *Priddle*.[39] The accused shot two deer while he was on a farmer's land with the farmer's permission to hunt deer on his land. The first deer was hit after it had left the farm and had run on to land where the accused had no permission to hunt; the second deer was hit on the farm, but lived long enough to run

[37] 'The Territorial Principle in Criminal Law', 1160.
[38] Ibid. 1162.   [39] [1972] 1 All Eng. Rep. 539, [1972] 1 Q.B. 489.

off the farm; both deer died on the neighbouring land. The accused was charged and convicted of the offence of killing deer without a game licence (the statutory exception, allowing unlicensed killing on one's own land or on the land of another with consent, being held not applicable).

The accused's conviction was reversed, but only as to the second deer, not the first. The implicit metaphysics of action presupposed by this result is like that adopted by the common law for jurisdictional and venue purpose: defendant's act (of killing a deer) is located neither at the place where the basic act is done (on the farm where he had permission) nor at the place where death of the deer takes place. Rather, the act must be thought to be located when the dying process in the deer began (by its being struck by the bullet).[40]

Clearly, some of what motivates courts to the curious metaphysics of action revealed by an odd spatial location of actions is simply confusion. Courts confuse what are parts of act-types with parts of act-tokens, so that some consequence-type that is part of the act-type killing is also taken to be a part of that act-token of the killing done by the defendant. Courts don't see, in other words, that an act-token done in place s_1 can acquire the property being-a-cause-of-event-e (which takes place at s_2 or s_3), without that act-token taking place anywhere but at s_1. Courts don't see that this remains true even when what makes the act-token illegal is precisely that it acquires just this property, so that death as a type of event is part of killing as a type of prohibited act.

Equally clearly, however, sometimes courts and legislatures have been willing to create self-consciously some metaphysics of action they know to be false in order to arrive at results that on policy grounds they take to be desirable. Consider the much discussed case of *Simpson* v. *State*, [41] where the Georgia Supreme Court upheld jurisdiction of the Georgia courts to try a defendant, in Georgia and under Georgia law, for attempted murder. In *Simpson*, the defendant had fired across the water from where he was standing, which was in South Carolina, missing his intended victim but striking the latter's boat, which was in

---

[40] I thus do not construe this case as compatible with a coarse-grained metaphysics of action individuation. For a contrasting interpretation, see Arnold, 'Conceptions of Action', 91–6.

[41] *Simpson* v. *State*, 92 Ga. 41, 17 S.E. 984 (1893).

Georgia waters. According to the Georgia Supreme Court, 'the law regards him [the shooter] as accompanying the ball, and as being represented by it, up to the point where it strikes'. The actor 'started across the river with his leaden messenger . . . and was, therefore, in a legal sense . . . in Georgia'.[42]

This kind of constructive presence is too transparently fictional to be taken as the reason for the Georgia court's conclusions—that Georgia courts had jurisdiction and that Georgia law applied. Surely the court was motivated by a sense that Georgia had at least as much interest in trying this defendant under its law as South Carolina would have in trying him under its attempted-murder statute. Surely the Georgia court was anticipating what came to be known as the 'injured forum' principle, supplementing the territorial principle in order to achieve a plausible result.

If one thinks that the policy judgement here is correct,[43] then the metaphysically correct, spatial location of actions becomes only one sufficient basis for finding jurisdiction to adjudicate, venue to try, and jurisdiction to legislate; equally sufficient would be where the salient consequences prohibited by criminal statutes take place, a quite different metaphysical enquiry. If one has this bifurcated view of these matters, then one should supplement the territorial principle with an injured-forum principle, as has South Dakota, for example, by statute: 'when the commission of a public offence commenced without this state is consummated within its boundaries, the defendant is liable to punishment therefore in this state'.[44] An act of killing is *consummated* (in the sense of 'becomes a killing') wherever and whenever the victim dies (even if *the* killing takes place earlier and elsewhere),[45] and it may

[42] *Simpson* v. *State*, 17 S.E., at 985.

[43] Glanville Williams argues that it is not. See Williams, 'Venue and the Ambit of Criminal Law', *Law Quarterly Review*, 81 (1965), 518–38.

[44] So. Dak. C.L., §23–9–10. German law is similarly explicit: 'An offence has been committed at that place at which the perpetrator has acted . . . or at which the consequences that are necessary for the crime have occurred . . .' (Strafgesetzbuch, §9).

[45] Jonathan Bennett develops a notion of *consummation* without any apparent realization that lawyers had already used the notion to his exact purpose in this context. Compare Bennett, *Events and Their Names*, 197, with Williams, 'Venue and the Ambit of Criminal Law', 520–1. This notion of consummation cannot be criticized on the ground that 'consummation . . . of actions take place at the same time, when and where the basic action itself takes place' (Arnold, 'Conceptions of

THE IDENTITY OF COMPLEX WITH BASIC ACTS   297

make as good a sense to predicate jurisdiction, venue, and choice-of-law issues on consummation as on action. This is to supplement the territorial principle, with its metaphysical presuppositions about actions, with an injured-forum principle, with its metaphysical presuppositions about the *effects* of actions.

One might think that with respect to interpretive issues such as that regarding the two deer, sometimes at least the courts' curious metaphysics of action is here too in reality a fictional cover for a policy-based interpretation not requiring the location of action. At issue in the case of the two deer, for example, is the scope of the exception created for unlicensed killing of deer. Although the exception might be put in terms of 'killing *on* one's own land or *on* the land of another but with permission', one might well construe such a statute to be unconcerned with the metaphysical question of where the killing took place. After all, why should one be able to shoot from one's own land and kill deer on someone else's land, when one has neither a licence nor permission to kill deer on the latter's land? And if one agrees with the intended rhetorical force of this last question, then it does not matter where the act of killing took place; it matters where one of the *effects* of that act took place, namely, the hitting of the deer with the bullet.

On the other hand, sometimes the best interpretation of criminal statutes will be to require an enquiry into where the prohibited act took place. Consider my seemingly similar hypothetical statute in Chapter 10, prohibiting the killing of a California correctional officer *in* a California correctional institution. If an inmate shoots two guards, hitting one while in the prison and the other outside the prison, both guards dying outside the prison but the inmate doing his shooting from inside the prison, I take it to be the best interpretation of the statute to convict the inmate of two killings in the prison. Such interpretation might be justified, for example, on grounds of prison discipline requiring punishment of inmate actors even if the consequences of their actions may be outside the prison. If this is the best interpretation of this not so hypothetical statute, then courts must take the 'killing in the correctional institution' literally, i.e. they must locate where the act of killing actually took place. For such

Action', 103). For one is not talking of consummation of action-*tokens* but of action-*types*.

statutes, the correct metaphysics of action is indispensable.

The upshot is that one cannot take the court's apparent location of actions at face value. Sometimes a court's apparent metaphysics of action betokens, not a metaphysical confusion, but a policy disagreement with the various rules requiring it to locate actions. Once one gets one's policy preferences clear, however, there is no excuse for the fictional metaphysics; it is just what Bentham said of all legal fictions, the theft of power by hands that durst not steal it openly.[46] If courts think that the best interpretation of some statute does not require it to locate actions, or if courts disagree with the territorial principle, they should say so and adjudicate openly under the alternative interpretation or principle.

Sometimes the best interpretation of a statute does require courts to locate actions. In addition, if one retains the territorial principle, either by itself or as supplemented by an injured-forum principle, courts will also have to locate actions for jurisdictional, choice-of-law, and venue purposes. In such cases courts should discard the unintelligible common-law metaphysics of action in favour of the correct view, namely, the coarse-grained metaphysics defended throughout this book.

2.2 The Temporal Locations of Actions in Law

The law has shown greater good sense with regard to the temporal location of an action; where the legal principle does not require that courts get the metaphysics of action right in order to serve the principle in question correctly, the courts quite sensibly abandon the search for the correct metaphysics of action and look instead for the admitted *effects* of the actions. In applying the statute of limitations, for example, it matters when a killing takes place in order to see if the killer has been prosecuted for it within the time allowed for the bringing of such prosecutions. The courts date killings for these purposes from the time death occurs, not from the time of the basic act nor from the time of that basic act's first effects on the victim's body. If courts did not do this a defendant could give another a poison which takes a period in excess of the statute of limitations for it to work, and escape prosecution for murder entirely.

[46] See C. K. Ogden, *Bentham's Theory of Fictions* (Paterson, NJ, 1959), pp. cxiii–cxviii, 141–50.

Consider next the legal doctrine allowing a defendant to be punished for being an 'accessory after the fact'. One is an accessory after the fact to murder if one gives aid to the murderer 'after the fact', that is, after the killing that is the murder. Is one who gives aid to Jones guilty if the aid is given after Jones has clubbed Smith but before Smith has died? Our courts usually answer no, dating the killing from the death of the victim.[47] Yet this isn't because the courts think that killings occur then; rather, it is thought unfair to such accused not to so date the killings, because only after the victim has died can the aider know he is aiding a killer. One can easily concede that we cannot know that an act is a killing before death of the victim ensues, without for a moment conceding that the death is a part of the killing.

Despite these examples, sometimes the law demands that the courts get it right on the time of a killing. For these kinds of doctrines, only the correct metaphysics of actions can yield the correct legal decisions. Consider first the simultaneity principle of criminal law. As is well known, it is not enough that an accused kill another in order to make out the prima-facie case for murder; it is not even enough that, in addition, the accused at some time formed the intent to kill; in addition, the simultaneity principle requires that the intent to kill be simultaneous with the act of killing. Assuming we know when an intention is present in the mind of an accused, under the simultaneity principle courts need to know *when* the actor killed.

The reason that the law requires that the metaphysics be correct here is because the simultaneity principle is not just a legal principle. It is also a moral principle, a principle of culpability. And morality requires that there be a true simultaneity, not a fictional or constructive simultaneity, and not some statutory substitute. To be so culpable that one deserves the severe punishment reserved for murder, one must kill *while* one intended to kill.

It being important that the metaphysics of action be correctly decided, the courts have in fact accurately located actions temporally—which is to say, they have adopted the coarse-grained view of the matter: 'When there is a difference in time between the initial act and either the result of that act or the combination of the initial act and its result, the courts would insist that the time of the *mens rea* must be that of the act and not of the result or the act plus the result.'[48]

Consider another example, that involving the *ex post facto* ban on retroactive criminal legislation. That constitutional principle bans punishment for any act done when the law that forbids it was not already in place. Underlying this legal principle is of course a moral principle of substantive fairness: it is unfair to surprise individuals with criminal punishment for acts not criminalized when they were done. Such moral principle again demands that judges get the metaphysics right as they judge when an action was done. And here again, judges have perceived the true metaphysics: for these purposes, a killing is deemed to be done when the defendant performs the basic act that causes death. The killing is not dated at the time the accused's act first impacts upon the victim's body (the common-law position for jurisdictional purposes) nor at the time of the victim's death (the extremely fine-grained view), nor is the killing conceived of as occurring over the course of time required for the basic act to cause death (the moderate fine-grained view). Rather, for these purposes 'the crime is committed on the date at which the deed, the original act, is performed, and not on the date of the victim's death'.[49]

Properly understood, the doctrines of the criminal law locating actions temporally give some evidence of the correctness of the coarse-grained metaphysics of action. Where the policies behind legal doctrines require the asking of the metaphysical questions of when an action was done, the courts have located complex actions at their source, in the corresponding basic actions. Such

[48] White, *Grounds of Liability*, 46. Given his rejection of my volitional theory of action, White means something different by his notion of an 'initial act' than I do by my notion of a 'basic act'. White thus attributes a minor mistake to courts, who adopt the time of a killing as being when the *basic* act took place. This is no mistake, as is seen clearly in G. Marston, 'Contemporaneity of Act and Intention in Crime', *Law Quarterly Review*, 86 (1970), 208–38.
[49] *State* v. *Masino*, 216 La. 352, 43 S. 2d 685, 686 (1949). Said of the crime of negligent homicide, a crime added to the Louisiana Code after the defendant negligently failed to encase gas pipes in concrete during construction but before the explosion and deaths (resulting from such failure) four years later. German law reaches a like result with §8 of its Strafgesetzbuch: 'An offence has been committed at the time at which the perpetrator . . . has acted . . . The time of the consequence is irrelevant.'

location presupposes the coarse-grained metaphysics identifying complex actions with basic ones. Such intuitive metaphysical theorizing, done by those with no philosophical theories to defend, gives some evidence of where the truth of the matter lies.

PART III

THE IDENTITY CONDITIONS OF ACTIONS AND THE DOUBLE-JEOPARDY REQUIREMENT

12

THE DOCTRINAL AND NORMATIVE BASIS OF THE DOUBLE-JEOPARDY REQUIREMENT

Having completed our explication of both the act and the *actus reus* requirements, it remains to examine a third (act-related) requirement for criminal punishment under our law. This is the requirement that the offence done by an accused not be one for which he has already been put 'in jeopardy' by prior criminal proceedings. That is, it is not enough that the accused perform a volitionally caused bodily movement (thus satisfying the act requirement) and that that movement$_T$ have the (causal and circumstantial) properties described by some complex action description contained in some statute in effect when the action took place (thus satisfying the *actus reus* requirement); in addition, the particular act done and its relevant properties cannot in some sense(s) be 'the same' as some other act and its properties for which the accused has already been prosecuted and/or punished.[1]

We have the same reasons for being interested in this third pre-condition of punishment as we had for examining the first two requirements. For scepticism about the first two requirements has certainly infected the third as well. As before, that scepticism operates at three levels: the level of legal doctrine, the level of morality, and the level of metaphysics. The first kind of scepticism questions whether there is any non-vacuous double-jeopardy requirement and whether there is *one* such requirement

---

[1] In my discussion of double jeopardy I shall ignore those aspects of double jeopardy having to do with *mens rea* rather than *actus reus*. In judging sameness of both offences and actions, courts sometime enquire into whether the 'same intent' is required by two or more offences, or whether but one intention is manifested in some defendant's course of conduct. Except in so far as these are presented as criteria for sameness of action-types or action-tokens, I shall ignore all the problems lurking in the proper individuation of intentions. For a start on such problems, see Moore, 'Intention and *Mens Rea*'.

rather than many. The second kind of scepticism questions whether there is any justification for having such a requirement, or at least whether there is any set of justifications capable of justifying one, unitary requirement. Both the doctrinal and the moral scepticisms are the kinds Peter Westen raises when he urges that there is no sameness to double jeopardy's 'same offence' requirement.[2] The third kind of scepticism questions whether the identity of actions are such that there could be any sense to asking the question whether one action was the same as some other—conceding *arguendo* that we both do and should ask such a question. It was this kind of scepticism that Salmond raised when he analogized the location of actions to the place where one lives[3]—while there is a place where one lives, the boundaries of it are sufficiently imprecise that there is little sense in asking if one lives in the *same place* as someone else without stipulating the artificial boundaries (country, state, county, city, borough, block, house, etc.) that mark out one's concern in asking the question.

In examining these scepticisms I shall proceed much as before, examining first the doctrinal and normative questions about the requirement, and then asking the metaphysical question of whether actions have sufficiently precise identity conditions that the requirement can meaningfully be imposed.

1. THE DOUBLE-JEOPARDY DOCTRINES AND THEIR RATIONALE

Sir William Blackstone held it to be a 'universal maxim of the common law of England that no man is to be brought into jeopardy of his life more than once for the same offence'.[4] Despite this seemingly straightforward statement, there is room for the sceptical view that there is no coherent or unitary double-jeopardy requirement. For there are several legal sources of the requirement, several differing contexts in which the double-jeopardy question is asked, several different policies justifying the

[2] Westen, 'The Three Faces of Double Jeopardy', 1004.
[3] See Salmond, *Jurisprudence*, 401–2.
[4] Sir William Blackstone, *Commentaries on the Laws of England*, Sharwood edn. (Philadelphia, 1859), iv. 334.

requirement, several doctrinal formulations of the test for double jeopardy, and several different questions of action individuation asked by the different tests. Enough values to enough variables to give pause to any rational reconstructor of the requirement.

One source of the requirement in American law lies in the federal constitution, which provides that no personal shall 'be subject for the same offense to be twice put in jeopardy of life or limb'.[5] Another source lies in similar state constitutional provisions, which prohibit any person to be twice put in jeopardy for the same offence.[6] A third source lies in the statutes of several states, which variously prohibit multiple punishment for any single 'act or omission'[7] or for the 'same conduct' if it constitutes two or more offences of 'similar import'.[8] A fourth source of the requirement lies in the common-law merger doctrine, which is a court-created rule of statutory construction whereby multiple punishment under two or more offences is banned where such offences are committed by a single 'act, transaction, or episode'.[9]

The last two of these sorts of provisions apply only to the question of whether a defendant may be multiply punished for two or more offences. The federal and state constitutional provisions against double jeopardy speak to this issue as well, but in addition they bar multiple prosecutions for the same offence. As the US Supreme Court has often said about the double-jeopardy clause: 'It protects against a second prosecution for the same offence after acquittal. It protects against a second prosecution for the same offence after conviction. And it protects against multiple punishments for the same offence.'[10]

The justification for this now entrenched tripartite division of double jeopardy is functional. It is thought that these three contexts must be separated because the values that justify the

[5] United States Constitution, Amend. v.
[6] See e.g. Pennsylvania Constitution, Art. 1, §10. See generally Note, 'Twice in Jeopardy', 262–3 n. 3.
[7] See e.g. California Penal Code, §654; Arizona Revised Statutes, §13–116.
[8] Ohio Revised Code, §2941.25(A). See generally Note, 'Twice in Jeopardy', 270 n. 34.
[9] See the Pennsylvania Superior Court's full exposition of the common-law merger doctrine in *Commonwealth* v. *Williams*, 344 Pa. Super. 1087, 496 A. 2d 31 (1985). Despite the different doctrinal formulation of the merger test, 'the doctrine is to a large extent coterminous with the double jeopardy protection against multiple punishments'. 496 A. 2d, at 39.
[10] *North Carolina* v. *Pearce*, 395 U.S. 711, 717 (1969).

double-jeopardy ban are different depending on which of these three contexts is involved. Because these differences are the source of one of the scepticisms about double jeopardy that we wish to examine, we should look at such contexts and their differing values with some care.

The fundamental distinction in context is between the ban on multiple *punishments* for the same offence versus the bans on multiple *prosecutions* (whether after conviction or acquittal) for the same offence. This is usually put as a distinction between a ban on multiple punishments for the same offence being meted out in a single trial, versus a ban on there being multiple trials for the same offence.[11] Although history may give some support to drawing the distinction in this way, a functional mode of dividing double jeopardy's bans does not. For multiple punishments inflicted for the same offence is bad for the same reasons no matter whether such cumulative punishments are given in one trial or in several successive trials. Functionally, therefore, it is better to distinguish a ban on multiply *punishing* anyone for the same offence (no matter how many trials it took to set such punishments) from a ban on *trying* such defendants on such repetitive charges at all.

The functional justification for further subdividing the ban on multiple prosecutions (into those that follow convictions versus those that follow acquittals) is also that the values justifying such bans are different. Granting that this is so, the proper way to subdivide here is not exactly between acquittals and convictions.[12] Rather, the division should be between a ban against successive prosecutions on any issue on which the defendant prevailed in the first trial, and a ban against successive prosecutions on issues on which the defendant did not prevail. While questions of ultimate guilt or innocence are certainly the most significant issues on which a defendant either does or does not

[11] See e.g. the highly influential student note by Larry Simon, 'Twice in Jeopardy', 265–6, 281–3. Simon's note was the only authority cited by the US Supreme Court in its laying down of the tripartite division of double jeopardy in *Pearce*, 359 U.S., at 717 n. 8. This note's influence in entrenching this way of drawing the distinction between the multiple-punishment and multiple-prosecution contexts can also be seen in the US Supreme Court's opinion in *Brown* v. *Ohio*, 432 U.S. 161, 164–6 (1977).

[12] Again, the traditional way to draw the distinction is between prosecutions after acquittal versus prosecutions after conviction. See e.g. *Brown* v. *Ohio*, 432 U.S., at 166–7.

prevail in the first prosecution, there are other issues, such as severity of sentence, and the subsidiary issues within the guilt phase of trial, such as the elements of the prima-facie case or of defences. If the defendant has prevailed on any of these issues, the same policy argues that he not be retried; such policy does not apply where the defendant did not prevail in the earlier trial, so that his interest in a ban on reprosecutions must be different.

It is easier to see the separate values justifying these three bans of double jeopardy once they are redrawn in these two ways. Consider first the ban against multiple punishments for the same offence. The ban on multiple punishments is motivated by a proportionality worry: granted, the legislature could severely punish any given criminal act, but when potentially severe punishment is only possible because multiple provisions converge on a single act, such punishment is probably excessive by the legislature's own measures if the provisions punish essentially the same thing. The premiss here is that legislatures set the punishment for each offence on the assumption that that offence would be punished by itself. Given this assumption, when statutory provisions overlap in their coverage, the rational legislator would not wish for judges to multiply punish crimes that are in some sense the same, for to do so would be to cumulate excessively the legislatively set punishments for each offence.

It is important to see how this proportionality worry both resembles, and differs from, the proportionality worry under the Eighth Amendment's cruel-and-unusual-punishments clause. The difference lies in the degree to which a court is to review legislative determinations of appropriate punishment. Under the cruel-and-unusual-punishments clause, the review is straightforward: if the legislature has punished too much for a given offence, the punishment is unconstitutional. The multiple-punishments ban of double jeopardy does not operate this way. It may be assumed for double-jeopardy purposes that a legislature could cumulate punishments for a given offence if it wanted to, because it could simply increase the punishment for that offence to the desired level. What is required for double-jeopardy purposes is a rebuttable presumption that for offences that are in some sense essentially the same, a rational legislature would not want to cumulate punishments. For if legislatures set the level of punishment for each offence *without regard to the question of how overlapping this offence may be with others*, then multiple punishments for cases falling in the overlap will be too much punishment—too much punishment by the legislature's own standard, not by some court-imposed idea of proportionality.

Despite this difference of constitutional force, when the legislature has not clearly indicated the appropriateness of multiple punishments for similar offences, the multiple-punishments ban of double jeopardy is justified by the same value that justifies proportionality review under the Eighth Amendment: both are concerned with excessive punishment. To apply either provision requires a court to judge whether a defendant deserves punishment of a certain severity. Such question of desert has two components: how much wrong did the defendant do? And, how culpable was he in the doing of that wrong? A negligent killer does a great wrong, but with minimal culpability; an intentional thief does a lesser wrong, but with greater culpability; an insane thief does the same lesser wrong, but with no (or perhaps diminished) culpability.

Let us now turn to double jeopardy's ban on multiple prosecutions for the same offence, when the defendant did not prevail in the prior prosecution. The primary purpose of this ban is what I shall call its 'handmaiden' purpose, so named because this ban is largely the handmaiden of the multiple-punishments ban. The idea is that because multiple punishments are barred (by the first ban), then successive trials must also be banned as a means of preventing multiple punishments. After all, a good way to prevent multiple *punishments* of a defendant for one offence is to prevent multiple *trials* of that defendant for that offence. As the US Supreme Court stated in its leading nineteenth-century double-jeopardy decision:

Why is it that, having once been tried and found guilty, he can never be tried again for that offence? Manifestly, it is not the danger or jeopardy of a second time being found guilty. It is the punishment that would legally follow the second conviction which is the real danger guarded against by the Constitution.[13]

---

[13] *Ex parte* Lange, 85 U.S. (18 Wall.) 173, 173 (1873). See also Note, 'Twice in Jeopardy', 266 nn. 12–13 ('It is clear that preventing multiple punishment for the same offence was foremost in the minds of the framers of the double jeopardy clause').

On this 'handmaiden' interpretation, the ban on successive prosecutions after conviction should apply wherever the ban against multiple punishments should apply. This means that the value of securing punishments proportional to desert should also guide the application of this ban.

It is often said that a second purpose to the multiple-prosecution-after-conviction ban is to give peace of mind to those who have already been convicted and sentenced for an offence. As Peter Westen describes this value, which he calls 'finality': 'the interest in finality . . . is easy to articulate: it is a need for "repose", a desire to know the exact extent of one's liability, an interest in knowing "once and for all" how many years one will have to spend in prison.'[14]

Whether this is truly a purpose served by the ban on multiple prosecution after conviction depends on how one conceptualizes the ban on multiple prosecutions. If, as I urged, the multiple-punishments ban bars multiple punishments imposed in *successive prosecutions* as much as it bars multiple punishments imposed on separate charges *in the same trial*, and if the ban on successive prosecutions after 'acquittal' applies as much to sentence severity as to guilt or innocence, then there is no 'quieting of mind' work to be done by the ban here in question. For whatever 'quieting of mind' work is to be done here is already done by these other two bans, so construed.

Another purpose commonly said to be served by the ban or multiple prosecution after conviction is the value of efficient criminal adjudication. As in civil trials, it is inefficient to try criminal charges separately when such charges are largely overlapping in the mode of their proof and defence. Barring the second prosecution, like the barring of second civil trials by the *res judicata* and compulsory-joinder doctrines, is thought to give prosecutors every incentive to bring all closely related charges in one prosecution.

Yet on closer examination this can be seen not to be a value justifying the ban on multiple prosecutions after conviction, either. Again, assume my reconceptualization of the three bans. The incentive to prosecutors to bring all closely related charges in one trial is already given by the ban against reprosecution when

---

[14] Westen, 'The Three Faces of Double Jeopardy', 1051.

the prior prosecution resolved the issue(s) in the defendant's favour. For the latter ban already deals with the prosecutor's ability to retry issues, or offences, or sentences, on which he did not prevail to his satisfaction before. Prosecutors cannot care about reprosecutions where there is no hope of an increased sentence, a more favourable jury verdict in total, or a more favourable finding on a particular issue. So no additional incentive to bring all charges in one trial is given by a reprosecution ban where the prosecution has *prevailed* in the earlier prosecution; the only incentive is given when the prosecutor has *failed* in the earlier prosecution, which is the subject of the other ban.

A second purpose that *is* served by the ban on successive prosecutions after conviction is the purpose of avoiding the expense and embarrassment attendant upon being tried on criminal charges when there is no point in doing so. After all, if there can be no successive punishment (because of the multiple punishments ban), what is the point of a second prosecution? Not only does this cause the defendant unnecessary suffering and expense, but such repetitive prosecutions could also be a potential tool for harassment of particular citizens by the executive branches of government. Let us call this the embarrassment–harassment value served by barring successive prosecutions after the defendant has already lost all relevant issues in a prior criminal proceeding.[15]

Double jeopardy's third ban is the ban against successive prosecutions where the defendant has prevailed in ultimate verdict, on subordinate issues, or in lighter sentence. This ban does not serve as the handmaiden to double jeopardy's main ban, that against multiple punishments, for the situation with which this ban deals is one where the defendant will not be punished for his first prosecution (because he earlier *prevailed*). To justify this third ban will accordingly be to do something other than show how it serves as a vehicle to avoid multiple punishment.

[15] The 'harassment' value is often given a broader reading than that I intend, for often included as 'harassment' by multiple prosecution are situations where the prosecutor is seeking a stiffer sentence or where the prosecutor reserves charges in order to correct possible jury error on issues or charges. See e.g. Note, 'Twice in Jeopardy', 286–92; *North Carolina* v. *Pearce*, 395 U.S., at 733–5. What I mean by harassment is much more limited: where a prosecutor brings successive prosecutions without hope of getting more favourable determinations but simply to harass the defendant with the expense and embarrassment of trial. The broader 'harassments' I take to be relevant to justifying double jeopardy's third ban, not the second one here under consideration.

The values served by this third ban are three. First, there is the increased risk of punishing the innocent attendant upon successive prosecutions when the defendant has earlier prevailed. There are two sets of assumptions that can ground this fear. One is that successive criminal trials operate differently on prosecutors versus defendants and their counsel: prosecutors hone their cases, making their prosecutions more effective with each repetition, whereas defendants are worn down, making their defences and rebuttals less effective with each successive prosecution. Alternatively, simply the fact that prosecutors could (without double jeopardy's third ban) retry any verdict, issue, or sentence they did not like, but defendants could not retry any verdict, issue, or sentence they didn't like, asymmetrically increases the statistical likelihood of conviction with each prosecution. In either case, under either or both sets of assumptions, the evil that double jeopardy's third ban is aimed at preventing is the evil of innocence being punished.

A second value served here is what might be called the 'trauma' value. Criminal trials are not only time-consuming, expensive, and embarrassing to a defendant; where guilt and punishment are at stake, they are also very traumatizing to many individuals precisely because of the risk of conviction and punishment. Defendants of course must suffer such expense, embarrassment, and trauma whenever sufficient evidence of criminal conduct to warrant a trial exists, but they should not have to undergo this expense, embarrassment, and trauma more than once for that same offence. If they have (fully or partly) prevailed once, they should not have to undergo it all again.[16]

A third value here is efficiency in criminal adjudication. There is an apt analogy to be made to civil litigation with its doctrines of *res judicata*, collateral estoppel, and compulsive joinder,[17] but

[16] These fairness concerns were summarized by the US Supreme Court in *Green* v. *United States*, 355 U.S. 184, 187–8 (1957): 'The underlying idea . . . is that the State with all its resources and power should not be allowed to make repeated attempts to convict an individual for an alleged offence, thereby subjecting him to embarrassment, expense and ordeal and compelling him to live in a continuing state of anxiety and insecurity, as well as enhancing the possibility that even though innocent he may be found guilty.'

[17] See Justice Brennan's concurrence in *Ashe* v. *Swenson*, 397 U.S. 436, 454 (1970), where Brennan approvingly endorses 'the increasingly widespread recognition that the consolidation in one lawsuit of all issues arising out of a single transaction or occurence best promotes justice, economy, and convenience'. As

it is to be made in defence of this third ban of double jeopardy, not the second. For it is here that prosecutors can truly be given incentives to join all closely related counts into one trial. The incentive, of course, is that if they do not join every viable theory they have in the first trial, they will lose it, for there will not be a second trial. This ban has the incentive effects it does here because it deals precisely with the situation where prosecutors might want a second trial, namely, where they did not prevail to their satisfaction on guilt or sentence.

Not surprisingly, given this divergence in the values served by the different double-jeopardy prohibitions, the doctrines used in these differing prohibitions differ. Under the American federal constitution, so long as only multiple punishments are at issue, the exclusive doctrinal test is the *Blockburger* test: 'The applicable rule is that where the same act or transaction constitutes a violation of two distinct statutory provisions, the test to be applied to determine whether there are two offences or only one, is whether each provision requires proof of a fact which the other does not.'[18]

The seemingly univocal statement of this test, together with its name (the 'same evidence' test), should not be allowed to obscure the test's true nature. To begin with, the test is given to answer two distinct questions of action identity.[19] One is the question of when there are two or more 'acts or transactions', and when two putatively different acts or transactions are in reality the same. Only multiple punishments for acts or transactions that are the same give rise to double-jeopardy problems; if they are not the same acts then of course there is no double-jeopardy problem about multiple punishments, even if such punishments are for qualitatively identical doings of the same kind of crime. A serial killer who murders one victim on Tuesday, another on Wednes-

---

Brennan went on to observe: 'The considerations of justice, economy, and convenience that have propelled the movement for consolidation of civil cases apply with even greater force in the criminal context because of the constitutional principle that no man shall be vexed more than once by trial for the same offence.' 397 U.S., at 456.

[18] *Blockburger* v. *United States*, 284 U.S. 299, 304 (1932). The 'same evidence' test of identity for offences is much older than the *Blockburger* decision, even though the test bears its name. English common-law judges originated the test in *King* v. *Vandercomb*, 2 Leach 708, 720, 168 Eng. Rep. 455, 461 (1796).

[19] I will argue for this shortly, at the end of this chapter.

day, and a third on Thursday, all by strangulation, of course has no double-jeopardy objection for multiple punishment for his three acts of murder, even though he has done the same (kind of) offence. The 'same act' requirement thus raises the question of identity between *particular* actions ('act-tokens').

The 'same evidence' test also raises the question of identity between *types* of action. The question is whether, for example, kidnapping is a distinct type of act from false imprisonment. This calls for a comparison of the statutory elements of these two types of action to see how much overlap there might be. In such cases there need be no doubt about whether the defendant performed a single particular act—say, of holding another physically while transporting that person elsewhere; the relevant identity question is whether the two *types* of action made criminal by different statutes—false imprisonment and kidnapping—are the same or not.

The same-evidence test is the exclusive test used in multiple-punishment contexts to answer both of these questions under the federal double-jeopardy clause. Yet the very name of the test misleads as to what questions the test is to answer. Justice Brennan has seen this clearly with respect to the type-identity question answered by the test: 'Commentators and judges alike have referred to the *Blockburger* test as a "same evidence" test. This is a misnomer. The *Blockburger* test has nothing to do with the *evidence* presented at trial. It is concerned solely with the statutory elements of the offences charged.'[20] Not seen as clearly by Brennan or others on the court is that the 'same evidence' test is equally a misnomer for the 'same act or transaction' requirement of *Blockburger*. Although proof of different acts requires different evidence, that differential mode of proof is not what makes them different acts.[21]

---

[20] *Grady* v. *Corbin*, 495 U.S. 508, 521 n. 12 (1990) (citations omitted). See also *Commonwealth* v. *Williams*, 496 A. 2d, at 36 ('The *Blockburger* "same evidence" test . . . depends solely on a comparison of the elements of the crimes charged, not on the similarity or even the identity of the evidence introduced at trial to establish their commission.')

[21] Brennan has recognized that the 'same evidence' test has been applied to the same-act question. See Brennan's concurrence in *Ashe* v. *Swenson*, 397 U.S., at 451. But what neither Brennan nor the Court has recognized is that plainly a 'same act' question has to be asked and answered negatively before one gets to the same *type* of act question. The Court was particularly confused about this in *Brown* v. *Ohio*, 431 U.S. 161 (1977), which I shall discuss shortly.

Under the current interpretation of the federal double-jeopardy provision, the two-prong *Blockburger* test is not the exclusive test for when two offences are the same when multiple punishment is *not* the issue. When the issue is multiple prosecution, not multiple punishment, the *Blockburger* test for the same offence is supplemented by a more stringent test. As Brennan recently stated this more stringent test: 'The Double Jeopardy Clause bars any subsequent prosecution in which the government, to establish an essential element of an offence charged in that prosecution, will prove conduct that constitutes an offence for which the defendant has already been prosecuted.'[22] Thus, in *Grady*, the prosecutors could not prosecute the defendant for any form of homicide, given that the defendant had already been prosecuted for driving while intoxicated and driving on the wrong side of the road, and given that the prosecutor would seek to prove culpable negligence with respect to death in the second trial by proving just those types of act. The prosecutor could, however, prosecute the homicide if the proof of wrongdoing were to be based solely on the fact that the defendant was speeding.

Although Justice Scalia's dissent in *Grady* disputes it, this more restrictive test does not raise a question of token-identity; it does not ask whether there was only one particular act (of driving by the defendant). Rather, the question is one of type-identity, even if the types are of somewhat finer gradations than those used under the *Blockburger* test. The relevant identity question in *Grady* was whether any of the types of action that the defendant's particular act instanced, which types would be used to show that defendant's act was also of another type, negligent homicide, was the same as the types of action for which he had already been prosecuted. Was *Grady*'s particular act of driving of the type of act 'negligent killing' only because it also was of the same types of act (drunk driving and wrong-side-of-the-road driving) for which he had already been prosecuted? Or was it a negligent killing also because it was of the type 'excessively speedy driving'? which is also an offence but for which he had not been prosecuted? Brennan's test does not ask about the defendant's particular act(s), but only compares those types of act contained in the criminal code (drunk driving, wrong-side-of-the-road driving, speeding) to ask whether they are the same. This is a more

---

[22] *Grady* v. *Corbin*, 495 U.S., at 521.

stringent requirement of double jeopardy than that used by *Blockburger*, which compares the type 'homicide' with the types 'drunk driving' and 'wrong-side-of-the-road driving'; still, this is a type-identity question that is asked by the more stringent, federal, multiple-prosecution test for double jeopardy.

Most American state constitutional double-jeopardy provisions follow this federal pattern.[23] A substantial minority, however, adopt only a one-prong test. This asks whether the various offences charged arise out of a single 'act, transaction, or episode'—in which case, multiple prosecution is barred.[24] This was Justices Brennan and Marshall's preferred test for assessing multiple prosecutions under the federal double-jeopardy clause,[25] but such standard has never been adopted.

I doubt that any state jurisdiction adopts the 'single act, transaction, or episode' test for assessing double-jeopardy claims having to do with multiple punishment. Even where the language of state constitutional, statutory, or common-law double-jeopardy tests is that of 'single act', the test actually employed has two prongs, only one of which asks whether the defendant performed one or several particular acts. Typically, a second prong allows punishment for multiple offences (even if there was but one particular act) so long as that act constituted several *injuries* to distinct *interests*.[26] One might well think, for example, that an act of shooting a sawed-off shotgun at a police officer who was trying to arrest the shooter is but a single act, yet also think that act injures distinct interests protected by statutes forbidding: assault with a deadly weapon; resisting arrest; and possession of a sawed-off shotgun.[27]

2. AN INTRODUCTION TO THE DOCTRINAL, NORMATIVE, AND METAPHYSICAL SCEPTICISMS ABOUT DOUBLE JEOPARDY

This tripartite division of double jeopardy into three separate contexts, together with the differing values justifying somewhat

[23] See Note, 'Twice in Jeopardy', 273 n. 52.

[24] See Stephen Schulhofer, 'Double Jeopardy', in S. Kadish (ed.), *Encyclopedia of Crime and Justice* (New York, 1983), 630.

[25] *Ashe* v. *Swenson*, 397 U.S. 436 (1970).

[26] See e.g. *Commonwealth* v. *Williams*, 344 Pa. Super. 108, 396 A. 2d 31 (1985). See also *Ashe* v. *Swenson*, 397 U.S., at 460 n. 14; and Otto Kirchheimer, 'The Act, the Offence, and Double Jeopardy', *Yale Law Journal*, 58 (1949), 513–44.

[27] So held by the court in *Commonwealth* v. *Williams*, 344 Pa. Super. 108, 496 A. 2d 31 (1985).

different tests in each of the three contexts, gives rise to one form of scepticism about the viability of there being a coherent or univocal double-jeopardy requirement. As we saw in Chapter 1, these attributes of double-jeopardy doctrine and its justification convinced sceptics like Peter Westen that it is hopeless 'to try to formulate a single definition of "same offence"'.[28] Indeed, three definitions of what should be taken to be the 'same offence' may well not satisfy Westen, who has a much more Legal Realist, *ad hoc* approach to double-jeopardy cases in mind: 'the way one approaches all complex double jeopardy cases . . . [is to] identify the separate values of double jeopardy, ascribe to them their respective weights, and then apply the resulting standards to the particular issue in dispute.'[29]

In the two chapters that follow we shall want to examine these doctrinal and normative scepticisms. Given the plausibly different values served by the plausibly separate three bans of double jeopardy, there appears to be much in favour of such scepticism.

The other scepticism which we shall wish to examine is not doctrinal or moral, but metaphysical. Here the worry is not that the values justifying the double-jeopardy prohibition(s) are too diverse to yield a univocal or coherent doctrine; rather, the worry is that acts are so chameleon-like in their nature that there will be no things there to answer to such coherent doctrine, if we had one. As we saw in Chapter 1, John Salmond was sceptical in this way. Also typical is an influential study of double jeopardy, which cites and quotes J. L. Austin for the view that:

any sequence of conduct can be defined as an 'act' or a 'transaction.' An act or transaction test itself determines nothing. . . . Whether any span of conduct is an act depends entirely upon the verb in the question we ask. A man is shaving. How many acts is he doing? Is shaving an act? Yes. Is changing the blade in one's razor an act? Yes. Is applying lather to one's face an act? . . . Yes, yes, yes.[30]

We shall wish to examine each of these scepticisms. Since the scepticisms are different, depending on whether one is individuating *types* of action ('offences') or act-*tokens*, I shall organize my discussion of these scepticisms around each of these two issues in the next two chapters. In Chapter 13, I shall discuss the issue of

---

[28] Westen, 'The Three Faces of Double Jeopardy', 1004 n. 12.
[29] Ibid. 1046.   [30] Note, 'Twice in Jeopardy', 276.

act-type individuation. This sort of case arises where there need be no doubt that an accused has performed but one (particular) act; the doubt is whether the *types* of act that this act instantiates, and which are made criminal by various statutes, are the same or not. *Woodward* v. *United States*[31] illustrates this kind of case. Woodward seemingly performed but one act in checking off the 'no' box on a customs declaration form asking about introduction of large amounts of cash into the United States. The question was whether he could be multiply punished for the crime of making a false statement to a federal agency as well as for the crime of failing to report currency, arguably two types of act instanced by his single act-token.

The second kind of case that I shall examine in Chapter 14 arises where doubt *is* felt about whether the defendant has performed a single act-token or not. Doing the identical *kind* of act several times unquestionably allows for multiple prosecutions and punishments, but only if the accused truly did that kind of act more than once. Usually this involves doubt about the spatio-temporal boundaries of an act. As the Supreme Court of Nebraska observed, 'the ability to determine when a criminal act commenced and when it terminated appears essential in order to determine when a subsequent crime occurs'.[32] Sometimes, however, questions of act-token individuation do not turn on spatio-temporal locations. As we saw in Chapter 4, events are unlike physical objects in that they occupy their spatio-temporal zones *non*-exclusively. Suppose it were criminal to 'hold a convention' in a certain city. Suppose the American Association of Law Schools and the American Society for Political and Legal Philosophy decided to hold their annual conventions at the same hotel at the same time, with many of the same people. If there is one meeting in one room with all people present, holding diverse discussions, is there one criminal act of convention-holding or two?[33] One might well conclude that there are two separate illegal meetings going on, even though they occupy the same spatio-temporal zone.

Unfortunately, double-jeopardy adjudication in the United

---

[31] 469 U.S. 105 (1985).

[32] *State* v. *Williams*, 211 Neb. 650, 319 N.W. 2d 748 (1982).

[33] Jonathan Bennett's sort of example of this kind of counting-of-events problem. See Bennett, *Events and Their Names*, 124.

States has not always separated the type-identity question from the token-identity question. Since that distinction is crucial to the organization of the next two chapters, we should examine it in more detail before proceeding to either individuation question.

In Chapter 13 we will be concerned with the question of when two *types* of act, made criminal by different statutes, are none the less the same for double-jeopardy purposes. Types of act, as we saw in Chapter 4, are complex universals; they consist of a constellation of properties, which are themselves universals. To answer a question of sameness of action-types, accordingly, is to answer the question of the identity and individuation of these sorts of universals.

The question asked by Chapter 14 is different. Here, we will seek the identity conditions of act-*tokens*. To seek to identify and to individuate act-tokens will be to do something different from seeking to identify and to individuate universals; rather, it is to seek the identity conditions of a kind of particular, an event-particular.

Once one sees this distinction, it is blindingly obvious. If I practise the violin every morning for a week, there are two very different answers to the question 'How many acts did I do?' For there are two very different questions of identity and individuation that could be asked by such a sentence: 'How many *kinds* of acts did I do?' (Answer: one.) And 'How many *particular* acts did I do?' (Answer: seven.)

That the words and sentences of ordinary, idiomatic English are so systematically ambiguous in whether they refer to act-types or act-tokens[34] helps to explain why courts often confuse the two sorts of identity question. Consider Justice Powell's majority opinion in *Brown* v. *Ohio*.[35] In *Brown* the defendant

---

[34] It is not of course just the English nouns that refer to actions that are systematically ambiguous in this way. Consider the noun 'word': 'There will probably be about twenty *the's* on a page, and of course they count as twenty words. In another sense of the word "word," however, there is but one word "the" in the English language; and it is impossible that this word should lie visibly on a page or be heard in any voice, for the reason that it is not a single thing or single event. . . . Such a definitely significant Form I propose to term a Type.' Peirce, *Collected Works*, iv. 537. For numerous other examples of the type–token ambiguity in ordinary usage of the nouns of English, see N. Wolterstorff, 'On the Nature of Universals', in M. Loux (ed.), *Universals and Particulars* (Garden City, NY, 1970), 162–4.

[35] *Brown* v. *Ohio*, 432 U.S. 161 (1977).

stole a car in one city in Ohio and was caught driving the car nine days later in another city in Ohio. He was first prosecuted, pleaded guilty, and sentenced in the latter town for the offence of joy-riding. He was then prosecuted in the first town for the offence of car theft.

The Ohio courts rejected the defendant's plea of double jeopardy as a bar to the second prosecution. In doing so the Ohio courts nicely kept separate the questions of type versus token-identity. The Ohio Court of Appeals found there to be (what I shall call in Chapter 13) a partial type-identity between the offences of joy-riding and car theft:

Every element of the crime of operating a motor vehicle without the consent of the owner is also an element of the crime of auto theft. . . . The crime of operating a motor vehicle without the consent of the owner is a lesser included offence of auto theft . . . [and therefore] for purposes of double jeopardy the two prosecutions involve the same statutory offence.[36]

None the less that court held there to be two different act-tokens of that same act-type, and thus disallowed defendant's double-jeopardy claim:

The two prosecutions are based on two separate acts of the appellant, one which occurred on November 29th and one which occurred on December 8th. Since appellant has not shown that both prosecutions are based on the same act or transaction, the second prosecution is not barred by the double jeopardy clause.[37]

Despite the admirable clarity with which the Ohio courts distinguished the token-identity question from the type-identity question, Justice Powell managed less than complete clarity about the distinction. Beginning with the erroneous supposition that 'the principal question in this case is whether auto theft and joy-riding, a greater and lesser included offence under Ohio law, constitute the "same offence" under the Double Jeopardy Clause',[38] Justice Powell's opinion spends almost all of its time berating the obvious: so construed, these are of course the 'same offence' for

[36] The relevant portions of the opinion of the Ohio Court of Appeals are quoted in the US Supreme Court's opinion, 432 U.S., at 163–4.
[37] 432 U.S., at 164.   [38] *Brown* v. *Ohio*, 432 U.S., at 164.

double-jeopardy purposes. Powell's only notice of the actual basis for the Ohio courts' holding was to say that:

> After correctly holding that joy-riding and auto theft are the same offence under the Double Jeopardy Clause, the Ohio Court of Appeals nevertheless concluded that Nathaniel Brown could be convicted of both crimes because the charges against him focused on different parts of his 9-day joyride. We hold a different view. The Double Jeopardy Clause is not such a fragile guarantee that prosecutors can avoid its limitations by the simple expedient of dividing a single crime into a series of temporal or spatial units. The applicable Ohio statutes, as written and construed in this case, make the theft and operation of a single car a single offence. Although the Wickliffe and East Cleveland authorities may have had different perspectives on Brown's offence, it was still only one offence under Ohio law. Accordingly, the specification of different dates in the two charges on which Brown was convicted cannot alter the fact that he was placed twice in jeopardy for the same offence in violation of the Fifth and Fourteenth Amendments.[39]

This misses entirely the question—of act-token identity—so nicely distinguished by the Ohio courts. That 'it was still only one offence under Ohio law' is completely irrelevant to the actual issue in the case, which was whether there were one or two instances of such offence performed by the defendant. On the other hand, the facts Powell finds irrelevant—the different dates and places of defendant's taking versus his driving—are (as we shall see) highly relevant to the actual issue in the case, the issue of act-token identity. Bringing forward such facts is not, as Powell supposed, an illegitimate attempt at 'dividing a single *crime* into a series of temporal or spatial units'; it is rather an attempt to individuate *instances* of such crime by their different spatio-temporal locations. However one comes out on such a question of individuation of act-tokens, one first has to see the question in order to answer it.

It is very clear that the law on double jeopardy does not generally elide the token-identity question into the type-identity question in this way. Federal double-jeopardy doctrine, followed by state constitutional provisions too, plainly separates the token-identity question from the type-identity question. The former question is commonly dubbed the 'unit of prosecution' or 'unit of

---

[39] *Brown* v. *Ohio*, 169–70 (citations omitted).

crime' question under both the federal and the state doctrines. This Justice Blackmun recognized in his dissent in *Brown*: 'Only if the [Double Jeopardy] clause requires the Ohio courts to hold that the allowable unit of prosecution is the [defendant's entire] course of conduct would the Court's result here be correct.'[40] As Blackmun concluded, the unit of prosecution is not so broad because the act-tokens done by defendant were 'separate and distinct'.[41]

If it were not the case that the defendant's act-tokens must be identical (in addition to the offences for which he is charged being the same), then boringly repetitive criminals would have valid double-jeopardy objections to their successive prosecution and punishment. If Smith assaults (or robs, or steals from, or burglarizes, or rapes, etc.) Jones on Wednesday, and does so again on Thursday and again on Friday, then on Powell's seeming view of double jeopardy there can be but one prosecution or punishment; for these are even clearer cases (than in *Brown*) where the offences (types of act) are the same, and for Powell no token-identity question is to be asked. Allowing the type-identity question to dominate over the token-identity question is this way would severely flout double jeopardy's purpose of securing each defendant a level of punishment proportionate to his wrongdoing.

Not surprisingly, therefore, despite the seemingly univocal nature of the *Blockburger* plus *Grady* double-jeopardy test, what is in fact being tested is not only the question of whether a defendant did one or more offences but also the question of whether that defendant did each of those offences more than once. Thus, even if the offences of distributing heroin and possessing heroin are the same for double-jeopardy purposes, where the particular acts of distributing and possessing did not occur 'at the same time, in the same place, and with the involvement of the same participants', defendants may be multiply prosecuted and punished.[42] Likewise, even if the offence of possessing counterfeit money is the same for double-jeopardy purposes as the offence of transferring counterfeit money, as long as it is not the same counterfeit money, then the acts of possession and transfer 'were clearly separate and distinct acts occurring at different times for

---

[40] Ibid. 172.   [41] Ibid. 171.
[42] *United States* v. *Rodriguez-Ramirez*, 777 F. 2d 454, 457–8 (9th Cir. 1985).

which separate punishments were applicable'.[43] Similarly, conceding that it is just one offence to make a false statement material to the lawfulness of the sale of firearms, to make thirty-seven such statements on thirty-seven separate forms is to make one liable for thirty-seven separate prosecutions and punishments—for 'each firearms transaction form was a proper unit of prosecution'.[44] Similarly, conceding that it is just one offence to possess cocaine, where 'for each possession the facts as to location, quantity, and manner of possession are different', the defendant may be separately punished for each such distinct possession-token.[45] Finally, while of course contempt of court is but one offence, 'a witness refusing to testify at each of two trials can twice be adjudged in criminal contempt and separately punished' because these are two distinct contempt-tokens.[46]

The upshot is that for double jeopardy to bar a second prosecution or punishment, the defendant must show *both* that his offence-type was the same in both prosecutions or punishments, *and* that his offending act-token was the same in both prosecutions or punishments. This is what I take to be meant by the often-repeated mantra that 'offences must be the same in law and in fact.[47] before double jeopardy bars multiple prosecutions or punishments. The mantra should not be taken to be, as some commentators have done,[48] a particularly strict sort of test for when offences are the same; it is, rather, a way of noting that there must be two distinct questions of identity that *any* double jeopardy test must ask and answer.

---

[43] *United States* v. *Shear*, 825 F. 2d 783, 787 (4th Cir. 1987).

[44] *United States* v. *Hawkins*, 794 F. 2d 589, 590 (11th Cir. 1986).

[45] *United States* v. *Rich*, 795 F. 2d 680, 683 (8th Cir. 1986).

[46] *United States* v. *Coachman*, 752 F. 2d 685, 691 (D.C. Cir. 1985). The state law on the separateness of the 'unit of prosecution' question from the type-identity question is the same. See the summary of state cases in C. Dix and M. Sharlot (eds.), *Criminal Law, Cases and Materials*, 3rd edn. (St Paul, Minn., 1987), 259–60.

[47] e.g. *People* v. *Helbing*, 61 Cal. 620 (1882); *Commonwealth* v. *Roby*, 12 Pick. (Mass.) 496 (1832); *Winn* v. *State*, 82 Wisc. 571, 52 N.W. 775 (1892).

[48] See Note, 'Twice in Jeopardy', 268–9.

13

LEGAL, MORAL, AND METAPHYSICAL NOTIONS OF 'SAMENESS' OF ACTION-TYPES

1. THREE APPROACHES TO ACT-TYPE INDIVIDUATION

As we have seen, the American federal double-jeopardy provision asks two sorts of type-identity questions with respect to the ban on multiple *prosecutions* for the same offence. The first is the *Blockburger* question,[1] where one asks if the statutorily prohibited type of act for which the accused is now being prosecuted is the same as the statutorily prohibited type of act for which he has already been prosecuted. If the accused was previously prosecuted for manslaughter, and is now being prosecuted for murder, the question is whether these two types of act are the same.

The second type-identity question is that asked by *Vitale*[2] and *Grady*:[3] does the statutorily prohibited type of act for which the accused is now being prosecuted include as one of its elements (or as one of the ways of proving one of its elements) an act-type for which the accused has already been prosecuted? If the accused was previously prosecuted for shooting at an occupied building, and if he is now being prosecuted for murder on the theory that a death occurred during the commission of a felony, namely, possession of a firearm by a convicted felon,[4] the question is whether, as types of action, shooting at an occupied building, and possession of a firearm by a convicted felon, are the same.

It is important to see that these are both type-identity questions; in particular, the second is not a question of whether the

---

[1] *Blockburger* v. *United States*, 284 U.S. 299 (1932).
[2] *Illinois* v. *Vitale*, 447 U.S. 410 (1980).
[3] *Grady* v. *Corbin*, 495 U.S. 508 (1990).
[4] The facts of *Davis* v. *Herring*, 800 F. 2d 513 (5th Cir. 1986).

accused performed one particular act or several. Moreover, under both double-jeopardy tests, the types of action to be compared for their identity are fixed by the statutory descriptions used in the special part of the criminal law. The only difference in the two type-identity questions lies in the different statutory sources of the descriptions: for the *Blockburger* question, one uses only the statutory descriptions of the act-types for which the accused was or is now being prosecuted, whereas for the *Vitale–Grady* question, one uses any statutory description of an act-type which the prosecutor will use to prove that the accused did an act of the type for which he is now being prosecuted. All types of action considered under both questions are fixed by statutory descriptions, but the *Grady* question is none the less finer-grained because of its more particular enquiry into statutory descriptions other than the one defining the offence of which the accused is charged in the second prosecution.

Although legally which of these two questions one asks can make a good deal of difference, the metaphysics on which they rely is the same; for both presuppose we have some way of individuating types of action. I accordingly will intermingle both sorts of cases in my discussion of the type-identity question. I shall also intermingle both state statutory decisions, and state common-law decisions under the earlier-mentioned merger doctrine, even though on their face such doctrines explicitly ask only a same act-token question; for as discussed earlier, such state law decisions implicitly ask a type-identity question as well when they examine whether an accused has twice been punished for 'the same act'.

There are easy cases of type-identity for double-jeopardy purposes. If Jones is prosecuted for murdering Smith and either acquitted or convicted, and if Jones is again prosecuted for the murder of Smith, it is uncontroversially correct to say that the second prosecution is barred. For not only is this the same particular act, but the type of act for which Jones is again prosecuted—murder—is obviously the same as that for which he was earlier prosecuted.[5]

---

[5] As the US Supreme Court said in its first major double-jeopardy opinion, *Ex parte* Lange, 85 U.S. (18 Wall.) 163, 168 (1874): 'no man can be twice lawfully punished for the same offence. And though there have been nice questions in the application of the rule to cases in which the act charged was such as to come

Although such easy cases rarely find their way into appellate decisions, it is important to separate two dimensions in which they are easy. Recall that a statutorily prohibited type of complex action consists of elements. Take common-law burglary. The overall act-type, burglary, consists of five elements: (1) a breaking and an entering; (2) of a dwelling house; (3) belonging to someone other than the accused; (4) at night; (5) with the intent to commit a felony therein. If Jones is prosecuted twice for burglary, it is easy to say that the act-types are identical, first, because the two offences share all the same elements and, secondly, because those elements themselves are identically described.

Suppose that Jones, having already been prosecuted for burglary, is then prosecuted: (1) for breaking and entering the dwelling house of another; or (2) for breaking and entering an unoccupied structure belonging to another at night with the intent to commit a felony therein. In neither case can we as easily say that the types of act for which Jones is later being prosecuted are identical to the type of act for which he was first prosecuted, burglary. For in (1), only three of the five elements of burglary are replicated in the breaking and entering offence; and in (2), while there are five elements (to what might be second- or third-degree burglary), one of these elements is not identical to the other (as types of act, entering a dwelling house is not identical to entering an unoccupied structure).

If the sameness in the prohibition against twice being put in jeopardy for the 'same offence' meant full identity of act-types, these last two cases would not be easy applications of the prohibition, for there plainly is no full identity of act-types in either of them. Double-jeopardy 'sameness' has never meant full identity of act-types, however, nor (as we shall see) should it be so construed. Some kind of partial identity of act-types has always sufficed for double-jeopardy purposes.[6]

within the definition of more than one statutory offence . . . there has never been any doubt of its entire and complete protection of the party when a second punishment is proposed in the same court, on the same facts, for the same statutory offence.'

[6] As the US Supreme Court has also noted. See *Brown* v. *Ohio*, 432 U.S. 161, 164 (1977) ('It has long been understood that separate statutory crimes need not be identical . . . in order to be the same within the meaning of the constitutional prohibition.')

To make sense of such partial identity two problems must be overcome: (1) How many elements of two offences must be the same for those offences themselves to be the same for double-jeopardy purposes? (2) When are two elements of different offences the same even though their statutory descriptions differ in some respect? These two problems correspond to the two dimensions of easiness earlier distinguished. I shall call the first the problem of degree vagueness, and the second, the problem of language-independent individuation.

There are three approaches to these two problems. The first is what I shall call the sceptical approach. On this view there is no answer to the 'same offence' question that is not arbitrary once one leaves the only easy case there is of sameness, namely, where two offences are defined in terms of identical elements that are identically described. The exact statutory language used to describe offences governs, on this view, so that any difference in description—either in numbers of elements, or in how each element is described—betokens a non-'sameness' of offences.

The second approach to legal 'sameness of offences' is unabashedly and unrestrictedly metaphysical. On this view there is an answer to the problems of degree vagueness and language-independent individuation, an answer that lies in the nature of the act-types themselves. Action-tokens, as we have seen, are instances of properties, and an act-type is simply a constellation of such properties. Properties, like relations, are universals. Thus, however one individuates universals is how one will individuate act-types. On the approach here considered, universals have a nature that is independent of the language used to refer to them. They exist no matter how we describe them, and, indeed, whether we refer to them at all.

The third approach to the two problems I shall call the moral approach. It may or may not be metaphysical, depending on one's view of morality. I shall defer a description of this last approach until after we have explored the first two.

I shall first explore the unrestrictedly metaphysical approach, and then examine the sceptical approach's response. To get started we first need to understand the notions of full and of partial identity. Full identity ('identity' hereinafter) between act-types is the simpler notion, so let me begin with it. It is commonly thought that 'the identity belonging to a universal . . . is a different sort of thing from what is involved in specifying wherein the identity of an individual consists'.[7] Whether this is so depends on where one comes out on Leibniz's two Principles of Identity, the Indiscernibility of Identicals, and the Identity of Indiscernibles. The first principle, often called Leibniz's Law, asserts that if any two things x and y are in reality one thing only, then any property that x has y must have as well, and vice versa. More simply: identicals must share all the same properties. The second principle asserts that if any two things do share all the same properties, then they are in reality one and the same thing. Put in other language: if two potentially different things are truly qualitatively identical, then they are numerically identical as well.

If one takes either the first principle alone, or both together, to define identity, and if (as seems plausible) properties can themselves have properties,[8] then the notion identity for properties need be no different than the notion of identity for particulars. Venus is the Morning Star only if they have all the same properties; spinning is the same as rotating only if they have all the same properties.[9]

This is fine as far as it goes, but it doesn't go very far. One might shore it up a bit by focusing on what it is about properties that gives us reason to think that they exist. As I mentioned in Chapter 4, the prevailing scheme for deciding whether anything exists is to ask whether such a thing has a necessary place in our best explanation of everything else we believe to be true. To ask whether numbers, zoological species, moral qualities, or mental states exist is to hold constant the rest of our ontological commitments and to ask whether the items in question do any explanatory work for us. To argue for the existence of properties in this way is to show where language-independent properties seem to do necessary explanatory work for us as we explain the rest of the world to ourselves. If such an argument is successful, then we have all the reason we need (or can have, for that matter) to believe in the independent existence of such properties.

[7] M. K. Munitz (ed.), *Identity and Individuation* (New York, 1971), p. iv.
[8] On the question of whether properties can themselves have properties (and thus be capable of being qualitatively identical to one another), see Armstrong, *A Theory of Universals*, 133–47.
[9] Baruch Brody for one takes Leibniz's second notion (of the Identity of Indiscernibles) to define identity, and has no problem accommodating the identity of universals within this general view. Brody, *Identity and Essence*, 60–5.

To accept this mode of arguing for the language-independent existence of properties is also to accept a certain mode of arguing for the correct individuation of such properties. As David Armstrong shows in detail, once one accepts that the existence of a property depends on its explanatory work (and, thus, upon that property's causal powers), then: (1) one and the same property must bestow the very same causal powers upon every particular of which it is a property; and (2) every different property must bestow different causal powers on each particular possessing each such different property.[10] True, as Armstrong notes, we can judge identity of causal powers only in terms of the identity of effects, and we can judge the identity of effects only if we can already individuate properties, so that these identity conditions are as circular as are the more general Leibnizian criteria.[11] But they are not viciously circular, because the properties by virtue of which we individuate the effects will not be the properties which we are seeking to individuate by their effects. As long as we know how to individuate some properties, we can individuate others. This is no more than an application of our non-foundationalist epistemology generally, where we hold unquestioned part of what we know (here, about the individuation of properties) in order to test other of our beliefs.

With regard to the more difficult notion of the partial identity of properties, a natural way to proceed would be to say that one property is partially identical to another when, but only when, they share one or more properties. As David Armstrong shows in detail,[12] this way of proceeding has serious disadvantages. Among other things: one might think that the property of redness is partly identical to blueness by virtue of sharing the property of being coloured. Yet: (1) being coloured is not a property of either redness or blueness so much as it is a property of the same particulars as those of which red and blue are properties (as well as other particulars); (2) and, even if being coloured is a common property of being red and being blue, 'the property in which they are supposed to resemble each other [being coloured] is the very thing in which they differ', and 'this seems impossible'.[13]

---

[10] Armstrong, *A Theory of Universals*, 43–7.   [11] Ibid. 46–7.
[12] Ibid. 105–8.   [13] Ibid. 106.

It is preferable to think of properties like red and blue as being related to the property of being coloured by the 'mereological' notions of part and whole. This hints at the direction in which to move in explicating the idea of partial identity of action-types and other complex properties.

There are in fact two sorts of partial identity (these will correspond to the two dimensions of easiness of double-jeopardy cases earlier put). The first and simplest sort is the partial identity existing between a conjunctive property and one or more of the properties that are its conjuncts. A conjunctive property is, as the name suggests, a property whose nature consists of the conjunction of other properties.[14] The property of bachelorhood is a conjunction of the properties of being male, being a person, and being unmarried. Likewise, the property of being a burglary might be thought to be a conjunction of the properties of being a breaking and an entering, being an act done at night, being an act done to the (legal) property of another, etc.

It is important to be clear here as to the difference between *properties* and *classes* (where the latter term is understood in a set-theoretical way). The class of things male, the class of things unmarried, and the class of things that are persons each include as a part the class of things that are bachelors. Such set inclusion of the relevant particulars is not what is meant by the part–whole relationship of properties. Rather, in the example given, the conjunctive property is bachelorhood, part of which is the property of being male, part of which is the property of being unmarried, and part of which is the property of being a person. A conjunctive property, in other words, includes in its extension a set of particulars that is usually smaller, not larger, than the sets of particulars included in the extensions of each of the conjunct properties.[15]

Since the notion of partial identity most clearly involves the notion of part to whole, the conceptually clearest sort of partial identity is the sort involving conjunctive properties—for it is here that we can most easily see that one property is *part* of another

---

[14] On conjunctive properties, see ibid. 30–42.

[15] The general difference between properties and sets is usually put in terms of the extensionality of the latter but not the former: two sets are identical when they share all the same members; two properties need not be identical just because they share all the same instances. See Wolterstorff, 'On the Nature of Universals', 164–6. Brody, *Identity and Essence*, 60–5.

(whole) property. Yet the second not-easy double-jeopardy case is not of this sort. Even though being a burglary might be a conjunctive property, in the second case (of breaking and entering an unoccupied structure) we cannot simply say that a clearly separable part—one of the conjuncts—is or is not absent. There seems to be some relation between being a dwelling house, and being an unoccupied structure, but it is not clear that the relation is part to whole. The relation might more intuitively be thought of as a kind of *overlap*, where part of the one is also part of the other.[16] That is, part of what it is for a thing to be a dwelling house is also part of what it is for a thing to be an unoccupied structure. Then two complex properties share this common part, much as red and blue share (as part of what makes them the properties they are) the part coloured. Overlap of common part(s) constitutes a second sort of partial identity.

One might object to the unpacking of double jeopardy's 'sameness' in terms of either kind of partial identity on the ground that identity is not a scalar affair (that is, that it is not a relation that admits of degrees). 'Venus either is identical to the Evening Star, or it isn't, but it makes no sense to say it is partly (or "sort of") identical to the Evening Star' might be the slogan for this objection.

The proper response is to deny the incoherence of partial (numerical) identity. Even for particulars like the planet Venus, there are partial identities. Venus is partially identical to the solar system; Venus is also partially identical with the part of Venus now visible to us on earth, even more fully identical to its entire surface area. Every (whole) particular is partially identical to one of its parts. The same is true for universals such as properties (and for those kinds of property that are act-types). Property P is partially identical to the conjunctive property (P and R); property S is partially identical to property T when they overlap, that is, when part of S is part of T.

So there is no metaphysical objection to be made here stemming from the allegedly non-scalar nature of numerical identity. The notions of parts to wholes, or of overlap of parts, can make sense of the idea of partial identity of act-types and other complex properties.

[16] See Armstrong, *A Theory of Universals*, 38, 123.

This, then, is a sketch of the metaphysical approach to the individuation of act-types for double-jeopardy purposes. The sceptic that I earlier described would of course deny the possibility of this metaphysical enterprise succeeding. The sceptic's view is that there is no real basis for individuation of act-types or other properties. Universals, on this nominalist view of the world, lack a sufficiently robust existence to have identity conditions that are in any sense grounded in reality. Rather, to the extent universals can be said to exist at all, it is only by virtue of some predicate of a language creating them. 'For every predicate a property, and for every property a predicate' might be the slogan for this view. Such a view spawns a purely language-relative, syntactic mode of individuating universals: if a different form of symbols is used to refer to a universal, it is a different universal, and if not, not.

As a general view about all universals, this nominalist view is surely false. There are four situations in which we need language-independent universals to explain the world as we explain it.[17] (1) There are predicates that do not refer to any property, e.g. 'accelerates through the speed of light' or 'is a winged horse' or 'is raining or is not raining' or 'is omitting to save the drowning man'.[18] These are not instances of meaningless predicates, so they cannot be written off in that way; yet they do seem to be cases where a predicate does not correspond to any existing property. (2) It seems very intuitive to think, conversely, that there are many properties of things in the world that we have not yet discovered and thus that we have not yet named. Since we haven't yet discovered them, I can't of course give contemporary examples. The history of science, however, is littered with examples (positive ionization potential, for example) of properties that were unknown to some generation of language-users and thus had no name in their language. (3) There seem to be many examples where the human community has stumbled on to the existence of some one property in different ways but was ignorant of the sameness for sufficient time to create different predicates to name the same property. The classical example with particulars are the names the 'Evening Star', the 'Morning Star', and

'Venus', which pre-Babylonian astronomers thought named three different particulars but which in reality name only one. An analogous example for properties is 'temperature' and 'mean kinetic energy of molecules', two different predicates referring to the same property. (4) Finally, there are also many examples where some linguistic community has ventured a guess (by creating one predicate) as to there being one distinct property when in fact the predicate names two or more distinct properties. Examples are 'jade', which in fact names two quite different sorts of stone; 'barbarian', as the ancient Greeks used the word (to name all non-Greeks);[19] perhaps 'multiple sclerosis', which may name several different kinds of disease.

These four kinds of counter-example do not of course decisively prove that properties must exist independently of the predicates in any language that might be used to name them. Such examples suggest the way that a more complete 'inference to the best explanation' argument in that direction would go, however. The best explanation of these four features of our experience is that there really are language-independent universals.[20]

The sceptical approach to double jeopardy should thus abandon a general nominalist scepticism about universals. The sceptic might none the less maintain his scepticism (about there being any language-independent way of individuating act-types) in the following way. Granted all that I say about the language-independent existence of *certain* properties, that has no impact upon the *sorts of properties needed by the statutes making up the special part of the criminal law.* For the causal powers that give reason to believe in the existence of certain properties, as well as a mode for their individuation, attach only to natural kinds. And, by my own earlier admission in Chapter 7, almost all the act-types used in the special part of the criminal law are not natural kinds but are what I called nominal kinds. More concretely: the act-type

---

[19] Plato's example: 'most people in this part of the world . . . separate the Greeks from all other nations making them a class apart; thus they group all other nations together as a class, ignoring the fact that it is an indeterminate class made up of people who have no intercourse with each other and speak different languages. Lumping all this non-Greek residue together, they think it must constitute one real class because they have a common name "barbarian" to attach to it.' *Politicus*, 262d.

[20] For examination of more traditional arguments for realism about universals, see Alan Donagan, 'Universals and Metaphysical Realism', in M. Loux (ed.), *Universals and Particulars* (Garden City, NY, 1970).

that is common-law (or now statutory first-degree) burglary is not plausibly thought to be a natural kind, for it is unlikely that such an act-type does any explanatory work for us in science. No true generalizations about the causal connections that may exist between events seems to require the act-type burglary. So how can it have a language-independent existence, even if other properties (including even those out of which burglary is constructed) do have such a language-independent existence?

The sceptic should be conceded the latter point, for almost all the properties explicitly created by criminal statutes are not, as such, natural kinds. It is thus a mistake to think that one can take each formally defined element of an offence such as burglary and treat it as if it were a *part* of burglary. For neither 'act done at night' nor 'burglary' are themselves natural kinds. The properties explicitly referred to are creatures of statutory creation, and as such 'they are mere shadows cast on the world by our predicates'.[21] In addition, even if there were a language-independent basis for individuating such act-types and their parts, that would not solve the problem of degree vagueness; for left unspecified would still be *how many* parts two offences must share in order for their partial identity to be considered an overall sameness of offence.

The problem for the sceptic is that his own approach does not begin to capture our practices with regard to double-jeopardy adjudication. Recall that, on the sceptic's view, every differently worded statutory description would refer to a different act-type, so that the only prosecutions or punishments that would be truly duplicative would be those under identically worded statutes. An accused who is once prosecuted for burglary could not again be prosecuted for the very same act as it constituted the offence of burglary, but she could be prosecuted for breaking and entering, daytime burglary, unoccupied-dwelling burglary, and criminal trespass, if the jurisdiction has statutes defining those offences in the ways earlier suggested.

This hardly squares with how we both do and should adjudicate double-jeopardy cases. Lesser included offences are thought to be easy applications of the bar against twice prosecuting or

---

[21] David Armstrong's conclusion about the properties of ordinary discourse, unlinked as many of them are to any real universals. Armstrong, *A Theory of Universals*, 18.

punishing for the 'same offence'. One prosecuted for burglary cannot be again prosecuted for breaking and entering, because the latter is a lesser included offence of the former. One 'joy-riding'—taking, operating, or keeping a motor vehicle without the consent of its owner—cannot again be prosecuted for car theft—taking a motor vehicle without the consent of its owner with the intent to deprive the owner permanently of it—because the former is a lesser included offence of the latter.[22] One prosecuted for kidnapping—unlawful confinement and asportation of another—may not later be prosecuted for false imprisonment—unlawful confinement of another—because the latter is a lesser included offence of the former.

Partial identity of act-types seems a very intuitive way of capturing these clear cases of 'same offence' under double jeopardy, but the sceptical view here considered would decide all such lesser included offence cases the other way.

Secondly, such a sceptical view makes a hash out of there being any *constitutional* basis for double-jeopardy review; for if the legislature chose to use separate predicates to describe separate offences, on the sceptical view those are by necessity different act-types and thus different offences. The legislature, in other words, would have the final word on when two offences were or were not the same for double-jeopardy purposes, for there would be no basis on which to judge its creation of two or more distinct offences. This would be like saying that legislative determinations of when a punishment is cruel or unusual are conclusive of when a punishment is cruel and unusual for Eighth Amendment purposes. In both cases this would just be a way of describing the absence of constitutional review.

Finally, even where double jeopardy is exclusively in the hands of the legislature—as it is for multiple-punishment purposes, although not for multiple-prosecution purposes under the federal scheme in America—it serves the legislature poorly to cast upon it the burden of considering all other possible punishments under all other possible statutes when it is setting the punishment for some class of acts covered by any particular statute. Yet this is the burden thrust upon it if only identity of description yields a judgement of 'same offence' in order to preclude multiple punish-

---

[22] The charges in *Brown* v. *Ohio*, 431 U.S. 161 (1977).

ment. For then the legislature, in setting the range of permissible punishments for, say, burglary, must also take into account the punishments that acts of burglary may receive if they are also acts of breaking and entering, acts of trespass, etc. How much easier to draft a criminal code if one can assume that the courts will preclude multiple punishments for offences that are, in some sense, essentially the same no matter how described. For then the legislature, in an effort to prohibit all forms of the conduct which it wishes to prohibit, can create many offences that plainly overlap in their coverage, without worrying that acts that fall within the overlapping coverage will be far too seriously punished.

Plainly needed is some third approach, one that fits our legal practices, is metaphysically possible, and is normatively desirable. Enter the moral approach to double-jeopardy individuation. The problem with the metaphysical approach, as we have seen, did not lie with the notions of full or partial identity of properties. We can make good sense of both such notions. Nor was the problem that nowhere are there language-independent properties. Rather, the problem was that the act-types used in criminal statutes did not seem very plausibly to be the kinds of property that have a language-independent existence.

The basic insight of the moral approach is to see that there are real properties about which we can seek the earlier-described metaphysical comparisons, but they are not the nominal properties each criminal statute creates. Rather, act-type descriptions in criminal statutes should be seen as referring to what I shall call morally salient act-types.

Suppose for the moment that morality consists at least in part of a set of norms either directing us to do or not to do certain act-types, or permitting us to do or not to do them. For criminal law, the most often relevant norms are those obligating us not to do certain act-types. It is these morally obligatory act-types to which statutory descriptions defining crimes like larceny, rape, arson, murder, etc. should be seen to refer. Such act-types have as real (as language-independent) an existence as do the norms of morality that make them morally salient.

To see how double-jeopardy adjudication under this approach works, consider instances of both full and partial identity of morally salient act-types. I take the crimes of first-degree murder, second-degree murder, voluntary manslaughter, involuntary manslaughter, and negligent homicide all to prohibit the same morally salient act-type, namely, the killing of another human being. Given that the same morally wrong act-type is referred to by each such statute, then no matter how such act-types may be described in the differing statutes, the offences are the same.

Now consider partial identity. As we have seen, there are two kinds of partial identity, the part–whole kind, and the overlapping kind. The second occurs when some morally salient act-type is part of several different (and more complex) statutorily created act-types. If killing another human being is such an act-type, then: all crimes of homicide, assassination of the President, killing of a policeman in the official performance of his duties, one form of deprivation of civil rights under colour of law (a federal crime under the Civil Rights Acts), killing of a prison official during the official performance of his duties, are all partially identical to one another in that they overlap, that is, they share a common part, the act-type of killing another human being. No matter how this common part is described in these various criminal statutes, we know that it is *one* part these more complex act-types share because we know that this is one type of act that morality forbids. That should be enough to preclude double jeopardy.

Now consider the part–whole kind of partial identity. Many *mala in se* crimes may plausibly be thought to name the same act-types that morality too forbids, without addition or subtraction. The various degrees of homicide, rape, and larceny, for example, may each name three of the distinct act-types contained in morality's distinct prohibitions. The crime of robbery may plausibly be added to this list, because it too seems to be an independent moral wrong. Yet robbery is also complex, since it consists of two act-types that are themselves morally wrong, when considered in isolation from one another: assault and larceny. This moral knowledge allows us to see that robbery and assault, and robbery and larceny, are partially identical act-types; for larceny is a whole moral wrong that is part of another whole moral wrong, robbery, and assault is likewise a whole moral wrong that is part of another whole moral wrong, robbery.

Sometimes the act-types contained in morality's prohibitions are complex without containing two independent moral wrongs. Burglary may be of this kind. It is an independent moral wrong to trespass (or 'enter') on another's property, and perhaps it is another independent moral wrong to use force against another's property (as in 'breaking'). But acts done at night, or acts done in another's dwelling, are not themselves independent moral wrongs. Yet one might think that burglary—breaking and entering into the dwelling house of another at night—is itself an independent moral wrong. If so, then burglary and trespass, and burglary and breaking and entering, are partially identical pairs of morally salient act-types. The latter of each pair is part of the former, which is true no matter how such act-types are described.

Contrast all of these cases of partial identity of morally salient act-types with a situation where there is no true partial identity, although there might appear to be one if one focuses on the nominal kinds of act-types created by statutes rather than on the act-types that are moral salient. Suppose there are four criminal statutes that variously prohibit: driving without a licence; driving while intoxicated; driving a vehicle without adequate smog devices; and driving an overweight vehicle. If one considers only the nominal act-types explicitly created by each of these statutes, there is a partial identity; each complex act-type overlaps with the other in that each shares a common part, namely, the act-type driving. Yet driving as such is not a moral wrong, so that on the moral approach this is not a case of partial identity. On the moral approach, these four statutes do not refer to morally salient act-types that share a common part that is itself morally salient. There is thus no identity between such offences on the moral approach (which intuitively seems the right result for double-jeopardy purposes).

This last example raises the question of statutorily created act-types, neither the whole of which, nor part(s) of which, corresponds with an act-type contained in morality's prohibitions. What, in other words, about crimes that are merely *mala prohibita*? Here at least must not the sceptic be conceded that there is no reality to the act-types (or to the nature of their individuation) except that created by the statutory predicates themselves? The truth still lies away from the sceptic. *Mala prohibita* crimes are created to prevent some bad state of affairs from occurring. Driving a motor vehicle without a driver's licence, for example, serves to prevent loss of life and other injuries to persons or property by keeping the unskilled, the ignorant, and the incapacitated off the roads. The morally ideal act-type here is the one that maximally achieves this intrinsically valuable state of affairs. Finding such a morally ideal act-type may be no easy matter, but that act-type is not simply a creation of the language of the statute. What ideally should constitute a 'motor vehicle', a 'driving', or a 'licence' may diverge considerably from what those terms ordinarily denote. There are thus still morally dictated act-types behind *mala prohibita* crimes even if such act-types are the product of instrumental calculation (rather than of the intrinsically wrongful nature of such act-types).

Seeing that we use instrumental wrongs as well as intrinsic wrongs when we make double-jeopardy comparisons allows us to complete the analysis of situations such as the four driving statutes above imagined. For each such statute, we must ask: what bad state of affairs does it seek to eradicate by its prohibition, given that the statute does not refer to intrinsically wrongful action? Such state of affairs for the driving without a licence statute will have to do with unsafe driving by those too unskilled, too ignorant, or too incapacitated to drive safely. Such states of affairs for the other statutes will have to do with safety (for the drunk-driving statute), environmental degradation (for the smog device statute), and road surface degradation (for the weight-limiting statute). However one reconstructs the maximally efficacious act-types that serve each such statutory purpose, driving as such will not be the instrumentally wrong act-type for any of them. Accordingly, driving is neither an intrinsically wrong act, nor is it instrumentally wrong; it cannot then be a *part* shared by each of the act-types referred to by these statutes making them partially identical, and thus 'the same', for double-jeopardy purposes.

The moral approach not only solves the problem of language-independent individuation, but it also points the way towards solving the problem of degree vagueness. As we have seen, partial identity of both the part–whole, and of the overlap kind, is a scalar relation, that is, a matter of degree. The conjunctive property (P and S), for example, is *more identical* to the conjunctive property (P and S and R) than is the property P. Similarly, the conjunctive property (P and R and S) is *more identical* to the conjunctive property (P and R and T) than is the property (P and Q and V). The degree vagueness worry is that even after one accepts that there are some language-independent properties that do bear the relation of partial identity, to use that notion for double-jeopardy purposes requires a court to take a position on *how identical* one act-type has to be to another before the act-types should be said to be 'the same'.

Despite the scalar nature of partial identity, there are two limiting cases: where full identity obtains and where no partial identity obtains. We have seen that the first of these extremes is too restrictive for double-jeopardy purposes, for it rules out some lesser included offences as being 'the same'. What about the other limiting case? If we were to say that *any* partial identity of act-types (including of course full identity) makes the two offences involving such act-types the same for double-jeopardy purposes, then we would have eliminated any need for judges to answer the question '*How much* partial identity?'

It is only because the act-types in question are those morality makes salient by its norms of obligation that there can be any plausibility to this way of avoiding the degree vagueness indeterminacy. Otherwise, *any* overlap, no matter how trivial, between the elements of different offences would make them partially identical and thus, the 'same offence' for double-jeopardy purposes. Rape-at-night, if there were such a crime, would be partially identical to burglary, because both would share the (nominal) 'part' an-act-done-at-night; 'driving', as we have seen, would be a common part of driving without a licence, driving while intoxicated, driving an overweight vehicle, and driving a vehicle without adequate smog devices. Whereas if the common part of criminal statutes has itself to be a (whole) moral wrong, then trivial partial identities between the nominal parts of the nominal act-types created by criminal statutes do not amount to the 'same offence' for double-jeopardy purposes. On the other hand, where the part is itself a whole moral wrong, either intrinsic or instrumental, then the overlapping is not a *trivial* overlapping, even if the overall act-types prohibited by different statutes differ considerably in other ways.

Consider the earlier example of common-law burglary, second-degree burglary, breaking and entering, and criminal trespass. Is there any partial identity between the underlying, morally salient act-types of these different crimes? The first three crimes share two morally salient act-types, use of force against another's property ('breaking') and trespass on to another's property ('entering'), whereas all four crimes share only the latter morally salient act-type. On the solution here proposed to the 'how much?' problem, all four crimes are none the less 'the same' for double-jeopardy purposes, so that for doing any given act-token that instantiates all four act-types an accused may be prosecuted or punished only once.

Consider by way of second example the crimes of robbery, assault, and larceny. I take all three of these *mala in se* crimes to name three distinct, intrinsic moral wrongs, even though one of them, robbery, is a conjunctive act-type of the other two. This gives the intuitively correct result that the pairs, robbery-assault, and robbery-larceny, are the same for double-jeopardy purposes, because the latter morally salient act-type of each pair is part of the former morally salient act-type of each pair. What about a second punishment or prosecution for larceny after there has been a punishment or prosecution for assault—or vice versa—in cases where the very same act-token is both (that is, the act is a robbery, although that is not charged)? It might seem that there is some partial identity between larceny and assault, namely, each is a part of the same act-type robbery. Yet there is no partial identity here between larceny and assault; they neither share a common (morally salient) part, nor are they (morally salient) parts of each other. Being each a part of a common whole is not to be partially identical.

The moral approach thus promises a solution to the two problems of language-independent individuation and of degree vagueness. It is not metaphysically indeterminate nor does it lead to results at odds with common intuitions about when two offences are or are not the same for double-jeopardy purposes. The moral solution to these problems does, however, obviously depend a good deal on morality having a certain nature. Morality must itself contain language-independent norms that allow the isolation of 'morally salient act-types' (moral wrongs), and we must be able to grasp such norms and wrongs well enough to identify when they are being referred to by some criminal statute. We must, in other words, not be either metaphysical or epistemological sceptics about morality in order to use the moral approach.

Before addressing each of these concerns about morality, one presupposition about morality that is *not* required by the moral approach is the idea that morality is deontological in nature. Because I believe (and have elsewhere defended) that morality in part consists of deontological norms,[23] and for ease of exposition, I earlier assumed this about morality. But if in fact consequentialism were exclusively correct about morality, the moral approach to act-type individuation would be viable none the less.

On the consequentialist view, morality does not contain norms making certain act-types intrinsically wrongful. Rather, morality consists of intrinsically good or bad states of affairs, so that a particular action becomes wrongful by virtue of its contribution to some bad state of affairs. This makes rules prohibiting certain *types* of action at best instrumentally true; shooting another person is prima facie wrong because such an act-type causes pain, risk of death, and bodily violation of that other, all intrinsically bad things, a consequentialist may suppose.

Even if morality consists ultimately only of intrinsically good or bad states of affairs, it will thus contain (instrumental) norms prohibiting (instrumentally) wrong action-types. It is to these act-types that the moral approach to double jeopardy will repair, on this consequentialist view of morality. There is nothing more problematic about such act-types than there was about the instrumental act-types underlying *mala prohibita* crimes. (Indeed, on the consequentialist view of morality, all crimes are *mala prohibita* crimes, since no act-types are intrinsically wrong.

Turning to the question of metaphysical scepticism about morality—I shall not here mount even a summary of the answer to the sceptic about moral reality. This I have done elsewhere, and in detail.[24] What I shall do is sketch how the moral anti-realist must do his double-jeopardy adjudication. There are two kinds of anti-realists about morality: idealists and sceptics. The most prominent form of idealism is conventionalism about morals, according to which moral norms exist but only as social conventions. The most prominent form of moral scepticism is non-cognitivism, according to which moral norms do not exist except as projections on to the world by each individual evaluator as she expresses her approval or disapproval of various acts or states of affairs.

---

[23] Moore, 'Torture and the Balance of Evils'.
[24] Moore, 'Moral Reality'; id., 'Moral Reality Revisited', *Michigan Law Review*, 90 (1992), 2424–2533.

Double-jeopardy adjudication would change little in the hands of a moral non-cognitivist using the moral approach. The moral non-cognitivist would purport to find moral norms of the deontological or consequentist kind, and would use the act-types made salient by such norms in order to assess whether there is full or partial identity. It is only in quiet moments that the non-cognitivist would admit to herself that these moral norms, like all moral norms, do not really exist except as her own projections on to the world. Then she would go on as before, treating them as real. Although this bifurcated consciousness is a problem, it is not a problem with the moral approach to act-type individuation; it is a problem for non-cognitivist metaethics generally.[25]

The moral conventionalist would practise the moral approach to double-jeopardy adjudication differently. He would have to find his language-independent act-types in his society's moral conventions. While those conventions will no doubt differ from the statutory law in certain places, allowing some substitution of 'moral' act-types for the nominal act-types created by statutes, the former act-types themselves would not be language-independent. For although moral conventions do not have canonical formulations like statutes, they do have historical formulations that individuate them as one convention.[26] The result for the moral conventionalist is that his determinations of full or partial identity of act-types will collapse into a language-independent enterprise, even if the language is not statutory. He will accordingly end up counting act-types as the same or different depending on the accident of their conventional formulation.

For the moral realist, by contrast, the act-types used by the moral approach to double jeopardy will be as real as any other act-types. Indeed, on the going version of moral realism that I and others have defended, moral properties like the badness of state of affairs or the wrongness of actions themselves possess causal powers and are to be individuated by the causal test earlier given for properties generally.[27] For the moral realist, there-

---

[25] See M. Moore, 'Note: Waldron on Realism', in R. George (ed.), *Natural Law Theories* (Oxford, 1992).

[26] On the necessary tie of all social rules to some kind of historical formulation, see M. Moore, 'Three Concepts of Rules', *Harvard Journal of Law and Public Policy*, 14 (1991), 771–95.

[27] Moore, 'Moral Reality Revisited'; David Brink, *Moral Realism and the Foundations of Ethics* (Cambridge, 1989); Nicholas Sturgeon, 'Moral Explana-

fore, the moral approach shares the same metaphysics with what
I called the metaphysical approach, with the crucial restriction
that only some properties count (the moral ones) in individuating
act-types for double-jeopardy purposes.

Coming to the epistemological sceptic about morality—this
position questions whether there is any basis for identifying *which*
moral wrong is referred to by any given criminal statute.
Granted, there are lots of separate moral wrongs, but how do we
know when they are separate? There are also lots of statutes, and
how do we know to match them to the separate moral wrongs
that we may be able to identify?

Neither of these questions should be unfamiliar to a criminal
lawyer, because he asks and answers them all the time in two
contexts other than double-jeopardy adjudication. One such context is the interpretation of the *actus reus* prohibition of a criminal statute. As I have argued elsewhere in detail, the best way for
a judge to interpret such prohibitions is to seek the nature of the
intrinsic moral wrong referred to, if there is one, or to identify
the act-type that ideally serves to eliminate the evil such statute
should be seen as aiming to eradicate.[28] In their interpretation of
criminal statutes, judges seek intrinsic or instrumental moral
wrongs all the time in order to match them up to the statute
being interpreted.

The second context has to do with the *mens rea* portion of
criminal statutes, where it is a familiar problem as to how much
of the *actus reus* of a given crime should be included in the
object of the *mens rea* prohibition. To illustrate the problem,
consider a statute making it criminal for one who is the husband
of a pregnant woman to leave her with the intent to abandon
such pregnant woman.[29] Does a defendant have to know that his
wife is pregnant in order to have the requisite *mens rea*, or is it
enough that he knows that she is a woman, that she is his wife,
and that he is abandoning her for good? The old common-law
'moral wrong' doctrine requires that a court ask whether the act
a defendant thought he was doing (e.g. abandoning his wife, but

tion', in D. Copp and D. Zimmerman (eds.), *Morality, Reason and Truth*
(Totowa, NJ, 1984); Richard Boyd, 'How to Be a Moral Realist', in G. Sayre-
McCord (ed.), *Essays on Moral Realism* (Ithaca, NY, 1988).
[28] Moore, 'A Natural Law Theory of Interpretation'.
[29] The facts of *White* v. *State*, 44 Ohio App. 331, 185 N.E. 64 (1933).

not abandoning his *pregnant* wife) was a moral wrong.[30] The more modern Model Penal Code solution to the problem is to require a court to ask whether the element of pregnancy is or is not a 'material' element, i.e. whether that element is or is not 'unconnected with the harm or evil . . . sought to be prevented by the law defining the offence'.[31] Both the common law and the Model Penal Code have thus long required courts to identity the moral wrong underlying each criminal offence (the only differences being that the Model Penal Code requires the identification of instrumental as well as of intrinsic moral wrongs, and the Code contains a bit more clarity that it is only the specific moral wrong referred to by the statute in question that is relevant to determining materiality).

None of this is to deny that there are not hard cases where we are genuinely perplexed as to what the moral wrong is that may be referred to by some statute, or whether two act-types are two variations of the same wrong or, alternatively, whether they are not in reality two separate moral wrongs. Despite such perplexities, we do have considerable moral knowledge on these topics, using that knowledge constantly to interpret the *actus reus* and *mens rea* portions of criminal statutes. The moral approach only requires that we adjudicate double-jeopardy claims on the same basis.[32]

It remains to enquire how well the moral approach to act-type individuation squares with the doctrines (and decisions under them) of double jeopardy we summarized in the last chapter. Consider in this regard a recent Pennsylvania court opinion applying the common-law merger doctrine to the question of whether multiple punishment was permissible for one act that was an instantiation of four offences: aggravated assault, resisting arrest, carrying a firearm on the public streets of Philadelphia, and illegal possession of a sawn-off shotgun.[33] The court nicely separated the token-identity question from the type-identity question, and interpreted the latter question as asking how many separate moral wrongs were done by defendant's admittedly single act:

[30] *Regina* v. *Prince*, L.R. 2 Cr. Cas. Rg. 154 (1875).

[31] Model Penal Code, §1.13(10).

[32] The closer analogy for double jeopardy is the courts' contrual of what elements of crimes are material for *mens rea* purposes.

[33] *Commonwealth* v. *Williams*, 344 Pa. Super. 108, 496 A.2d 31 (1985).

In order to find that separate statutory offences merge, we must . . . determine not only that the crimes arose out of the same criminal act, transaction, or episode, but also that the statutes defining the crimes charged were directed to substantially the same harm or evil.[34]

The court's general recipe for individuating moral wrongs, was:

In determining how many different 'evils' are present in a given criminal act, the sentencing court should devote close attention to the language the Legislature has used and the scheme it has followed in defining offences . . . The court must also approach the question with a heavy dose of common sense.[35]

If one reads 'common sense' to mean 'moral sense', this is surely right. Just as in ascertaining which of the elements of a given offence are morally material for purposes of interpreting the *mens rea* requirement, so, for merger and double jeopardy, courts should ignore formal elements of offences that are unconnected to the wrongfulness of the acts they prohibit. Further, for those elements of offences that are connected to the wrongfulness of the actions prohibited, courts should redescribe those elements so as better to capture the aspects of them that are morally relevant.

Thus, the court in *Williams* ignored the 'public streets or public property' element of the charge of carrying a firearm, because this element seemed incidental to the evil the statute was seeking to eradicate, the potential use of firearms in acts of violence. Likewise, the court ignored the element of the firearm being a sawn-off shotgun (or other prohibited type of firearm), thinking presumably that any firearm was potentially dangerous in the relevant way. Furthermore, the court also ignored the differing descriptions of the act-types that were material—'carrying' under the first statute, 'possessing' under the second—thinking any difference between the two to be immaterial to the type of instrumental moral wrong each statute aimed to prohibit. Not surprisingly, given these alterations, the court found the morally salient act-types to be at least partially identical and thus prohibited multiple punishment.

---

[34] *Commonwealth* v. *Williams*, 496 A.2d, at 41–2.
[35] Ibid., at 50.

To some extent the Pennsylvania court's analysis could be duplicated under current US federal court doctrine (the *Blockburger* and *Grady–Vitale* tests for sameness of offences). The US Supreme Court, for example, has held: that the offence of joy-riding contains no elements not contained in the offence of car theft—even though the *actus reus* of joy-riding is to 'take, *operate, or keep* any motor vehicle' while the *actus reus* of car theft is to take into one's possession, and to carry away, a motor vehicle;[36] that the offence of robbery with a firearm contains no new elements not contained in felony-murder (when predicated on robbery)—despite the fact that armed robbery requires use of a firearm, while robbery sufficient for felony-murder only requires the use of force or fear but not necessarily use of a weapon;[37] that the offence of adultery contains no elements not contained in the offence of illicit cohabitation—despite the fact that adultery requires sexual intercourse with someone who is married to another, whereas illicit cohabitation was accomplished by living together with two or more members of the opposite sex irrespective of their marital status with others;[38] that the crime of possessing a prohibited firearm shipped in interstate commerce contains no element not contained in the crime of receiving a prohibited firearm shipped in interstate commerce—despite the possibility that one might possess something without having received it.[39] Without saying so, in each instance the US Supreme Court slid over differences in the nominal act-types prohibited by different statutes, presumably because the court saw a moral sameness beneath the formal difference.

Still, it remains true that both the *Blockburger* and the *Grady–Vitale* tests invite a more formal approach, taking elements of offences at face value and holding there to be a difference even when morally that difference is trivial. Thus, courts have held: that the act-type transporting in interstate or foreign commerce a motor vehicle that has been stolen is distinct from

[36] *Brown* v. *Ohio*, 431 U.S. 161 (1977). Since the two offences were prosecuted under Ohio state law, Ohio's definitions of the offences governed. Ohio Revised Code Annotated, §4549.04(D) provided at the time: 'No person shall purposely take, operate, or keep any motor vehicle without the consent of its owner.' §4549.04(A) on car theft provided: 'No person shall steal any motor vehicle.'
[37] *Harris* v. *Oklahoma*, 433 U.S. 682 (1977).
[38] *In re Nielsen*, 131 U.S. 176 (1889).
[39] *Ball* v. *United States*, 470 U.S. 856 (1985).

the act-type possessing a motor vehicle which has crossed a state or United States boundary after being stolen;[40] that the act-type making of a false statement material to the lawfulness of the sale of firearms is distinct from the act-type making of a false statement with respect to information required by law to be kept in the records of a federally licensed firearms dealer;[41] that the act-type possessing counterfeit money is distinct from the act-type transferring counterfeit money;[42] that the act-type failing to report currency contains an element that the act-type making a false statement to a federal agency does not, and vice versa.[43] Indeed, the Pennsylvania court in *Williams* for federal-law purposes also adopted the more formal approach invited by the federal tests, holding that the act-types carrying a firearm on a public street, and possessing a prohibited type of firearm, each contained an element the other did not.[44]

The upshot is that if the moral approach to sameness of offence for double-jeopardy purposes is the right approach, these federal tests and their state law counterparts are too restrictive. The moral approach would demand that courts do regularly and explicitly what they now do only irregularly and implicitly: ask not about the nominal act-types each statute creates, but ask instead whether the morally wrong act-types to which each statute refers are or are not 'the same'—that is, are or are not fully or partially identical.

---

[40] *United States* v. *Wolf*, 813 F.2d 970 (9th Cir. 1987). 18 U.S.C. §2312 ('Transportation of Stolen Vehicles') provides that: 'whoever transports in interstate or foreign commerce a motor vehicle or aircraft, knowing the same to have been stolen, shall be fined not more than $5,000 or imprisoned not more than five years, or both'. §2312 ('Sale or Receipt of Stolen Vehicles') provides that: 'whoever receives, possesses, conceals, stores, barters, sells, or disposes of any motor vehicle or aircraft, which has crossed a State or United States boundary after being stolen, knowing the same to have been stolen, shall be fined not more than $5,000 or imprisoned not more than five years, or both'.

[41] *United States* v. *Hawkins*, 794 F.2d 589 (11th Cir. 1986). The offences are defined by §§922(*a*)(6) and 924(*a*) of 18 U.S.C.

[42] *United States* v. *Shear*, 825 F.2d 783 (4th Cir. 1987). Possession and transfer of counterfeit money are separately defined in 18 U.S.C., §§472, 473, respectively.

[43] *United States* v. *Woodward*, 469 U.S. 105 (1985).

[44] *Commonwealth* v. *Williams*, 496 A.2d, at 37–9.

2. ADJUDICATING BETWEEN THE THREE APPROACHES TO SERVE THE PURPOSES OF DOUBLE JEOPARDY

As we have seen, there can be a direct relevance of metaphysical analyses of action to the meaning and application of legal doctrine. When the general part of the criminal law calls for a voluntary act, the best interpretation of that doctrine is to require what metaphysically is an act (on my theory, a volitionally caused bodily movement). Similarly, when the special part of the criminal law prohibits killing, the best interpretation of that prohibition is to prohibit what metaphysically are killings (on my theory, those volitionally caused bodily movements that proximately cause deaths). It may thus be tempting to think that the metaphysics of morally salient action above described can play a similarly direct role in unpacking the various legal doctrines that prohibit multiple punishment or prosecution for the 'same act' constituting the 'same offence'.

Whether such straightforward legal applications of the morally restricted metaphysics of action individuation is possible or not depends on the moral theory that justifies such doctrines. The reason why the metaphysics of basic acts so directly fills out the meaning of the general part's voluntary-act requirement is because of the morality that justifies the latter requirement. Voluntary acts are required as a prerequisite to legal liability only because such acts are a prerequisite of moral responsibility. And morality requires that someone *act*, as the correct metaphysics would unpack that notion. Similarly, the reason why the metaphysics of complex actions so directly fills out the various prohibitions of the special part of the criminal law is because of the morality that justifies the *actus reus* requirement. Both for legality reasons, and because criminal prohibitions generally follow moral prohibitions, when a criminal code prohibits 'killings', the correct metaphysical analysis of the nature of killings is what is needed to understand the meaning of the law's prohibition. Morality itself prohibits acts that are (really) killings, and the law that would mirror that morality should abide by its metaphysics.

Surprisingly, perhaps, we have not yet shown whether the morally restricted metaphysics of act-type individuation above described can be an appropriate answer to double jeopardy's 'same offence' requirement—the possible surprise, because we have already built a considerable amount of morality into act-type individuation for double-jeopardy purposes with our third, 'moral' approach. Yet the morality we used in order to surmount the two problems of language-independence and degree vagueness was the morality that determines whether an act-type is morally wrongful. What we have not yet consulted is the morality that determines whether the moral wrongfulness of an act-type is relevant for double-jeopardy purposes. What we haven't done, in other words, is shown that the moral approach to act-type individuation serves the purposes of the double-jeopardy requirement.

It will be recalled from the previous chapter that those purposes are somewhat diverse. With regard to the multiple-punishments question, the purpose of the double-jeopardy prohibition is to prevent too much—that is, undeserved—punishment. Such a purpose is not to set an outer limit to permissible punishment that the legislature cannot exceed, as it is for the Eighth Amendment's prohibition against disproportionate punishment. Rather, the purpose of the double-jeopardy ban on multiple punishment is to make it easier for legislatures to set the appropriate level of punishment for each offence, since the presumption against multiple punishment for offences that are 'the same' allows legislative selections of punishments without worrying about the overlapping of offences (and consequent risk of excessive punishment for acts that fall within the overlaps). Despite this difference, the phrase 'same offence' should be interpreted to as to serve the same value that animates interpretations of the cruel and unusual punishment ban, namely to prevent disproportionate punishment. Only by doing so can courts give a content so the double-jeopardy presumption that allows legislatures not to worry about the overlapping-prohibition problem.

Given that the (mediate) purpose of the multiple-punishments ban of the clause is to prevent undeserved punishment, it becomes a straightforward matter to show how this purpose is served by utilizing the moral approach above described. For there are two variables that determine how much punishment anyone deserves: first, what wrong they did, and secondly, the culpability with which they did it. The person who does a less serious wrong deserves less punishment than one who does a more serious wrong; the person who does but one wrong (of a greater degree of seriousness) deserves less punishment than another who accomplishes several wrongs (of similar seriousness) with a single act.

One thus must individuate wrongs in order to assess the degree of deserved punishment. Adopting the moral approach to individuating act-types does precisely this. Such an approach thus maximally serves the purpose of the multiple-punishments ban of double jeopardy. Indeed, neither of the other two approaches that we examined—the sceptic's linguistic approach, and the pure metaphysical approach—even begins to serve this proportionality purpose. They thus would have to be rejected on this ground alone, even if they did not possess other infirmities.

There are two purposes behind the second of double jeopardy's bans, that against multiple prosecution for the same offence after prior conviction for that offence. One purpose makes the ban on multiple prosecution after conviction the handmaiden of the multiple-punishments ban. After all, one way to prevent multiple punishment for the same offence is to prevent multiple prosecutions for that offence. Indeed this is an efficient way of carrying out the multiple-punishments ban. If there could be no additional punishment at the end of a second prosecution (because of the multiple-punishments ban), it surely is a waste of everyone's time to go through a second trial. Better to ban the multiple punishment as soon as it can be determined that the already convicted accused is being reprosecuted for the same offence.

To determine whether or not the second prosecution of a defendant is for the *same offence* as that for which he was earlier convicted, one must (on this handmaiden purpose) interpret 'same offence' identically to the way it is interpreted in the multiple-punishments ban. That is, one should interpret the phrase in a way that prevents excessive and undeserved punishment. This again mandates the moral approach to interpreting 'same offence'.

The second value served by the ban on successive prosecutions after conviction is that of preventing unnecessary embarrassment and expense to defendants and preventing purely harassing prosecutions by prosecutors. This value certainly sounds distinct from any concern with preventing excessive punishment, and, thus, would seem to demand its own definition of what makes two offences 'the same'. Yet this would be to miss the common thread here. The time, expense, and embarrassment of a criminal trial are harms to every criminal defendant, but they are harms we obviously are and should be willing to inflict on a defendant sometimes, else we'd never prosecute anyone. Similarly, the knowing infliction of these harms on a defendant by a prosecutor cannot always amount to harassment; it is only where these harms are inflicted on a defendant for no good reason that a prosecutor is harassing a defendant. When offences are not truly the same, there is a good reason for a defendant to have to suffer the harms of criminal trials more than once, and there is a good reason for a prosecutor to knowingly inflict them more than once. That reason is 'insuring that the guilty are punished'.[45] If a defendant has truly done more than one wrong, he deserves more than one punishment. Hence, conviction for an earlier wrong should not bar a second prosecution for the second wrong. The embarrassment, time, and expense of a second trial are no reason to bar the second prosecution, nor can such a prosecution be called harassment.

The upshot is that what limits the domain in which the harassment–embarrassment value operates is the need not to give too *light* a punishment. This need, like the need not to give too *heavy* a punishment, demands the moral approach to where offences are the same. For as we have seen, it is the identity of wrongs that determines when multiple punishments are excessive; it must be equally true that this same variable determines when single punishment is too lenient. Thus, although the mix of values justifying these first two of double jeopardy's bans do differ somewhat, the test for 'same offence' that these values justify differs not at all.

The third ban of federal double jeopardy is that against multiple prosecutions for an offence of which the defendant has earlier been acquitted. There is no 'handmaiden' purpose to this ban, since the defendant has not been previously punished (this, because he was acquitted). Rather, the purposes here are different and are three in number. One is the danger of convicting the innocent. The second purpose of this ban is to prevent unfairness to defendants of having to go through the trauma of more than one criminal trial for the same offence. Quite apart from any

---

[45] *Burks* v. *United States.*, 437 U.S. 1, 15 (1978).

increased chance of conviction with each successive prosecution, it is unfair to demand of the successful defendant that he be at risk again against the same charges. The third purpose is the promotion of efficient criminal adjudication. This ban encourages prosecutors to try all related charges in one proceeding and thus to avoid the duplicative inefficiencies of separate trials.

These three values seem even more removed from the proportionality worry that motivates the ban against multiple punishment for the same offence. Yet here too, as with the second ban, these three values are limited in their operation by the paramount value of seeing that the guilty are punished in proportion to their wrongdoing. Increased familiarity with the character of the defendant, the specifics of his act, or the tactics of his counsel undoubtedly comes from a previous trial, and perhaps increases the likelihood of conviction in a second trial; yet if the second trial is for a different moral wrong from that for which the defendant could have been punished had he been convicted of the charges brought in the first trial, there can be no objection on this ground. It is only where new wrongdoing is not at issue that the worry stemming from increasingly better prosecutorial attempts to convict has sufficient weight to bar the second prosecution.

Similarly, the domains in which the values of unfair trauma and efficiency can do their work are marked by the identity of wrongdoing. It is not sufficiently unfair or inefficient to bar a second prosecution when that prosecution is for a distinct moral wrong. Those for whom there is good reason to think that they did a second moral wrong can't complain that they have to experience the trauma of a second trial for that moral wrong. Such defendants have no more complaint than do those who do separate moral wrongs less efficiently, i.e. with separate acts. In both cases, the additional trauma of more than one trial is not an unfair price to be paid in order that wrongdoing receive its just desert.

The borders defining 'same offence' for this third ban are thus, again, patrolled by the overriding purpose of achieving proportionate punishment of the guilty. Such a purpose of achieving sufficient punishment for wrongdoing, like the purpose of avoiding excessive punishment for wrongdoing, demands an individuation of 'same offence' by morally wrong act-types. Here as elsewhere in double jeopardy, the 'same offence' means partial or full identity of those act-types that morality makes, intrinsically or instrumentally, wrong.

By this analysis, we have completed our case for the use of the metaphysics of morally salient act-types in double-jeopardy adjudication. With the same analysis, we have also answered the sceptic who takes the multiplicity of contexts of, and policies behind, double jeopardy to deny any univocality to the sameness of offences for double-jeopardy purposes. Divergent as those contexts and purposes may be, running through each is the constant theme that punishments be in proportion to wrongs done. Such a constant theme justifies a univocal meaning to 'same offence' in terms of the identity between morally salient act-types. Such univocal meaning rules out the US Supreme Court's current division, applying one test of sameness to multiple-punishment contexts, and another, to multiple-prosecution contexts. Also ruled out is the much more context-sensitive criterion of sameness of those Legal Realist-like sceptics who call for 'same offence' to be no more than the legal name put on the conclusion of an *ad hoc* balancing of values done in each double-jeopardy case.

14

LEGAL, MORAL, AND METAPHYSICAL NOTIONS OF THE 'SAMENESS' OF ACT-TOKENS

1. AGENT-RELATIVE AND VICTIM-RELATIVE INDIVIDUATION OF MORALLY WRONGFUL ACT-TYPES

We now turn to the other 'same offence' question under double jeopardy, the question of whether what is seemingly one (morally salient) act-*type* has been instanced by one or more than one act-*token*. Suppose that we are clear that killing of another human being is one morally salient act-type (a wrong), so that when defendant D shoots and kills victim B there is only one liability under all statutes that include this wrong in their prohibitions. This conclusion about how many wrongful act-*types* the defendant has done does not conclude the question of how many act-*tokens* the defendant did in order to accomplish this one wrong. If D shot V twice, are there two acts of killing even though there is but one death? This is the kind of question with which we shall be concerned in this chapter.[1]

Because this question is often confused with a hitherto ignored aspect of the act-*type* individuation question we explored in Chapter 13, it will be useful to defer consideration of act-token individuation until after we have disposed of the classes of cases that give rise to the confusion. There are two classes of such cases. The first occurs when two or more defendants independently (that is, without prior arrangement or knowledge of the other) cause one legally prohibited result. For example D_1 and D_2 each shoot V, who dies from the loss of blood from both

---

[1] As I noted in Chapter 12, our question here is what is often termed the 'unit of prosecution' problem. For a history of the phrase, see Peter Westen and Richard Drubel, 'Toward a General Theory of Double Jeopardy', *The Supreme Court Review* (1978), 81–169: 111 n. 147.

wounds. Are there two murders despite the one death, or only one? The second class of case arises when there is only one defendant but that defendant by one action manages to kill two victims. For example, D shoots one shot into V_1, thereby killing him, and the bullet continues through V_1's body into the body of V_2, also killing him. Is D guilty of one homicide or of two?

It is easy (if incorrect) to think that these two classes of cases involve questions of act-token individuation. Yet if we were to so categorize them, we would be forced to legal results that either contradict how we individuate acts elsewhere, or are both contrary to established law and are morally very counter-intuitive. We would have to conclude of the first class of cases (two defendants, one victim) that there are two acts of homicide. While this is both the morally intuitive result and the actual result under current law, it does not square very well with the equally intuitive and equally well-established conclusion that two shots by one defendant that kill one victim (by loss of blood from both wounds) is only one homicide. If we are individuating act-tokens, surely each shot by one defendant is as much a distinct act-token as is each shot by two different defendants, D_1 and D_2.

With regard to the second class of cases (one defendant, one shot, two dead victims), by hypothesis there is but one act here, causing the one bullet to be fired. By the arguments given in Chapter 11, the consequences of that one act are not *parts* of that act, as they would be on the 'componential' or 'moderately fine-grained' mode of act-token individuation that we examined and rejected. By the same analysis in Chapter 11, we are also foreclosed from the extremely fine-grained mode of act-token individuation, according to which we could conclude that the defendant moving his finger, pulling the trigger, shooting the bullet, killing V_1, and killing V_2 are all separate acts that he did all at once. Our earlier-defended coarse-grained metaphysics forces us to the conclusion that there is only one act here, even if it has two very bad consequences.

The upshot is that we should reconsider whether both classes of cases do not present problems of act-*type* individuation, not act-*token* individuation. Consider first the two-defendants, two-acts, one-victim situation. How many wrongful act-types did *each* defendant's act instantiate? With regard to homicide-like statutes, 'one' is the answer I defended in Chapter 13. How many wrongful act-types did *both* defendant's acts instantiate? The answer to this question depends on whether we individuate wrongs not only by the type of action done but also by the identity of the agent who does them. Why not say that each defendant's act in the scenario imagined instantiates a separate wrong, on the view that moral wrongs are 'agent-relative?'[2] On this view, moral norms like that prohibiting the unjustified killing of another human being speak to us individually even if the content of the message is the same. This means, on the standard account, that I obey such moral norms by not killing myself; I do not obey such norms by killing whenever that killing minimizes killings by myself or others. Even if I am a KGB officer who can save five Russian diplomats (who will otherwise be killed by terrorists) by myself killing an innocent relative of the terrorists, I may not minimize killings by killing myself.[3] The moral norm is 'Don't *you* kill', not 'By your acts, minimize killings'.

On such an agent-relative view of morality, it matters who is doing an act causing death in the sense that each of us has a moral ledger unaffected by what others do. On such a view it thus makes sense to say that D$_1$ does one wrong in *his* act causing V's death, and D$_2$ does another wrong in *his* act causing V's death, even though there is only one death of V. On this view, there are as many moral norms prohibiting unjustified killings of human beings as there are moral agents to whom they are addressed.

Putting things this way saves us the embarrassment of concluding that where one defendant kills one victim with two causally efficacious shots, he has done two moral wrongs and thus two homicides. Moral norms may be agent-relative, but they are not 'act-relative' in the sense that a different norm is involved each time a given agent kills. On the agent-relative view about individuating moral wrongs, therefore, we can leave this sort of case to be dealt with as a case of act-token individuation.[4] If we think

---

[2] The phrase is Derek Parfit's. See Parfit, *Reasons and Persons* (Oxford, 1985).

[3] The example is one from the recent history of the Middle East. See Moore, 'Torture and the Balance of Evils', when I use this and other examples to defend a complex agent-relative view of morality.

[4] This sort of case is dealt with in the succeeding sections of this chapter, where I shall conclude that there is properly held to be only one homicide in such cases no matter how many metaphysically distinct acts of a given defendant there may be jointly causing one victim's death.

that the two-shots, one-victim defendant is guilty of two homicides, that will be because he did two act (tokens) of killing, not because he did two wrongs of killing one human being.

Now consider the second class of conceptually troublesome cases, where we have one act by one defendant but where we have two victims hurt by that one act. Since the law is here less uniform than with respect to the first class of conceptually troublesome cases, consider three sorts of examples. First example: suppose Smith shoots at Jones with a powerful gun, intending to kill him. He does kill Jones with the single bullet, but also kills Long, a bystander who was standing behind Jones. Is Smith guilty of one murder (or of any other degree of homicide) or of two? Second example: same as the first, except that Smith did not intend to shoot or kill anyone; he was negligent (*vis-à-vis* Jones although not *vis-à-vis* Long) in that he did not weigh adequately the risk of Jones's being hit while he was target-shooting with an inadequate backstop. Third example: Smith without threat or force against any person, takes a bag containing two valuable watches from a single locker; the watches belong to two people, one to Jones and the other to Long. Is Smith guilty of two thefts, or only of one?

In results as well as in reasons, the courts have come out pretty well across the board on such 'multiple consequences' cases. With large splits of authority, the courts mostly say of the first sort of example that there are two murders; of the second, that there is one negligent homicide (or involuntary manslaughter); and of the third, that there is one theft.[5]

In the present context what is troubling about such cases is not so much the checkerboard results reached by different courts on these three sorts of examples. It is rather the conceptual confusion that underlies such inconsistency of result. Consider the first example in isolation for a moment. I take it that what motivates those minority of courts that hold there to be but one murder here are two insights: (1) that there is only one offence (type of act) at issue here; and (2) that there is only one instance of that type of act committed by Smith—there was, in other words, only one act (token) of killing even though there were two deaths.

---

[5] See the summaries of cases in Frank Horack, 'The Multiple Consequences of a Single Criminal Act', *Minnesota Law Review*, 21 (1937), 805–22; Dix and Sharlot, *Criminal Law, Cases and Materials*, 258–60.

The second insight is, metaphysically, correct. As we saw in Chapter 11, the consequences of a basic act are not parts of that act, as they are considered to be by the moderately fine-grained (or 'componential') view that we rejected. Nor does the possession of different properties (say, the causing of the death of Jones, and the causing of the death of Long) make for two different acts entirely, as would be the case on the extremely fine-grained view we also rejected. Rather, on the coarse-grained view of act-token individuation that I defended, Smith performed but one act here, even if there are two different (complex) ways of describing it ('killing of Jones', 'killing of Long').

The majority of courts that come out the other way on cases like my first example do not do so because they reject the coarse-grained metaphysics of act-token individuation that I defended in Chapter 11. For that mode of individuating act-tokens is, I believe, by far the most in accord with common sense. Smith's pulling the trigger once, sending one bullet on its way, is by common sense but one act, no matter what happened after that. Rather, such courts have had a moral insight (that they have not known how to conceptualize in terms of action-identity): Smith did much more wrong in killing both Jones and Long than he would have done if he had killed Jones only. As the West Virginia Supreme Court put it, in deciding that in such situations there are two homicides: 'Certainly, the degree of culpability, and as a consequence the degree of punishment, must bear some proportion not only to the magnitude of the crime but also to the number of victims involved. These are fundamental considerations that society expects from a criminal justice system.'[6]

The problem is how to give voice to such a moral insight in light of our two questions of action-identity. To begin with, we should be clear what the moral insight is. It is not exactly that killers like Smith of my first example have greater *culpability* when a bullet intended to kill Jones also kills another. For culpability is a matter of *mens rea* and excuses, and Smith had but one intent to kill, namely, an intent to kill Jones.[7] Rather, the moral

---

[6] *State ex rel. Watson* v. *Ferguson*, 166 W. Va. 377, 274 S.E. 2d 440, 446 (1980).

[7] That is, his actual, psychological intent was to kill Jones, not to kill Jones and Long. That the law will impute an intent to Smith to kill Long under its doctrine of 'transferred intent' does not alter the morally relevant fact that Smith had an intent to kill only one person, not two.

insight is that Smith does *more wrong* when he kills Long as well as Jones. Smith deserves more punishment, on retributive grounds, because with the same culpability he in fact accomplished more wrongdoing than would have been the case had he only killed Jones.[8]

The problem is how to accommodate this seemingly valid moral insight—of greater wrongdoing—within the confines of our two questions of action-identity. Smith didn't seem to do more wrongdoing by our first measure of that—namely, how many morally wrong act-types his act-token instantiates—because his act seemingly instantiates but one morally salient act-type, namely, that of killing another human being. That suggests that Smith did this one moral wrong (killing) more than once, which would be to use our second measure of degree of wrongdoing, namely, how many times did a defendant do a given wrong? Yet Smith did but one act here, on the common-sense, coarse-grained metaphysics we earlier defended.

The common conceptual manœuvre given by courts and commentators in order to accommodate the above moral insight is to attempt to introduce a third measure of wrongdoing.[9] In addition to measuring degree of wrongdoing by how many wrongs of what degree of significance a defendant accomplished, and by how many times a defendant did each of those wrongs, one measures it by how many harmful consequences the defendant's act (token) causes. If Smith's act causes two deaths, then even if it is but one act (token) that instantiates but one morally wrong act (type), Smith has done more wrong. This manœuvre, unfortunately, is incoherent, for there is no room for this supposed third measure of degree of wrongdoing. If there is but one moral wrong done here, and it is done only once, it makes no sense to claim that none the less there is greater wrongdoing by people

---

[8] I here assume without argument, as I have throughout this book, that there is such a thing as 'moral luck'. Moral luck here means that one can deserve more punishment (because one has done more wrong) solely because of factors like causation, over which one has at best tenuous control. If one were to reject moral luck, that would not mean that desert was a function only of culpability; it could not mean this, because what intentions one forms, and what risks one takes, is a function of factors over which one has tenuous control too. Whether an opportunity presents itself in which to form a culpable intention is also a matter of moral luck. The true implication of rejecting moral luck entirely is that desert disappears entirely.

[9] See e.g. Horack, 'The Multiple Consequences of a Single Criminal Act', 821.

like Smith. The degree of wrongfulness of Smith's action(s) is fixed once we admit he did one wrong of a certain significance, once.

A preferable conceptualization is to withdraw the concession that Smith did but one wrong. Rather, we should say that Smith by his single act-token did two distinct moral wrongs, a killing of Jones, and a killing of Long. True, these two moral wrongs are similar in that they are both *killings*. Yet one should individuate moral wrongs not only by their character (killing, for example), but also by the victim of them. It is one moral wrong to kill Jones, another to kill Long. Indeed, the norm against 'killing another person' is in reality a related family of moral norms prohibiting the killing of each actual or potential person.

This rather more populous map of moral norms should sound familiar, at least to legal philosophers. Wesley Hohfeld made the same observation years ago about legal rights and duties: Hohfeld held that a duty not to kill Jones was distinct from a duty not to kill Long, for all duties, like all correlative rights, were relative to particular people.[10] It is only convenience of exposition that allows us to talk of 'a' duty not to kill, or of 'a' right not to be killed; in reality, what we are referring to is a duty on A not to kill C, a duty on B not to kill D, etc.

The upshot is that we should amend yet again our analysis of act-type individuation. Underlying each nominal act-type created by each criminal statute is not one morally salient act-type; there is in reality a family of morally salient act-types, as numerous as there are potential agents *and* potential victims.[11]

Why should we suppose that morality is 'victim-relative' as well as 'agent-relative'? Because: (1) I take the moral insight of the courts in this kind of case to be a compelling one: to cause by a single act the same kind of harm to more than one victim increases the wrongdoer's desert because it increases the degree of wrongdoing he has done; and (2) there is no alternative way of conceptualizing there being increased wrongdoing in such cases except by individuating moral norms by their victims as well as by both their agents and the moral significance of the sort of act-type instantiated by an agent's action on a given occasion.

[10] Wesley Hohfeld, *Fundamental Legal Conceptions* (New Haven, Conn., 1919).
[11] For conjunctive act-types like robbery and burglary, there will be two or more families of such related act-types.

If this is correct for my first example, then to the extent that the courts have decided my second and third examples the other way, they are incorrect. Take the second sort of example, where the actor is only negligent or reckless with respect to death. It is no doubt correct that 'the apparent difference in the criminality of D's conduct may provide the explanation'[12] for why those courts that count these to be two murders in my first example count only one manslaughter in my second. Yet the greater 'criminality' of murder over manslaughter is *not* a difference in the wrongfulness of their respective act-types, for these are identical: killing. Rather the difference is one of culpability, a murderer either intending to kill, knowing his act will kill, or consciously running the risk of death for absolutely no reason, whereas a person guilty of manslaughter only risks death by his activities. Courts that count murders differently from manslaughters may be confusing the greater culpability of an actor with a supposed greater wrongfulness of his action. In reality, there is no difference in the increment of wrongfulness in these two classes of cases. He who negligently or recklessly kills two has done more wrong than one who negligently or recklessly kills one, and the increment of his increased wrongdoing is the same as he who intentionally kills two rather than one. Multiple killers of either sort deserve multiple punishments to match the increased wrongfulness of their actions.

The only qualification to this last conclusion might be along the following lines. Suppose it to be the case that degree of culpability is at least as important an ingredient in determining degree of deserved punishment as is degree of wrongdoing. Then perhaps the killer who in fact kills two but only intended to kill one (my first hypothetical), and the killer who in fact kills two but only negligently risked death to one (my second hypothetical), only deserve the punishments of one murder and one manslaughter, respectively—for they each had only that amount of culpability. Only the killer who not only in fact kills two, but intends (or negligently risks) the death of two, deserves the punishment of two murders (or two manslaughters)—for only in such cases is there *both* increased culpability as well as increased wrongdoing.[13]

---

[12] Horack, 'The Multiple Consequences of a Single Criminal Act', 809.
[13] It is because of such added degree of complexity that I put aside culpability

Such a qualification, if true, would indeed decide the multiple-consequence cases differently from my simple recommendation (two deaths, two homicides, irrespective of culpability with respect to each death). Yet such qualification could not, of course, rationalize the actual pattern of decisions, which counts two deaths as two murders, but as one manslaughter, irrespective of whether there is the required level of culpability with respect to each death.

My third example was one of decreased wrongdoing compared to homicide, because property crimes generally are thought to be of less seriousness than crimes that cause violence against a person. Thus, courts that would decide there to be two murders in my first example and two manslaughters in my second will none the less decide that there was but one theft in my third hypothetical.[14]

This distinction, too, cannot be maintained. If there are two murders in the case of two deaths resulting from a single act, then there are two thefts when the property stolen by a single act belonged to two different victims. To think otherwise would be to think that morality has a very peculiar shape. It would be to think that, although the norms of morality separately enjoin each of us not to harm each of the rest of us, such norms do not separately enjoin us not to steal from (or otherwise harm the property interests of) others. Rather, morality has one norm here, which is not to steal, period. Such a view about morality's norms would in turn be backed by a view holding that the amount of wrongdoing done in theft is measured by the value of the property taken, and by the number of times one separately takes it—but not by the number of victims from whom it is taken. Yet is not the interest of each of us in maintaining the possession and ownership of our property as separate and distinct as is the interest of each of us in maintaining our lives and our bodily

measurement in double-jeopardy contexts, focusing exclusively on wrongdoing measurement. See Ch. 12 n. 1.

[14] See e.g. *Commonwealth* v. *Donovan*, 395 Mass. 20, 478 N.E. 2d 727 (1985) where defendants picked up a fake deposit box containing seven different depositors' money. The court held that 'only a single crime has been committed even though the larcenous scheme involves the taking of property from a number of owners', even while the court adhered to the rule that 'whenever a single criminal transaction gives rise to crimes of violence which are committed against several victims, then multiple indictments (and punishments) are appropriate'. 395 Mass., at 28, 478 N.E.2d, at 734–5.

integrity? If so, is not the wrong done greater when the thief steals by one act the property of two of us rather than one? If so, isn't the duty not to steal from me as distinct from the duty not to steal from you, so that stealing from each of us instantiates a separate wrong for which separate punishment is deserved? My own answers to these questions are all affirmative, so that the single act that relieves two victims of their property is two wrongs, and should be punished accordingly.

2. THE METAPHYSICS OF ACT-TOKEN INDIVIDUATION

We are now ready to come to the topic of this chapter proper, which is how to decide whether a defendant has done one act-token or several. As the discussion of the multiple-consequence cases illustrates, we have already done some of the metaphysics of act-token individuation. In Chapter 11 we defended the coarse-grained view of individuation against both the extreme and moderate fine-grained views. This metaphysical work on the individuation of act-tokens has had the pay-off we have seen in the multiple-consequence cases that there is but one act-token in such cases, no matter how numerous the bad consequences. That insight led us to characterize such cases as presenting problems of act-type identity, not act-token identity.

Yet much of the work on the metaphysics of act-token individ-uation we have not yet done. We have thus far only established that where there is one basic act, there is but one act, no matter how many complex descriptions of that act there may be. We have not yet explored how to establish when there is one basic act-token (as opposed to two or more). Are five finger move-ments on a trigger that result in five shots being fired into a house five basic acts or one?[15] Does it matter whether we ask this question in the context of asking how many offences of 'shooting into a building that is usually occupied' have occurred, as opposed to the context of asking how many offences of homicide have occurred when all five bullets strike a single victim in the house, who then dies from the loss of blood from all five wounds?

---

[15] *Davis* v. *Herring*, 800 F. 2d 513 (5th Cir. 1986).

How many act-tokens of driving are there when an accused over a period of nine days drives a stolen car to various locales, over various distances?[16] Does it make any difference whether we ask this question in the context of assessing how many offences of joy-riding the accused is guilty of, as opposed to asking it in the context of assessing how many offences of driving without a licence the accused is guilty of?

These questions cannot be answered by recourse to the work on act-token individuation we did in Chapter 11. They can only be answered by developing a theory for how to individuate act-tokens *vis-à-vis* each other, not *vis-à-vis* more complex actions. It is to developing such a theory that I now turn.

As we saw in Chapters 4–6, human actions are a kind of event. It accordingly makes some sense to seek the identity conditions of event-tokens before we see what might be changed in those conditions when the events in question are human actions. Giving identity conditions for events generally (including bodily movements), and then amending those conditions to suit human-action events (bodily movements caused by volitions), are the two tasks that will occupy us in this section.

2.1. The Individuation of Bodily Movements and Other Events Generally

What we here seek is some general formula for what counts in determining when we have just one particular event (versus when we have two or more). When do we have one flash of lightning, one avalanche, one arm movement, and when do we have two, several, or many?

As with types of events in the last chapter, the very general criteria of identity of Leibniz are the place to start although here too they will not be enough. It is of course true that for all events x and y, if x is identical to y, then x and y must have all the same properties (Leibniz's Law, or the Principle of the Indiscernibility of Identicals). It may also be true that for all events x and y, if they share all the same properties, then they are one and the same event (the Principle of the Identity of Indiscernibles). Yet these general Leibnizian principles do not give identity conditions for events. Rather, they tell us what *identity* is

---

[16] *Brown* v. *Ohio*, 432 U.S. 161 (1977).

for any sorts of things. They are true of the identity of events, in other words, because they are true of the identity of any sorts of things. We need identity conditions that are more discriminating, conditions that tell us when *events* (but not necessarily any other things) are one or many.[17]

Such identity conditions will be related to our theory of the *nature* of events in an intimate way: the nature that particulars must have to be events determines the properties by virtue of which two putatively distinct events are in reality identical. It is this link to the essential nature of the kind events that Leibniz's formal criteria lack; the Leibnizian criteria are criteria of identity in general, whereas true identity conditions for events are criteria for event-identity more specifically. What the Leibnizian criteria can tell us is what follows (from the notion of identity itself) if 'two' events are in reality identical: namely, 'they' must share all of the same properties. Identity conditions for events utilize the *essential* properties of the kind events, in order to tell us when two putatively distinct events are identical: namely, whenever the 'two' share all the properties *essential to being an event*.

We have thus far distinguished the Leibnizian criteria of identity itself from the criteria of identity for events. We now need to distinguish a third sort of criteria relevant here, what Myles Brand helpfully calls 'identifying conditions'. As Brand notes:

Identity conditions are often confounded with identifying conditions. Identifying conditions are criteria for judging or telling how many of a certain kind of object there are . . . Identifying conditions serve epistemological or pragmatic functions. . . . Identifying conditions are true, general statements about the kind of evidence that provides good support for identity claims.[18]

---

[17] On the role of identity conditions for ontological kinds like events versus the Leibnizian criteria for identity, see *Brand, Intending and Acting*, 59–65, and Lombard, *Events*, 23–30. Compare Brody, *Identity and Essence*, who argues that Leibniz's Principle of the Identity of Indiscernibles is all one needs as identity conditions for anything. The disagreement between Brand, Lombard, and myself, on one hand, and Brody on the other, is not a disagreement about whether identity is relativized to ontological kinds (so that identity for events means one thing, identity for properties another, etc.). On relative identity, see David Wiggins, *Sameness and Substance* (Cambridge, Mass., 1980), 15–44. Rather, one should see Leibniz as decisive on the meaning of identity, the only issue being whether more powerful identity conditions are not both possible and desirable for different kinds of things.

[18] Brand, *Intending and Acting*, 59–60.

Identifying conditions are epistemically linked to identity conditions because the latter tell us what it is that the former give good evidence of.

Consider the literal formulation of the *Blockburger* test for identity of act-types. Where each act-type requires evidence for its proof not required by the other, they are not identical act-types, and where the evidence is the same, they are. As we saw in Chapter 12, this is an identifying criterion for act-type identity; it is not an identity condition because it in no way relies on the nature of act-types in order to specify when two are in reality the same. Moreover, such a criterion is itself only evidential: act-types will be identical *usually* if the evidence required for their proof is the same, but this is not necessarily always the case. The world could be arranged so that the same evidence is required to be offered to prove instantiation of two different act-types, when the evidence equally well evidences the existence of each.

We should be interested in all three sorts of criteria as we individuate act-tokens. Consider the analogous problem of individuating physical objects, say, mountains. If we know that Torrey's Peak and Grey's Peak—a pair of 14,000 (plus) feet bits of rock in the Colorado Rockies that are separated by a very high saddle— do not share some property, then by the Leibnizian criteria of identity alone we know that these are not names of a single peak. We can use, in other words, the Leibnizian criteria to a certain extent to individuate physical objects like mountains, even if we are ignorant of the identity conditions for physical objects.

As it happens, we do know the identity conditions for physical objects. Because physical objects have as their essential nature the exclusive occupation of a space at a time, the identity conditions for physical objects are standardly formulated as:

> For all things x and y, if x and y are physical objects, then x is identical to y if and only if x and y have the same spatio-temporal location.

If Grey's Peak and Torrey's Peak occupy the same spatial region, today, they are the same mountains; if they do not, they are not the same (or if, as is actually the case, *parts* of each peak occupy the same spatial region, then the parts are identical and the two peaks are partially identical).

Identifying conditions for physical objects give the conditions under which it would be rational to believe that 'two' physical objects occupy the same spatial region at the same time. For example, Chomolungma (the Tibetan name for Mt. Everest) is identical to Sagamartha (the Nepalese name for Mt. Everest) if and only if a Tibetan climber climbing from the north side would meet a Nepalese climber climbing from the south side, if both climbed their respective peak(s) at the same time.

We shall primarily focus on the identity conditions of events as we seek a basis for their individuation. As we saw in Chapter 4, the nature of events is given by three essential properties: for a particular to be an event, it must be an exemplifying of a *property*, by an *object*, at a *time*. The identity conditions for events should thus be:

> For all things x and y, if x and y are events, then x is identical to y if and only if x and y are exemplifyings of the same property, by the same object, at the same time.

We can usefully explicate this criterion for the identity of events by contrasting it to two other well-known proposals for identity conditions for events. One is the extremely fine-grained theory of Jaegwon Kim that we touched on in Chapter 11.[19] On Kim's view, two event-tokens are identical whenever they are identically (spatio-temporally) located instances of an identical property. The finger-moving$_T$ (that causes a gun to go off, that causes the death of someone) is, on this account, not identical with the shooting, not identical with the killing, because these are all three separate properties. This 'fine-grained' aspect of Kim's theory we rejected in Chapter 11, where we identified the finger-moving, the shooting, and the killing as all one action.

Kim's theory is none the less quite similar to my own in that the identity of events depends on the identities of objects, times, and properties. The difference lies in the sorts of property Kim counts as making a difference to event-identity. On Kim's view, almost any difference of property results in a non-identity of the instances of such properties.[20] On Kim's view, events cannot be

[19] Kim, 'On the Psycho-Physical Identity Theory'; id., 'Causation, Nomic Subsumption, and the Concept of Event'; id., 'Events as Property Exemplifications'.
[20] Not quite every difference in any property affects the identity of an event for Kim. For the refinements, see Brand, *Intending and Acting*, 71–2, Bennett, *Events and Their Names*, 79–82; Lombard, *Events*, 56–62.

instances of more than one property; an event e cannot be a finger-moving, and a shooting, and a killing, because these are three quite distinct properties.

The theory of the nature of events sketched in Chapter 4, and relied on in Chapter 11, is different from Kim's in this respect. Any event-token e not only can, but almost always is, an instance of numerous properties, and it is yet the same event. Saying this gets rid of the very counter-intuitive, fine-grained aspect of Kim's theory. It also unfortunately lands us in a very real difficulty of our own: of all the properties that an object possesses at any given time, *which* are the ones that constitute the identity conditions for events? From Chapter 4's discussion, one can of course answer 'the dynamic properties'. Yet the dynamic properties (of change) that we used in Chapter 4 to give the nature of events are just composed of some constellation of static properties either coming to be exemplified, or ceasing to be exemplified, by an object at a time. The question then recurs: *which*, of all the static properties possessed by an object during some interval of time and which either come to be exemplified or cease to be exemplified, are the ones constituting the identity conditions for events?

We need an example. Suppose that John kisses Mary tenderly on the cheek at noon of a given day.[21] How many events took place: (1) having the same objects (John, Mary) and (2) at the same interval of time (noon, or thereabouts)? On Kim's view, quite a few: there is the event of John kissing Mary tenderly on the cheek, the event of John tenderly kissing Mary, and the event of John kissing Mary on the cheek. This strikes most people as absurd; Mary got kissed only once by John, even if that event instantiated several distinct properties (of being a tender kiss, of being a kiss on the cheek, and of being a tender kiss on the cheek). Yet the opposite extreme—that there was only one event involving these objects at this time—seems also false. Suppose John had a cold, and that cold was transmitted to Mary by the kiss. It might be the case that the event (the transmittal of the cold from John to Mary) occupies exactly the same time, and involves exactly the same objects, as the event John kissing Mary, yet they remain distinct events.

---

[21] Jonathan Bennett's example. See Bennett, *Events and Their Names*, 79. See ibid. 119–28 for a very helpful discussion of the issue of property groupings for purposes of event-identity.

Given the possibility that more than one event can occur at the same place and at the same time, one cannot bypass the need of specifying which collections of properties make for one event as opposed to two or more. We sense that the warming of a sphere is a distinct event from the spinning of the sphere, even if both events involve the same object at the same internal of time; we sense that Esther Williams swimming the channel, and Esther Williams catching a cold, are distinct events, even if they occupy the same spatio-temporal zone.[22] Yet to specify in general how we individuate such events (by the differing constellations of properties that they instantiate) is a very tricky matter, for we seem to lack any principle for aggregating properties into separate 'constellations'.[23]

Fortunately for our purposes one can bypass this thorny issue of specifying adequately the identity conditions of events generally. For purposes of assessing when the same offence is done more than once, the relevant constellation of properties is already fixed. We are not counting events as such; we are counting burglaries, murders, and other act-types whose nature we have already fixed in the moral manner described in Chapter 13. Our problem is more like asking 'How many *kisses* did John give Mary, then?' than it is like asking 'How many *events* took place involving John and Mary then?' Once we know that we are only involved in counting kisses, we also know what sorts of property count in individuating kiss-type events: lip-touching etc. count, but tenderness and on-the-cheek-ness do not count in making an event a kiss. (Although they are properties of kisses, they are not kiss-making properties.)

Consider another example of individuating co-occupying event-tokens when we are interested in event-tokens that instantiate but one event-type. Suppose we have reason to count wedding-tokens (in order to ascertain how many instances of some wedding offence have taken place, perhaps). Suppose further a particularly

---

[22] The examples, respectively, of Donald Davidson and Myles Brand. See Davidson, *Essays on Actions and Events*, 178; and Brand, *Intending and Acting*, 66. See also the examples and discussion of Lombard, *Events*, 157–61.

[23] Thus Jonathan Bennett concludes that this issue has 'no determinate answer', meaning that 'our concept of a particular event has a large dimension of vagueness, letting it sprawl across much of the continuum of possible views running from Kim to the Quinean'. Bennett, *Events and Their Names*, 127.

passive ceremony is performed by one minister for two couples (x, y) and (w, z), the entire ceremony being a speech by the minister. How many illegal weddings have taken place? I take the answer to be 'two', but not because of any spatio-temporal separation of the two wedding events. Rather, what makes there be two instances of the type wedding event is the simultaneous coming into being of two marriage relations. The marriage relation Mxy is distinct from the marriage relation Mwz, and since the coming into being of a marriage relation is a wedding-constituting property, we have two wedding events, not one.

One other well-known position on the identity conditions for events focuses, not on the properties instanced by an event, but on the spatio-temporal locations of events. There are two variations here. On Quine's view, events are just like physical objects in that they *exclusively* occupy a unique spatio-temporal zone.[24] On such a view there is no need to worry what constellations of properties make for one event as opposed to another, for such properties have nothing to do with event-identity. If a sphere rotating occupies the same spatio-temporal region as that same sphere warming, then there is only one event-token, on Quine's view. The identity conditions for events on this view would be the same as that given for physical objects:

> For all things x and y, if x and y are events, then x is identical to y if and only if x and y have the same spatio-temporal location.

The second variation here is the view put forward by Myles Brand.[25] Brand, unlike Quine, agrees with common sense in thinking that events do not *exclusively* occupy their spatio-temporal zones. One cannot thus use actual spatio-temporal location as a sufficient condition of event-identity. This leads Brand to a modal test of event-identity:

---

[24] See Quine, *Word and Object*, 171. See also id., *Theories and Things*, 11–12. Others who have accepted this view include Jack Smart, 'Further Thoughts on the Identity Theory', *Monist*, 56 (1972), 149–62: 160, and E. J. Lemmon, 'Comments on D. Davidson's "The Logical Form of Action Sentences"', in N. Rescher (ed.), *The Logic of Decision and Action* (Pittsburgh, 1967), as well as (eventually) Donald Davidson ('Reply to Quine on Events', in E. LePore and B. McLaughlin (eds.), *Actions and Events: Perspectives on the Philosophy of Donald Davidson* (Oxford, 1985), 172–6).

[25] Brand, *Intending and Acting*, 65–7, 73–9.

For all things x and y, if x and y are events, then x is identical to y if and only if x and y *necessarily* have the same spatio-temporal location.[26]

On Brand's view, the spinning of a sphere and the warming of a sphere do not *necessarily* have the same spatio-temporal locations even if they do in fact coincide in their spatio-temporal locations. It is logically possible for a sphere to rotate without warming, and vice versa, so there is no (logical) necessity to these events coinciding in spatio-temporal location.

Brand's view has much the same difficulty in specifying how or in what sense a co-occupation of a spatio-temporal zone must be *necessary* as my own view does in specifying grouping of properties possession of which makes for distinct event-tokens. One suspects that the necessity Brand wants is a kind of analytic necessity that comes from certain ways of picking out events, an analytic necessity that gives us a warming grouping and a spinning grouping and that make these two separate groupings distinct (while identifying, say, rotatings with spinnings, etc.) If so, then Brand is stuck with the same problem just addressed, namely, how to pick the groupings of properties the possession of which makes for distinct events.

Quine's view is part of his highly revisionary metaphysics whereby we lose our grip on there being any distinction between objects and events. This view obliterates a feature of events that seems basic, namely, their *non*-exclusive occupation of spatio-temporal zones; it also obliterates a feature of objects that seems basic, namely, that they endure for long periods of time (rather than, on Quine's view, being a separate object for each slice of time). I put aside such radical revisions of our metaphysics not because I have good arguments that they are false, but because if they are true, a great deal changes besides our views on individuating co-occupying event-tokens for purposes of double jeopardy.

Whatever the differences between the views of Brand–Quine and my own, their criteria of event-identity would work out much the same as my own in their application to many cases. For most of the cases that puzzle us about event individuation involve issues of spatial or temporal location. The way in which we usually decide whether two putatively distinct instances of a

---

[26] *Intending and Acting*, 65.

given event-type are in reality one is by comparison of the spatio-temporal location(s) of the event(s).

Suppose two instances of finger movement x and y occur at the same time but at different locations. There being no co-occurrence of the two movements at the same place there can be no identity of the two on Quine's view. Likewise on Brand's view, if there is no *actual* sameness of spatial location of the two finger movements, there can be no *necessary* sameness of spatial location and thus no identity. Likewise on the view I hold: since the spatial location of an event (a property instance) is just the spatial location of the object possessing the property, identity of spatial location is as much a *necessary* condition of event-identity as it is on Quine's or Brand's views. This being so, the absence of identity of spatial location is as much a sufficient condition for the lack of identity between two events for me as it is for Quine and Brand. The difference between the views only shows up for those problems of event-identity where the two putatively distinct events occupy the same spatio-temporal zones. This latter kind of identity problem comes up infrequently, at least in double jeopardy contexts; usually where we have identity of spatio-temporal location, we have identity of events. We should thus regard Quine's criterion (of spatio-temporal coincidence) as an *identifying* condition of sameness: since *usually* sameness of spatio-temporal location is accompanied by sameness of events, *always* the former sameness is good evidence of the latter identity.

My own view on the identity conditions of events thus stands comfortably between the over-emphasis on properties by Kim and the over-emphasis on spatio-temporal location of Quine and Brand. Events are the same when but only when they have the same spatio-temporal locations *and* the same constellation of relevant properties. Such an identity condition for events is of course no more precise than are the identities out of which it is constructed. The indeterminacy in one of these underlying identities we have already touched upon—the difficulty of specifying what groupings of properties make for distinct events. This indeterminacy we bypassed for double jeopardy purposes, because in this context the relevant groupings of properties are fixed by the type analysis of the previous chapter. We now need to examine two other indeterminacies, involving the other two underlying identities out of which the identity conditions for events is constituted.

Consider first sameness of spatial location, which I hold to be the sameness of objects involved in events. In Chapter 11 we got rid of some irrelevant distractions on this topic. Once one sees that causing the death of another may be a property of some event e, but that neither the death nor the causal sequence of causing a death are *parts* of e, then we needn't worry about e being located where such effects of e take place. None the less, even without this distraction, a thorny issue remains: how do we know which, among the proper parts of an object, are the one(s) involved in any given event-token e?[27] When Jones's finger moves on the trigger, is the relevant object (and thus the relevant spatial location of e) Jones's entire body, Jones's arm (or so much of it as is required to flex in order to have the fingers move), just the particular finger, or just the part of the surface of the finger touching the trigger?

This problem—let us call it the spatial-sizing problem for events—can be exacerbated if we recognize the possibility of discontinuous objects, and thus of spatially discontinuous events. Lawrence Lombard, for example, asks whether there is *an* object all the world's snow, so that there could be *an* event the melting of all the world's snow.[28] Notice that if there is such an event, then it has multiple locations: at all the places where the world's snow might be.

Recognition of the spatial-sizing problem for events generally has led legal commentators on the identity of acts for double jeopardy purposes to undue scepticism about there being any metaphysical answers here. Phil Johnson, for example, concludes that 'any incident of physical conduct can be divided indefinitely into its component parts'.[29] In this he is seconded by Peter Westen and Richard Drubel, who despair of there being any 'way to make sense out of the notion that a course of conduct is "really" only one act, rather than two or three, or, indeed, as many as one likes'.[30] Yet I don't think the spatial aspect of the

---

[27] On the difficulties of isolating discrete spatial locations of events because events (and their corresponding objects) have parts that are themselves events (objects), see Bennett, *Events and Their Names*, 153–64; Lombard, *Events*, 120–31;' Davidson, *Essays on Actions and Events*, 175–7; and Thomson, *Acts and Other Events*, *passim*.   [28] Lombard, *Events*, 123–4.

[29] Philip Johnson, 'Multiple Punishment and Consecutive Sentences: Reflection on the Neal Doctrine', *California Law Review*, 58 (1970), 357–90: 363.

[30] Westen and Drubel, 'Toward a General Theory of Double Jeopardy', 114.

sizing problem is any more troublesome for events than it is for objects. Objects do have parts and those parts are themselves objects. It might be the case that there is such a thing as the smallest part of an object that is itself an object, and thus the smallest part of an event that is itself an event. One might then speak of 'atomic events' just as we speak (or used to speak) of indivisible atoms.[31] Recent physics should not make one sanguine about this possibility, however. It might alternatively be the case that one could make sense of objects being composed of *points*, points being things that have no spatial dimensions but only spatial locations. On this view, however, it is difficult to make sense of the idea of either objects or events *occupying spatial regions*— for if their smallest parts (points) do not, how do they? As a third alternative, it might be the case that there is no limit to the size of objects and thus of events. The universe is one object, and so is each of its parts, divisible into smaller parts without limit. At any given time there would thus be one event, and many other, smaller events, involving objects smaller than the universe as a whole.[32] This needn't be a very sceptical position, since it admits of the possibility that there are natural sizings of parts of objects and thus of events.

The spatial-sizing problem is somewhat alleviated by the fact that we rarely need to ask how many *events* took place. We usually have much more finely focused enquiries, as we do in double jeopardy contexts where we want to know how many *bodily-movement events* took place within some scenario. Still, even in such more finely focused enquiries, we face some sizing difficulties. Suppose someone drives by an occupied structure with a pistol in each hand. He simultaneously moves the trigger finger of each hand, sending two bullets into the structure, one of which kills its occupant. Has there been one instance of a bodily-movement event, the object being the two trigger fingers and the event being its movement? Or two, because there are two parts to the latter movement, one part for each finger? Or four, because after each finger-pulling there was a finger-releasing? Or many, because

---

[31] This is Lawrence Lombard's solution to the spatial-sizing problem for events. See Lombard, *Events*, 168.

[32] Judy Thomson's view of events and their parts (that are also events). Thomson, *Acts and Other Events*.

each finger moved over some distance d, and d is divisible into indefinitely many parts?

Absent some theory about there being natural sizings of objects and thus of events, we cannot yet answer such questions. There may be such a theory, as Lombard for one has recently proposed. I put aside the enquiry not out of scepticism about the possibility of such a theory. Rather, it will turn out that for double jeopardy purposes such a theory could not by itself be a sufficient basis to individuate act-tokens. As we shall see, we need to solve the sizing question relative to double jeopardy's purposes, and those purposes are such that we could not use a purely metaphysical answer to the problem.

The third indeterminacy in our identity conditions for events is of course the indeterminacy in the notion of sameness of temporal location. Again, this indeterminacy does not stem from any doubt about whether to include the temporal location of effects of an event e in e's overall temporal location; for again, in Chapter 11 we disposed of any such worry by showing that the effects of e are not proper parts of e. Deaths are not parts of killings, so we needn't worry about including the temporal locations of deaths as we temporally locate acts of killing.

Still, even with this clarification, we have enough problems of identity of temporal locations. For events, unlike objects, have temporal parts (as well as spatial parts) that are themselves events. We therefore seem to have some choice as to the temporal size of events when we are counting them up. Suppose the drive-by shooter in my earlier scenario behaved thus: while in a stationary car, he moved the trigger finger of his right hand five times, sending five bullets into the unoccupied structure, one of which killed its occupant. Is there one bodily movement occupying the entire duration t_1–t_{10} (the time from the beginning of the first movement to the end of the last)? Or five, t_1–t_2, t_2–t_3, etc. (each pair being the beginning of one movement and the end of that movement)? Or many, given the infinite divisibility of each interval, t_1–t_2, t_2–t_3, etc., into many temporal parts?

There are also three general responses to the temporal aspect of the sizing problem. One could think that there is such a thing as the smallest event-interval, a kind of atomic event-interval.[33]

---

[33] See Lombard, *Events*, 168 ff., where the shortest interval for events is given by the notion of an atomic quality space that must be 'crossed' by an atomic event.

Alternatively, one might think that all larger events have as their ultimate temporal parts instantaneous points in time, points having temporal location but no temporal duration. Or one might again adopt the promiscuous position, that the entire history of some spatial region is one event, and that each part of that temporal duration is itself an event of lesser duration, such division going on without limit.

I again put aside any defence of one of these three metaphysical positions as unnecessary for double jeopardy purposes. The purposes of double jeopardy demand an answer to the temporal-sizing problem that is different from any of these three metaphysical answers by themselves (although the legal answer might make use of one of these metaphysical answers).

2.2. The Individuation of those Bodily Movements that are Caused by Volitions

Before coming to the solution of the sizing problems in terms of the purposes of double jeopardy, we have one more purely metaphysical task to undertake. Actions, as we have seen in Chapters 5 and 6, are not identical with bodily-movement events. Rather, actions are identical to the causal sequence, volition–cause–bodily movement. We thus need to ask whether this more complex event should be treated any differently with respect to the identity conditions for events generally.

One difference is readily apparent: the spatio-temporal locations of actions are different from the spatio-temporal locations of the corresponding bodily movements. On the view of action defended in Chapters 5 and 6, actions are complex wholes, the parts of which are volitions and bodily movements. Given that each of those parts has a distinct location and a distinct temporal duration, the whole cannot have the same spatio-temporal location as either part, but only of both together.

To individuate actions, we must thus not only know how to individuate physical events (bodily movements) but also mental events (volitions). Volitions, like all events and states, are individuated by the *object* that possesses the relevant (volition-making) properties, the *time* at which such properties are instantiated, and by the relevant (volition-making) *properties*. But unlike non-mental events, among the properties individuating volitions are Intentional properties: different Intentional object, different volition.

Consider more familiar mental states like those of belief, desire, or intention. If different persons at some time t desire to eat the same piece of apple pie, there are two desire-tokens because there are two different persons involved. Also, if I desire at t_1 to eat a slice of apple pie, and I desire at t_2 to eat that same slice of apple pie, then the desires are distinct so long as t_1 is not identical to t_2. Finally, if I desire at t to go downtown, and I also desire at t to stay at home, I have two desires, not one, because the relevant property of those desires—the Intentional object—is different.

Individuating mental states by their Intentional objects is a very tricky business to which I have briefly sketched an approach elsewhere.[34] This much seems clear: despite the deceptive appearance of certain idiomatic English usages, we never intend, desire, predict, or will an act-token; rather, we form the objects of such mental states over *types* of acts (of which the act we do will then be an instance). Admittedly, we say things like 'He intended the action he did' where 'the action he did' seems to refer to an individual act-token. What we mean, however, is that he intended to do *some* act-token that instanced a particular type of action. When I intend to go downtown, I intend to do a going-downtown type of action (of which the act-token I do will then be an instance if I successfully execute my intention).

The implications of this feature of mental state (and hence, volitional) individuation is this: perhaps we can solve the sizing problem for actions even if not for events generally by this feature of volitional individuation. The objects of volitions might give us interval-sizings for actions that bodily movements by themselves lack.

The basic intuition goes like this. Volitions are the executors of our more general intentions and plans, which are themselves the executors of our belief–desire sets. The objects of volitions are the action-types we choose as appropriate to execute our intentions and motivations. While events like bodily movements may be infinitely divisible into ever smaller spatio-temporal parts that are themselves bodily-movement events, the objects of volitions are not. Persons have limited capacities to consider and to choose ever smaller bits of bodily movements as the executors of their

[34] Moore, 'Intentions and *Mens Rea*'.

intentions and motivations. Those smallest choosable bits are the simplest things we know how to do. It is these smallest bits that form the objects of volitions and by which we individuate volitions. Unless weakness of the will is much more pervasive than it seems to be, usually a volition with some type of (smallest bit of) bodily movement as its object will cause a movement-token to occur that has an instance of that type. Whatever sizing of movement is so caused is the appropriate sizing to count as one act-token.

Consider my example of a temporal-sizing problem, the drive-by shooter who shoots five quickly successive shots into a single building. Suppose it were true that the simplest bodily movements persons could learn to do to execute a plan to shoot a building or a person were just such a sequence of finger-movings. Then I take it there would be but one action event, the single volition having as its object the complex sequence of movements (as a type) and that volition causing an instance of such a type to occur.

This outcome is implausible, but only because the psychological suppositions on which it is based are implausible. More plausibly, we know how to move our trigger finger inward. More plausibly, therefore, the Intentional object of our volitions is an individual pulling inward of the finger. There are therefore at least five volitions, each having as its object this type of smaller movement and each causing an instance of this type to occur. There are thus at least five act-tokens here, not one.

Could there be more? Imagine that the defendant were trained to shoot in the following way. Those who suddenly pull their trigger finger usually jerk the gun sufficiently that they are inaccurate in their shooting. *Squeeeeeze*, the trigger, as my old Marine sergeant used to say. Suppose the squeezing movement is mastered only by mastering sequential movements: pull inward the smallest bit one can, and keep doing it until the gun goes off (the bits being small enough that *when* the gun goes off should be a surprise to you). For a person who is so trained and who exercises that training on the occasion in question, it might be the case that the object of his volitions is the smallest bit of movement that he can notice. That means he has several volitions with each trigger pull, each having such type of movement as its object, and each causing an instance of that type of movement to occur. Then we would have more than five act-tokens in the scenario suggested.

As we noticed before in Chapter 6, it is important not to confuse the volitions we actually have with the volitions we are aware of as we act. Even though volitions execute our intentions and motivations, and even though their objects are fixed by that executor's role, it need not be the case that we are aware of individual volitions as we act. My skilled sequential trigger-squeezer, like a skilled pianist, may well have the volitions needed to learn a routine recede from consciousness as he becomes better at the routine. It is none the less plausible to suppose that if the sequential movements persist in the way that he shoots or plays the piano, then the volitions that initially caused them are still causing such movements, even though the actor is no longer aware of them. We thus need not be embarrassed by our typical inability to find an introspective answer to Ryle's sort of question, 'how many acts of will [a person] executes in, say, reciting "Little Miss Muffet" backwards'.[35] For volitions and their objects are not fixed by our present phenomenological experiences of them.

As should be apparent from the foregoing discussion, act-tokens individuated in the way suggested will come in pretty small spatio-temporal sizings: at the largest, one trigger finger movement in each of my two drive-by shooting hypotheticals. There are good policy reasons why courts should not use this relatively fine-grained metaphysics when they individuate act-tokens for double jeopardy purposes; needed for double jeopardy purposes, I shall argue shortly, is a coarser-grained manner of individuation of act-tokens. Yet some courts have pretended to arrive at some such coarser-grained theory of individuation metaphysically, not as a matter of double jeopardy policy. Such courts seem to believe that the relevant items to consider in individuating act-tokens is not volitions but, rather, the intentions that cause such volitions.

This is the well-known 'same intent' test for individuating act-tokens. If one intent causes a variety of bodily movements by a defendant, those movements are all parts of one act, not several. For example, 'A series of shots may constitute one act . . . where they are fired with one volition.'[36]

[35] Ryle, *The Concept of Mind*, 65.
[36] *Spinnell* v. *State*, 83 Tex. Cr. App. 418, 203 S.W. 357 (1918).

To the extent that this is an attempt at metaphysics at all, it is bad metaphysics. To begin with, the 'single volition' causing the numerous movements is not a volition at all. As John Austin accurately observed, the object of a volition is the simplest movement we know how to do in order to accomplish something else.[37] Except on very implausible psychological suppositions, the sequence of movements needed to fire a non-automatic weapon several times does not satisfy this test for volitional objects.

Such a 'volition' is thus an intention causing more numerous volitions with more finely grained types of movements as their objects. The intent is typically an intent to shoot someone, or perhaps to kill them. It may well be one intention, but such an intention is not a *part* of the action (as are volitions). Such intentions are *causes* of actions. One could not think, therefore, that by individuating such intentions one was individuating a crucial part of actions, and that one could thus size such actions by their sharing a common intention-part. The most one could think is that one could individuate actions by individuating the immediate causes of actions. Since intentions are the immediate causes of those sequences of volition-movements we call actions, the same intent is indicative of there being the same action.

Donald Davidson once subscribed to a test for event-identity generally that would support this latter defence of the 'same intent' test. He urged that two events are identical if and only if they have all the same causes and all the same effects.[38] On the plausible suppositions that the same immediate causes will themselves have the same, more remote causes, and that the same immediate effects will themselves have the same further effects, Davidson's criterion can be tightened to:

> If x and y are both events, then x is identical to y if and only if x and y have the same immediate causes and the same immediate effects.[39]

There are two problems with using this Davidsonian criterion in defence of the same-intent approach to act-token individuation. One is that the criterion is not satisfied in the cases that the same-intent test counts as having but one act. Granted, five

---

[37] See the discussion in Chs. 5 and 6.
[38] Davidson, *Essays on Actions and Events*, 179.
[39] For this amendment of Davidson, see Lombard, *Events*, 74–5.

successive trigger finger movements done in execution of a single intent to kill have the same immediate cause, yet they do not have the same immediate effects. Even if all five bullets strike the victim, and he dies from the loss of blood from all five wounds, the *immediate* effect of each finger movement is different. Finger movement number one causes bullet number one to strike the victim at location number one, which causes wound number one, while finger movement number two causes bullet number two, which causes wound number two, etc.

The second problem lies with the Davidsonian criterion itself. Why cannot two distinct events have the same immediate causes and the same immediate effects?[40] Causally inert events, if there are any, might well be distinct even though having the same causes (none) and the same effects (none). At most, therefore, Davidson's conditions are identifying conditions for events, not identity conditions.[41] And the action examples we have been considering are cases of just the sort where part of those conditions break down, because the same immediate cause (an intention) is not much if any evidence that the acts caused are the same.

I shall come back to the 'same intent' test of act-token individuation as a test to be justified by the policies underlying double jeopardy. Although I shall there reject it as not serving those policies, my aim here has been more preliminary: such a test is a very bad metaphysical test for identity of act-tokens.

3. RESTRICTING THE METAPHYSICAL ENQUIRY SO AS TO SERVE THE PURPOSES OF DOUBLE JEOPARDY

It is worth reminding ourselves of the policies that justify double jeopardy's three bans. The dominant value served by the ban on multiple punishment and the ban on multiple prosecution after conviction is the prevention of excessive punishment, that is, the prevention of punishment that is disproportionate to wrongdoing. A secondary justification for the second of these two bans is

---

[40] For explorations of this criticism of Davidson, see Brand, *Intending and Acting*, 69; Thomson, *Acts and Other Events*, 70.
[41] The same can be said of David Armstrong's causal criteria of property-identity that I adopted in Ch. 13. That is because Armstrong, like most philosophers (myself included), lacks any very clear idea of identity conditions for properties.

what I called the embarrassment–harassment justification. This is the good of preventing both embarrassment to defendants arising out of unnecessarily duplicative criminal trials, and harassment of personal or political enemies by prosecutors by the use of such trials. The third of double jeopardy's bans, that against multiple prosecution after acquittal, is justified by three different considerations: first, the danger of convicting innocent defendants if prosecutors may 'try and try again'; secondly, the anxiety a defendant must suffer if he is again in jeopardy of conviction or increased sentence; and thirdly, the incentive that can be given to prosecutors to promote judicial efficiency by joining related counts against a defendant into a single trial.

As we saw in Chapter 13, none of these last four policies is so powerful that we allow it to justify a bar to multiple criminal trials of a given defendant when truly new wrongdoing is at stake in such trials. We (rightly) limit, in other words, the operation of these four values by the retributive value that those guilty of wrongdoing are not punished too lightly for their wrongdoing. It is this limit, together with the value of not punishing too severely, that creates the policy that guides questions of act-token and act-type individuation for double jeopardy purposes: punishment should be in proportion to wrongdoing.

As we saw in Chapter 13, this reduction of double jeopardy's policies to one—the achieving of punishment proportionate to wrongdoing—has the result that act-types are to be individuated for double jeopardy purposes by what I called the moral approach: one tests whether two offences are the same for double jeopardy purposes by seeing whether or not the (intrinsically or instrumentally) morally wrong act-types referred to by each offence are or are not (fully or partially) identical. The moral approach to act-type individuation thus requires that we isolate a morally salient act-type that is referred to by each statutory offence.

To serve the value of achieving punishment proportionate to wrongdoing, we again need to resort to such morally salient act-types as we individuate act-tokens. To see how and why, consider my earlier-mentioned drive-by shooter.[42] He squeezes off five shots into a building that is usually occupied, hitting with each

---

[42] Taken from the facts in *Davis* v. *Herring*, 800 F.2d 513 (5th Cir. 1986).

bullet the victim, who dies from the loss of blood caused by all five wounds. How many act-tokens has the defendant performed? Possible answers from our earlier discussion of the metaphysics of bodily-movement events: as many acts as there are 'atomic event intervals' crossed by atomic objects;[43] or, infinitely many, one for each spatio-temporal point, there being an infinite number of such points in the spatio-temporal zone consisting of the five finger movements;[44] or, one, several, or indefinitely many, depending on the level of subdivision (of whole events into their spatio-temporal parts, which are also events) on which one fastens.[45] Possible answers from our earlier discussion of the individuation of volitions: one, five, or many, depending on how large one takes to be the object(s) of the volition(s) involved.

I earlier refrained from defending any of these metaphysical answers to event and act individuation, not because I think there are no answers, but because double jeopardy's purposes demand that we eschew those answers in favour of what I shall now call the 'wrong-relative' manner of act-token individuation. On the wrong-relative mode of act-token individuation, there is no legally usable answer to the question 'How many acts?', until we know what morally salient *type* of act is involved with a given statutory offence. If the question is 'How many *murderous* acts were there?' the correct metaphysical and thus legal answer is 'One'.

We should individuate act-tokens this way for double jeopardy purposes because basic acts, as such, are not morally wrong. What makes a basic act-token wrong is not that it instantiates types of acts, such as moving the finger or moving the foot. Rather, morality is such that it is always further consequences of such movements that make them wrong. Since such wrong-making features of acts lie in such further consequences, we must repair to those consequences in order to assess how many times an accused has done a given wrong. This is true, not because such consequences are *parts* of such acts; rather, it is because the causing of such consequences is a *property* of such acts that makes them be morally wrong acts on this occasion.

---

[43] Lawrence Lombard's view. See Lombard, *Events*.
[44] A suggestion of Jonathan Bennett's. See Bennett, *Events and Their Names*.
[45] Judy Thomson's view. See Thomson, *Acts and Other Events*.

Taking such wrong-making properties into account is what we did when we were assessing what I called the multiple-consequence cases. Where a single act causes a single bullet to kill two victims, there are two wrongs done by this single act, and the perpetrator should be punished accordingly, i.e. twice. Here we face the converse case, where five bullets cause a single death. Only one wrongful killing was done, which means that we must lump five (or more) metaphysically distinct acts of killing together and call them one act of killing. We should say that not only was there but one morally salient act-type done here (killing); but it was done only once, meriting but one punishment.

Change the wrong (relative to which act-token individuation is done), however, and potentially you change the number of times the defendant acted. If no one was hit, so that the charge is shooting into an occupied building, our hypothetical drive-by shooter did this wrong five times. The morally salient act-type behind this offence is shooting into an occupied building, which act-type is instrumentally wrong because it risks serious injury to the occupants. This instrumental wrong is done five times by the defendant, who causes five bullets to enter the house; this, because it is the consequence of a bullet entering the house while it is occupied that is the wrong-making consequence—and there were five of these. No matter how spatio-temporally close the shootings might be, there were *at least* five of such instrumental wrongs done, not one; there were five, because five bullets create five separate occasions on which risk of serious injury was imposed on the occupants of the structure. No matter how many separate volitions it required to pull the trigger each time, there were *at most* only five instances of this wrongful act-type done, for each volitionally caused movement segment only jointly caused one wrongful state of affairs to exist.

Now consider the defendant who drives away with another's car in one city, and is caught driving that car in another city nine days later.[46] Conceding that the morally salient act-types that underlie the joy-riding and theft statutes are partially identical (both involving the taking of another car), how many times did the defendant violate this offence? We should again put aside the

[46] *Brown* v. *Ohio*, 431 U.S. 161 (1977).

metaphysical questions of how many bodily-movement events, and how many volitions, occurred; over this stretch of time, there were doubtlessly many such movements and volitions, but this does not answer to our purpose. Rather, we need first to identify the morally wrong act-type involved here in order to ask, how many times was *that* wrong done?

The morally salient act-type involved in the combined joy-riding–car theft offence is the taking of the property of another. To *take* another's property, one must move one's body in such a way as to cause that property to come into one's own possession. Our hypothetical defendant did this only once. Even though he had possession for the entire nine days, he caused himself to come into possession only once. This is true even if he stopped driving, got out of the car, and then after an interval drove off again in the car. The state he was prohibited from causing himself to be in—that of being in possession—he was in even when out of the car (so long as he had control of it). He could not cause himself to come into the state of possession when he already was in that state.

Now suppose the morally salient act-type is not *taking* another's car without their consent, but is *driving* that car without the owner's consent. Since this is arguably one of the morally salient act-types that underlie joy-riding statutes, we might imagine the defendant to be charged with two counts of joy-riding. (Alternatively, if one resists this interpretation of joy-riding, imagine that the defendant was unlicensed, and the charges are two counts of unlicensed driving.)

It may seem that with act-types like driving we have no consequence the causing of which the act-description 'driving', prohibits. Yet we have already established that this is not so in Chapter 8, where we argued for the causal version of the equivalence thesis. All morally salient act-types underlying the offences of Anglo-American criminal law include consequences. In the case of driving, that consequence is the state of being in control of a car's motion.

The problem with individuating bits of driving thus does not lie in the lack of any consequence by virtue of which we individuate driving-tokens. The problem rather lies with our individuating such consequences. Unlike deaths of persons, states of control over an automobile's motion are not readily individuated.

Granted, however many movements and volitions it takes to cause one of those states should count as one token of driving— yet when do we have *one* of those states?

This I take to be the problem known in criminal law as the problem of 'continuing offences'. Crimes of omission and possession are the usual examples, but crimes like the driving offences share the relevant feature of there being a long-term, unbroken state that seems determinative of the issue of individuation. The answer to such problems of individuation is to look for breaks in the continuity. It is to ask Justice Blackmun's question in dissent in *Brown* v. *Ohio*: did the defendant ever *stop* driving?[47] If he did, then when he was found driving again nine days after his initial driving in the first town, he was engaged in a distinct instance of driving and can again be prosecuted and punished. Unlike *taking* the car, if the relevant act-type is *driving*, he did that as many times as he got into the car and caused himself to be in control of its movements.[48]

The wrong-relative approach to act-token individuation is thus capable of solving the spatio-temporal-sizing problem,[49] and of doing so in a way that secures punishment proportional to wrongdoing. The solution lies in identifying as one act-token all those metaphysically distinct act-tokens that cause but one instance of some state of affairs to occur, where the causing of that state of affairs is the wrong-making property of each of the act-tokens to be identified as one. So long as we can individuate such states of affairs, then we can individuate instances of acts that cause them.

Once one sees the unifying function of morally salient consequences of actions, it is also easy to see both the temptations and

---

[47] *Brown* v. *Ohio*, 431 U.S., at 171.

[48] Thus, the ultimate disagreement between the majority and the dissenting opinions in Brown should have been about whether the morally salient act-type underlying the combined car theft–joyriding offence was a taking act-type, or a driving or possessing act-type. Since the offence of car theft is partially identical to the offence of joyriding by the overlapping of the act-type of taking, it is taking that is the relevant act-type.

[49] We also implicitly used the wrong-relative approach to solve the problem of deciding how many properties make up a separate 'constellation of properties' that individuates spatio-temporally co-present events. When we know that the morally salient act-type is kissing (our earlier example), then we know which properties make for one event (of this sort).

the error of the 'same intent' test for act-token individuation. One temptation stems from the fact that how many culpable intentions one has as one acts *is* relevant to one's overall desert. The error is to confuse culpability (for bad intentions) with wrongdoing (stemming from bad actions). That one was motivated to shoot five times into a building by a single intention—say to kill its inhabitant—is relevant in assessing the culpability with which the defendant acted over this interval of time. It is irrelevant to how much wrong he did in thus shooting whether that wrong be in terms of shooting into a building or in terms of killing.[50]

Other than this confusion of culpability with wrongdoing, the same-intent test may also seem tempting because it does often lead to intuitively appealing individuations for double jeopardy purposes. This is because defendants often do just the wrongs they intend to do, in which case the same-intent test individuates identically with wrong-relative individuation. In the drive-by shooting scenario: if the wrong relevant to individuating how many acts were done is killing (because that is what the defendant's shots accomplished), then there was one act-token for double jeopardy purposes on the wrong-relative approach. Analogously, if the defendant shot five times into the building in order to kill the inhabitant, there is also one act here using the same-intent test, because a single intention motivated all five shots. It is this coincidence in results between the two tests that accounts for the intuitively satisfactory individuations of the same-intent test.

Yet the latter is not a perfect proxy for the wrong-relative test, for the obvious reason that often the wrongs in fact done by a defendant differ from the wrongs he intended to do. If each of the shots of the drive-by shooter were motivated by a different intention, there would be five different acts under the same-intent test, yet if each of the bullets kills but one victim, there is but one act here if the charge is murder. Conversely, if each of the five shots was motivated by one intention, there would be but one act even if the charge is shooting into an unoccupied

[50] Again, my analysis of double jeopardy has been limited to proportioning punishment to wrongdoing only, not to wrongdoing *and* culpability. See Ch. 12 n. 1.

structure; whereas under the wrong-relative approach, there are five instances of this offence committed by the five shots.[51]

---

[51] It is interesting that courts often lump together tightly sequenced acts, like squeezing off five shots in quick succession, so as to treat them as but one instance of a wrong such as shooting into an occupied dwelling. To a large extent the confusion referenced in the text suffices to explain this phenomenon: courts see that there is but one intention in such cases and, confusing culpability with wrongdoing, assume that means that there is but one wrong done too. To the extent the culpability/wrongdoing confusion is not the explanation here, the courts must be imposing a kind of 'volume discount' on multiply done wrongs: it is significantly less wrongful to shoot the second, third, etc. times, once one has already shot into the building once. Yet such a volume discount is very hard to justify. Each bullet represents the same increment of risk to human life and well-being—or, at least, if it doesn't, the degree of risk each bullet poses is not related to its order in the temporal sequence of bullets. Moreover, a volume discount on a tightly sequenced series of shots is hard to square with the volume *penalty* we impose on those who are recidivists. As is well known, we often *increase* the penalty for each successive shooting into an occupied dwelling, if the separate shots are widely enough separated in time to be charged in separate trials.

REFERENCES

ALBRITTON, R., 'Freedom of Will and Freedom of Action', *Proceedings and Addresses of the American Philosophical Association*, 59 (1985), 239–51.

ALEXANDER, L., 'Reconsidering the Relationship among Voluntary Acts, Strict Liability, and Negligence in Criminal Law', *Social Philosophy and Policy*, 7 (1990), 84–104.

American Law Institute, *Model Penal Code* (Philadelphia, 1962).

—— *Model Penal Code and Commentaries* (Philadelphia, 1985).

—— *Restatement of the Law of Torts* (St Paul, Minn., 1934).

ANDERSON, J., *The Architecture of Cognition* (Cambridge, Mass., 1983).

ANNAS, J., 'How Basic Are Basic Actions?' *Proceedings of the Aristotelian Society*, 78 (1978), 195–213.

ANSCOMBE, G. E. M., *Intention*, 2nd edn. (Ithaca, NY, 1963).

ARMSTRONG, D. M., *A Theory of Universals* (Cambridge, 1978).

ARNELLA, P., 'Character, Choice and Moral Agency: The Relevance of Character to Our Moral Culpability Judgments', *Social Philosophy and Policy*, 7 (1990), 59–83.

ARNOLD, C., 'Conceptions of Action', *Sydney Law Review*, 8 (1977), 86–106.

AUDI, R., 'Intending', *Journal of Philosophy*, 70 (1973), 387–402.

—— 'Intending, Intentional Action, and Desire', in J. Marks (ed.), *The Ways of Desire* (Chicago, 1986).

—— *Practical Reasoning* (London, 1989).

AUNE, B., 'Can', in P. Edwards (ed.), *Encyclopedia of Philosophy* (New York, 1967).

AUSTIN, J., *Lectures on Jurisprudence*, 2 vols., 5th edn. (London, 1885).

AUSTIN, J. L., 'A Plea for Excuses', *Proceedings of the Aristotelian Society*, 57 (1956), 1–30.

—— *How to Do Things with Words*, 2nd edn. (Cambridge, Mass., 1975).

BALKIN, J. M., 'The Rhetoric of Responsibility', *Virginia Law Review*, 76 (1990), 197–263.

BAIER, A., 'The Search for Basic Actions', *American Philosophical Quarterly*, 8 (1971), 161–70.

BEARDSLEY, M., 'Actions and Events: The Problem of Individuation', *American Philosophical Quarterly*, 12 (1975), 263–76.

—— 'Intending', in A. Goldman and J. Kim (eds.), *Values and Morals* (Dordrecht, 1978).

BENNETT, J., 'Shooting, Killing, Dying', *Canadian Journal of Philosophy*, 2 (1973), 315–23.

—— 'Morality and Consequences', *The Tanner Lectures on Human Values*, 2 (1981), 45–116.

—— *Events and Their Names* (Indianapolis, 1988).

BENTHAM, J. *Introduction to the Principles of Morals and Legislation* (Buffalo, NY, 1988).

BIRO, J., and SHAHAN, R. (eds.), *Mind, Brain, and Function* (Norman, Okla., 1982).

BISHOP, J., *Natural Agency* (Cambridge, 1989).

BLACKSTONE, W., *Commentaries on the Laws of England*, Sharwood edn. (Philadelphia, 1859).

BOHLEN, F., 'The Moral Duty to Aid Others as the Basis of Tort Liability', *University of Pennsylvania Law Review*, 56 (1908), 217–44, 316–38.

BOYD, R., 'How to Be a Moral Realist', in G. Sayre-McCord (ed.), *Moral Realism* (Ithaca, NY, 1988).

BRACTON,. *De Legibus*, ed. Sir Travers Twiss (London, 1879).

BRAND, M., 'Danto on Basic Actions', *Nous*, 2 (1968), 187–90.

—— 'The Language of Not Doing', *American Philosophical Quarterly*, 8 (1971), 45–53.

—— 'Cognition and Intention', *Erkenntnis*, 18 (1982), 165–87.

—— *Intending and Acting* (Cambridge, Mass., 1984).

—— (ed.), *The Nature of Human Action* (Glenview, Ill., 1970).

—— and WALTON, D., (eds.), *Action Theory* (Dordrecht, 1976).

BRATMAN, M., 'Individuation and Action', *Philosophical Studies*, 33 (1978), 367–75.

—— 'Davidson's Theory of Intention', in B. Vermazen and M. Hintikka (eds.), *Essays on Davidson: Actions and Events* (Oxford, 1985).

—— *Intention, Plans, and Practical Reason* (Cambridge, Mass., 1987).

BRINK, D., *Moral Realism and the Foundations of Ethics* (Cambridge, 1989).

BRODY, B., *Identity and Essence* (Princeton, NJ, 1980).

BROWN, D. G., *Action* (Toronto, 1968).

BUCHANAN, J., *The Philosophy of Human Nature* (Richmond, Ky., 1812).

BUXTON, R., 'The Working Paper on Inchoate Offenses: Incitement and Attempt', *Criminal Law Review* (1973), 656–73.

—— 'Circumstances, Consequences, and Attempted Rape', *Criminal Law Review* (1984), 25–34.

CARE, N. S., and LANDESMAN, C. (eds.), *Readings in the Theory of Action* (Bloomington, Ind., 1968).

CHISHOLM, R., 'Law Statements and Counterfactual Inference', *Analysis*, 15 (1955), 97–105.

—— 'The Descriptive Element in the Concept of Action', *Journal of Philosophy*, 61 (1964), 613–25.

—— 'Freedom and Action', in K. Lehrer (ed.), *Freedom and Determinism* (New York, 1966).

—— 'Events and Propositions', *Nous*, 4 (1970), 15–24.

—— 'States of Affairs Again', *Nous*, 5 (1971), 179–89.

—— 'The Agent as Cause', in M. Brand and D. Walton (eds.), *Action Theory* (Dordrecht, 1976).

—— *Person and Object* (La Salle, Ill., 1976).,

CHOMSKY, N., *Studies on Semantics in Generative Grammar* (The Hague, 1972).

COOK, W. W., 'Act, Intention, and Motive', *Yale Law Journal*, 26 (1917), 645–63.

—— 'The Logical and Legal Bases of the Conflict of Laws', *Yale Law Journal*, 33 (1924), 457–88.

CORRADO, M., 'Automatism and the Theory of Action', *Emory Law Journal*, 39 (1990), 1191–1228.

DAN-COHEN, M., '*Actus Reus*', in S. Kadish (ed.), *Encyclopedia of Crime and Justice* (New York, 1983).

DANTO, A., 'What We Can Do', *Journal of Philosophy*, 60 (1963), 435–45.

—— 'Basic Actions', *Philosophical Quarterly*, 2 (1965), 141–8.

—— 'Freedom and Forebearance', in K. Lehrer (ed.), *Freedom and Determinism* (New York, 1966).

—— *Analytical Philosophy of Action* (Cambridge, 1973).

—— 'Consciousness and Motor Control', *The Behavioral and Brain Sciences*, 8 (1985), 540–1.

D'ARCY, E., *Human Acts* (Oxford, 1963).

DAVIDSON, D., *Essays on Actions and Events* (Oxford, 1980).

—— 'Reply to Quine on Events', in E. LePore and B. McLaughlin (eds.), *Actions and Events: Perspectives on the Philosophy of Donald Davidson* (Oxford, 1985).

—— 'Reply to Bruce Vermazen', in B. Vermazen and M. Hintikka (eds.), *Essays on Davidson 'Action and Events'* (Oxford, 1985).

—— 'Reply to Michael Bratman', in B. Vermazen and M. Hintikka (eds.), *Essays on Davidson 'Action and Events'* (Oxford, 1985).

DAVIS, L., 'Individuation of Actions', *Journal of Philosophy*, 67 (1970), 520–30.

—— *Theory of Action* (Englewood Cliffs, NJ, 1979).

DENNETT, D., *Content and Consciousness* (London, 1969).

—— *Brainstorms* (Montgomery, Vt., 1978).

—— *Elbow Room* (Cambridge, Mass., 1984).

—— *The Intentional Stance* (Cambridge, Mass., 1987).

—— *Consciousness Explained* (Boston, 1991).

DEVITT, M., *Realism and Truth* (Cambridge, Mass., 1984).

DEWEY, J., *Experience and Nature* (New York, 1958).

DIX, C., and SHARLOT, M. (eds.), *Criminal Law, Cases and Materials*, 3rd edn. (St Paul, Minn., 1987).

DONAGAN, A., 'Universals and Metaphysical Realism', in M. Loux (ed.), *Universals and Particulars* (Garden City, NY, 1970).

—— *Choice: The Essential Element in Human Action* (London, 1987).

DONNELLAN, K., 'Reference and Definite Descriptions', *Philosophical Review*, 75 (1966), 281–304.

DRESSLER, J., *Understanding Criminal Law* (New York, 1987).

DRETSKE, F., 'Can Events Move?' *Mind*, 76 (1967), 479–92.

DUFF, R. A., *Intention, Agency, and Criminal Liability* (Oxford, 1990).

EPSTEIN, R., 'A Theory of Strict Liability', *Journal of Legal Studies*, 2 (1973), 151–204.

FALLON, R., 'A Constructivist-Coherence Theory of Constitutional Interpretation', *Harvard Law Review*, 100 (1987), 1189–1286.

FEINBERG, J., *Doing and Deserving* (Princeton, NJ, 1970).

—— *Harm to Others* (Oxford, 1984).

FINNIS, J., *Natural Law and Natural Rights* (Oxford, 1980).

FISH, S., 'Dennis Martinez and the Uses of Theory', *Yale Law Journal*, 96 (1987), 1773–1800.

FITZGERALD, P. J., 'Voluntary and Involuntary Acts', in A. Guest (ed.), *Oxford Essays in Jurisprudence* (Oxford, 1961).

—— 'Acting and Refraining', *Analysis*, 27 (1967), 133–9.

FLETCHER, G., *Rethinking Criminal Law* (Boston, 1978).

FODOR, J., *The Language of Thought* (Cambridge, Mass., 1975).

FOOT, P., 'The Problem of Abortion and the Doctrine of Double Effect', *Oxford Review*, 5 (1967), 5–15.

—— 'Killing and Letting Die', in J. Garfield (ed.), *Abortion: Moral and Legal Perspectives* (Amherst, Mass., 1984).

—— 'Morality, Action, and Outcome', in T. Honderich (ed.), *Morality and Objectivity* (London, 1985).

FREUD, S., *Dora: An Analysis of a Case of Hysteria* (New York, 1963).

—— *The Psychopathology of Everyday Life, Standard Edition of the Complete Psychological Works of Sigmund Freud* (London, 1960), vi.

—— 'Moral Responsibility for the Content of One's Dreams', in *Collected Papers*, ed. James Strachey (New York, 1959), v, 154–7.

FULLER, L., 'Positivism and Fidelity to Law', *Harvard Law Review*, 71 (1958), 630–72.

—— *The Morality of Law*, 2nd edn. (Stanford, 1969).

GEACH, P., *Mental Acts* (London, 1957).

—— 'Ascriptivism', *Philosophical Review*, 69 (1960), 221–5.

GIDE, ANDRÉ, *Lafcadio's Adventures* (New York, 1925).

GINET, C., *On Action*, (Cambridge, 1990).

GLOVER, J., *Causing Death and Saving Lives* (Harmondsworth, 1977).

GOLDBERG, G., 'Supplementary Motor Area Structure and Function:

Review and Hypotheses', *The Behavioral and Brain Sciences*, 8 (1985), 567–88.

GOLDMAN, A., *A Theory of Human Action* (Englewood Cliffs, NJ, 1970).

—— 'The Individuation of Actions', *Journal of Philosophy*, 18 (1971), 761–74.

—— 'The Volitional Theory Revisited', in M. Brand and D. Walton (eds.), *Action Theory* (Dordrecht, 1976).

GOODMAN, N., *Fact, Fiction, and Forecast*, 4th edn. (Indianapolis, 1983).

GREEN, O. H., 'Killing and Letting Die', *American Philosophical Quarterly*, 17 (1980), 195–204.

GREEN, L., 'Law, Co-ordination, and the Common Good', *Oxford Journal of Legal Studies*, 3 (1983), 299–324.

GREENAWALT, K., 'The Perplexing Borders of Justification and Excuse', *Columbia Law Review*, 84 (1984), 1897–1927.

GREENAWALD, A., 'Sensory Feedback Mechanisms in Performance Control: With Special Reference to the Ideo-Motor Mechanism', *Psychological Review*, 77 (1970), 73–101.

GRICE, H. P., 'Intention and Uncertainty', *Proceedings of the British Academy*, 57 (1971), 263–79.

GRIMM, R., 'Eventual Change and Action Identity', *American Philosophical Quarterly*, 14 (1977), 221–9.

GROSS, H., *A Theory of Criminal Justice* (New York, 1979).

HACKER, P. M. S., 'Events and Objects in Space and Time', *Mind*, 91 (1982), 1–19.

HALL, J., *General Principles of Criminal Law*, 2nd edn. (Indianapolis, 1960).

HAMLYN, D. W., 'Behaviour', *Philosophy*, 28 (1953), 132–45.

HAMPSHIRE, S., and HART, H. L. A., 'Decision, Intention, and Certainty', *Mind*, 67 (1958), 1–12.

HARDIE, W. F. R., 'Willing and Action', *Philosophical Quarterly*, 21 (1971), 193–206.

HARMAN, G., 'Willing and Intending', in R. Grandy and R. Warner (eds.), *Philosophical Grounds of Rationality* (Oxford, 1986).

HARRIS, J., 'The Marxist Conception of Violence', *Philosophy and Public Affairs*, 3 (1974), 192–220.

—— 'Bad Samaritans Cause Harm', *Philosophical Quarterly*, 32 (1982), 60–9.

HART, H. L. A., 'Ascription of Responsibility and Rights', *Proceedings of the Aristotelian Society*, 49 (1949), 171–94.

—— 'Positivism and the Separation of Law and Morals', *Harvard Law Review*, 71 (1958), 593–629.

—— *Punishment and Responsibility* (Oxford, 1968).

—— and HONORÉ, A. M., *Causation in the Law*, 2nd edn. (Oxford, 1985).

396   REFERENCES

Hogan, T., 'The Case against Events', *Philosophical Review*, 87 (1978), 28–47.

Hohfeld, W., *Fundamental Legal Conceptions* (New Haven, Conn., 1919).

Holmes, O. W., *The Common Law* (Boston, 1881).

Horack, F. E., 'The Multiple Consequences of a Single Criminal Act', *Minnesota Law Review*, 21 (1937), 805–22.

Hornsby, J., *Actions* (London, 1980).

Hughes, G., 'Criminal Omissions', *Yale Law Journal*, 67 (1958), 590–637.

Hurd, H. M., 'Sovereignty in Silence', *Yale Law Journal*, 99 (1990), 945–1028.

Husak, D., *Philosophy of the Criminal Law* (Totowa, NJ, 1989).

James, W., *Principles of Psychology,* 2 vols. (New York, 1890).

Jeffries, J., 'Legality, Vagueness, and the Construction of Penal Statutes', *Virginia Law Review*, 71 (1985), 189–245.

Johnson, P., 'Multiple Punishment and Consecutive Sentences: Reflection on the Neal Doctrine', *California Law Review*, 58 (1970), 357–90.

Jung, R., 'Voluntary Intention and Conscious Selection in Complex Learned Action', *Behavioral and Brain Sciences*, 8 (1985), 544–5.

Kadish, S., 'Codifiers of the Criminal Law: Wechsler's Predecessors', *Columbia Law Review*, 78 (1978), 1098–1144.

—— *Blame and Punishment* (New York, 1987).

—— 'Introduction', in *Indian Penal Code,* Legal Classics edn. (Birmingham, Ala., 1987).

—— 'The Model Penal Code's Historical Antecedents', *Rutgers Law Journal*, 19 (1988), 521–38.

—— 'Act and Omission, *Mens Rea*, and Complicity: Approaches to Codification', *Criminal Law Forum*, 1 (1989), 65–89.

—— and Schulhofer, S. (eds.), *Criminal Law and Its Processes*, 5th edn. (Boston, 1989).

Kamm, F., 'Harming, Not Aiding, and Positive Rights', *Philosophy and Public Affairs*, 15 (1986), 5–11.

Katz, L., *Bad Acts and Guilty Minds* (Chicago, 1987).

Kelman, M., 'Interpretive Construction in the Substantive Criminal Law', *Stanford Law Review*, 33 (1981), 591–673.

—— *A Guide to Critical Legal Studies* (Cambridge, Mass., 1987).

Kenny, A., *Action, Emotion, and Will* (London, 1963).

—— *Aristotle's Theory of the Will* (New Haven, Conn., 1979).

Kim, J., 'On the Psycho-Physical Identity Theory', *American Philosophical Quarterly*, 3 (1966), 227–35.

—— 'Events and Their Descriptions: Some Considerations', in N. Rescher (ed.), *Essays in Honor of Carl G. Hempel* (Dordrecht, 1969).

—— 'Causation, Nomic Subsumption, and the Concept of Event', *Journal of Philosophy*, 70 (1973), 217–36.

—— 'Events as Property Exemplifications', in M. Brand and D. Walton (eds.), *Action Theory* (Dordrecht, 1976).

—— 'Supervenience and Nomological Incommensurables', *American Philosophical Quarterly*, 15 (1978), 149–56.

—— 'Causality, Identity, and Supervenience in the Mind–Body Problem', *Midwest Studies in Philosophy*, 4 (1979), 31–49.

—— 'Supervenience and Supervenient Causation', *Southern Journal of Philosophy*, Suppl., 22 (1984), 45–56.

KIMBLE, G. A., and PERLMUTER, L. C., 'The Problem of Volition', *Psychological Review*, 77 (1970), 361–84.

KIRCHHEIMER, O., 'The Act, the Offense, and Double Jeopardy', *Yale Law Journal*, 58 (1949), 513–44.

LaFAVE, W., and SCOTT, A., *Criminal Law*, 2nd edn. (St Paul, Minn., 1986).

LAKOFF, G., 'On Generative Semantics', in D. Steinberg and L. Jakobovits (eds.), *Semantics: An Inter-Disciplinary Reader in Philosophy, Linguistics, and Psychology* (Cambridge, 1971).

LANGFORD, G., *Human Action* (Garden City, NY, 1971).

LEES, R., *The Grammar of English Nominalizations* (Bloomington, Ind., 1963).

LEMMON, E. J., 'Comments on D. Davidson's "The Logical Form of Action Sentences"', in N. Rescher (ed.), *The Logic of Decision and Action* (Pittsburgh, 1967).

LEWIS, D., 'Causation', *Journal of Philosophy*, 70 (1973), 556–67.

LIBET, B., 'Unconscious Cerebral Initiative and the Role of Conscious Will in Voluntary Action', *Behavioral and Brain Sciences*, 8 (1985), 529–39.

——, GLEASON, C. A., WRIGHT, E. W., and PEARL, D. K., 'Time of Conscious Intention to Act in Relation to Onset of Cerebral Activities (Readiness-Potential): The Unconscious Initiation of a Freely Voluntary Act', *Brain*, 106 (1983), 623–42.

——, WRIGHT, E. W., and GLEASON, C. A., 'Preparation or Intention to Act in Relation to Pre-event Potentials Recorded at the Vertex', *Electroencephalography and Clinical Neurophysiology*, 56 (1983), 367–72.

—— —— —— 'Readiness-Potentials Preceding Unrestricted "Spontaneous" vs. Pre-planned Voluntary Acts', *Electroencephalography and Clinical Neurophysiology*, 54 (1982), 322–35.

LLEWELLYN, K., *The Bramble Bush*, 3rd edn. (New York, 1960).

LOCKE, D., 'Action, Movement, and Neurophysiology', *Inquiry*, 17 (1974), 23–42.

LOCKE, J., *An Essay Concerning Human Understanding* (first pub. 1689; New York, 1965).

LOMBARD, L., *Events: A metaphysical study* (London, 1986).

LYCAN, W., 'On Intentionality and the Psychological', *American Philosophical Quarterly*, 6 (1969), 305–11.

LYCAN, W., *Judgement and Justification* (Cambridge, 1988).

LYNCH, A. C. E., 'The Mental Element in the *Actus Reus*', *Law Quarterly Review*, 98 (1982), 109–42.

McCANN, H., 'Volition and Basic Action', *Philosophical Review*, 83 (1974), 451–73.

—— 'Rationality and the Range of Intention', *Midwest Studies in Philosophy*, 10 (1986), 191–211.

MacCAULAY, T., 'Notes on the Indian Penal Code, 1837', in *Works*, vii (New York, 1897).

McCAWLEY, J. D., 'Prelexical Syntax', in P. A. M. Seuren (ed.), *Semantical Syntax* (London, 1974).

McDOUGALL, W., *Outline of Psychology*, 13th edn. (London, 1949).

MACK, E., 'Causing and Failing to Prevent', *Southwestern Journal of Philosophy*, 7 (1976), 83–9.

—— 'Bad Samaritanism and the Causation of Harm', *Philosophy and Public Affairs*, 9 (1980), 230–59.

MACKIE, J., *The Cement of the Universe* (Oxford, 1980).

MALCOLM, N., 'The Inconceivability of Mechanism', *Philosophical Review*, 77 (1968), 45–72.

MALM, H. M., 'Killing, Letting Die, and Simple Conflicts', *Philosophy and Public Affairs*, 18 (1989), 238–58.

MARSTON, G., 'Contemporaneity of Act and Intention in Crime', *Law Quarterly Review*, 86 (1970), 208–38.

MELDEN, A. I., 'Action', *Philosophical Review*, 65 (1956), 529–41.

—— *Free Action* (London, 1961).

—— 'Willing', *Philosophical Review*, 69 (1960), 475–84.

MILGRAM, S., *Obedience to Authority* (New York, 1974).

MILL, J. S., *A System of Logic*, 8th edn. (London, 1961).

MOORE, M., 'Responsibility and the Unconscious', *Southern California Law Review*, 53 (1980), 1563–1663.

—— 'The Nature of Psychoanalytic Explanation', *Psychoanalysis and Contemporary Thought*, 3 (1980), 459–543; rev. and repr. in L. Laudan (ed.), *Mind and Medicine* (Los Angeles, 1983).

—— 'The Semantics of Judging', *Southern California Law Review*, 54 (1981), 151–295.

—— 'Closet Retributivism', *USC Cites* (Spring–Summer, 1982), 9–16.

—— 'Moral Reality', *Wisconsin Law Review* (1982), 1061–1156.

—— 'Drunk-Driving Law: Precise Is Vague?', *Los Angeles Times*, 8 June 1983, pt. 2, p. 5.

—— *Law and Psychiatry: Rethinking the Relationship* (Cambridge, 1984).

—— 'Causation and the Excuses', *California Law Review*, 73 (1985), 201–59.

—— 'The Limits of Legislation', *USC Law* (Fall 1985), 23–32.

—— 'Intention and *Mens Rea*', in R. Gavison (ed.), *Issues in Contemporary Legal Philosophy* (Oxford, 1987).

—— 'A Natural Law Theory of Interpretation', *Southern California Law Review*, 58 (1985), 277–398.

—— 'The Moral Worth of Retribution', in F. Schoeman (ed.), *Character, Responsibility, and the Emotions* (Cambridge, 1987).

—— 'Thomson's Preliminaries about Causation and Rights', *Chicago–Kent Law Review*, 67 (1987), 497–521.

—— 'Mind, Brain, and the Unconscious', in P. Clark and C. Wright (eds.), *Mind, Psychoanalysis, and Science* (Oxford, 1988).

—— 'The Interpretive Turn in Modern Theory: A Turn for the Worse?', *Stanford Law Review*, 41 (1989), 871–957.

—— 'Law, Authority, and Razian Reasons', *Southern California Law Review*, 62 (1989), 827–96.

—— 'Sandelian Anti-liberalism', *California Law Review*, 77 (1989), 539–51.

—— 'Torture and the Balance of Evils', *Israel Law Review*, 23 (1989), 280–344.

—— 'Choice, Character, and Excuse', *Social Philosophy and Policy*, 7 (1990), 29–58.

—— *A Theory of Criminal Law Theories*, in D. Friedmann (ed.), *Tel Aviv Studies in Law* (Tel Aviv, 1990), x.

—— 'Three Concepts of Rules', *Harvard Journal of Law and Public Policy*, 14 (1991), 771–95.

—— 'Note: Waldron on Realism', in R. George (ed.), *Natural Law Theories* (Oxford, 1992).

—— *Michigan Law Review*, 90 (1992), 2424–2533.

—— 'Foreseeing Harm Opaquely', in J. Gardner, J. Horder, and S. Shute (eds.), *Action and Value in Criminal Law* (forthcoming, Oxford, 1993).

MORGAN, J. L., 'On Arguing about Semantics', *Papers in Linguistics*, 1 (1969), 49–70.

MORRIS, H., 'Dean Pound's Jurisprudence', *Stanford Law Review*, 13 (1960), 185–210.

—— (ed.), *Freedom and Responsibility* (Palo Alto, Calif., 1961).

MORRIS, N., 'Somnambulistic Homicide: Ghosts, Spiders, and North Koreans', *Res Judicatae*, 5 (1951), 29–33.

MORSE, S., 'The Guilty Mind: *Mens Rea*', in D. K. Kagehiro and W. S. Laufer (eds.), *Handbook of Psychology and Law* (New York, 1991).

MOYA, C. J., *The Philosophy of Action* (Oxford, 1990).

MUNITZ, M. K. (ed.), *Identity and Individuation* (New York, 1971).

MUNZER, S., 'Realistic Limits on Realist Interpretation', *Southern California Law Review*, 58 (1985), 459–75.

Note, 'Hypnotism and the Law', *Vanderbilt Law Review*, 14 (1961), 1509–24.

Note, 'Twice in Jeopardy', *Yale Law Journal*, 75 (1965), 262–321.

O'CONNER, D., 'The Voluntary Act', *Medical Science and the Law*, 15 (1975), 31–6.

OGDEN, C. K., *Bentham's Theory of Fictions* (Paterson, NJ, 1959).

O'SHAUGHNESSY, B., *The Will*, 2 vols. (Cambridge, 1980).

PACKER, H., *The Limits of the Criminal Sanction* (Stanford, Calif., 1968).

PARFIT, D., *Reasons and Persons* (Oxford, 1985).

PEARS, D. F., 'The Appropriate Causation of Intentional Basic Actions', *Critica*, 7 (1975), 39–69.

—— 'Intention and Belief', in B. Vermazen and M. Hintikka (eds.), *Essays on Davidson, 'Actions and Events'* (Oxford, 1985).

PEIRCE, C. S., *Collected Works of C. S. Peirce* (New York, 1935).

PERKINS, R., 'The Territorial Principle in Criminal Law', *Hastings Law Journal*, 22 (1971), 1155–72.

—— and BOYCE, R., *Criminal Law*, 3rd edn. (Mineola, NY, 1982).

PETERS, R., *The Concept of Motivation* (London, 1958).

POSTEMA, G., *Bentham and the Common Law Tradition* (Oxford, 1986).

PRICHARD, H. A., *Moral Obligation* (Oxford, 1949).

PUTNAM, H., 'The Meaning of "Meaning"', in H. Putnam, *Mind, Language, and Reality* (Cambridge, 1975).

QUINE, W. V., *From a Logical Point of View* (Cambridge, Mass., 1965).

—— *Word and Object* (Cambridge, Mass., 1960).

—— *Theories and Things* (New York, 1981).

—— 'Events and Reification', in E. LePore and B. McLaughlin (eds.), *Actions and Events: Perspectives on the Philosophy of Donald Davidson* (Oxford, 1985).

QUINN, W., 'Actions, Intentions, and Consequences: The Doctrine of Doing and Allowing', *Philosophical Review*, 98 (1989), 287–312.

QUINTON, A., 'Objects and Events', *Mind*, 88 (1979), 197–314.

RACHELS, J., 'Active and Passive Euthanasia', *New England Journal of Medicine*, 292 (1975), 78–80.

—— 'Killing and Starving to Death', *Philosophy*, 54 (1979), 159–71.

RAZ, J., *Practical Reason and Norms* (Oxford, 1975).

RESCHER, N., 'Belief-Contravening Suppositions', *Philosophical Review*, 60 (1961), 176–96.

—— 'On the Characterization of Actions', in M. Brand (ed.), *The Nature of Human Action* (Glenview, Ill., 1970).

RIPLEY, J., 'A Theory of Volition., *American Philosophical Quarterly*, 11 (1974), 141–7.

ROBINSON, P., 'Causing the Conditions of One's Own Defense: A Study in the Limits of Theory in Criminal Law Doctrine', *Virginia Law Review*, 71 (1985), 1–63.

—— 'A Theory of Justification: Societal Harm as a Prerequisite for Criminal Liability', *University of California Los Angeles Law Review*, 23 (1975), 266–92.

—— and GRALL, J., 'Element Analysis in Defining Criminal Liability: The Model Penal Code and Beyond', *Stanford Law Review*, 35 (1983), 681–762.

RUSSELL, B., *The Problems of Philosophy* (Oxford, 1912).

RYLE, G., *The Concept of Mind* (London, 1949).

—— 'Negative "Actions"', *Hermathena*, 81 (1973), 81–93.

SALMOND, J., *Jurisprudence*, 11th edn. (London, 1957).

SARTRE, J.-P., *Being and Nothingness*, trans. Hazel Barnes (New York, 1956).

SAUNDERS, K., 'Voluntary Acts and the Criminal Law: Justifying Culpability Based on the Existence of Volition', *University of Pittsburgh Law Review*, 49 (1988), 443–76.

SCHOLZ, FRANZ, *Sleep and Dreams*, trans. H. M. Jewett (New York, 1893).

SCHOPP, R., *Automatism, Insanity, and the Psychology of Criminal Responsibility* (Cambridge, 1991).

SCHULHOFER, S., 'Harm and Punishment: A Critique of Emphasis on the Results of Conduct in the Criminal Law', *University of Pennsylvania Law Review*, 122 (1974), 1497–1607.

—— 'Double Jeopardy', in S. Kadish (ed.), *Encyclopedia of Crime and Justice* (New York, 1983).

SCOTT, G. E., *Moral Personhood* (Albany, NY, 1990).

SEARLE, J., *Intentionality* (Cambridge, 1983).

SELLARS, W., 'Thought and Action', in K. Lehrer (ed.), *Freedom and Determinism* (New York, 1966).

—— 'Metaphysics and the Concept of a Person', in Karl Lambert (ed.), *The Logical Way of Doing Things* (New Haven, Conn., 1969).

—— 'Action and Events', *Nous*, 7 (1973), 179–202.

—— 'Volitions Reaffirmed', in M. Brand and D. Walton (eds.), *Action Theory* (Dordrecht, 1976).

SHER, G., *Desert*, (Princeton, NJ, 1987).

SHERRINGTON, C. S., *Integrative Action of the Nervous System* (New York, 1906).

SIDDONS, A. R., *Peachtree Road* (New York, 1988).

SIEGLER, F., 'Omissions', *Analysis*, 28 (1968), 99–106.

SILBER, J. R., 'Human Action and the Language of Volitions', *Proceedings of the Aristotelian Society*, 64 (1964), 199–220.

SMART, J. J. C., 'Further Thoughts on the Identity Theory', *Monist*, 56 (1972), 149–62.

SMITH, J. C., 'Comment', *Criminal Law Review* (1987), 480–5.

—— and HOGAN, B., *Criminal Law*, 6th edn. (London, 1988).

402 REFERENCES

SMITH, P., 'Recklessness, Omission, and Responsibility: Some Reflections on the Moral Significance of Causation', *Southern Journal of Philosophy*, 27 (1989), 569–83.

SOSA, E. (ed.), *Causation and Conditionals* (Oxford, 1975).

STALNAKER, R., 'A Theory of Conditionals', in N. Rescher (ed.), *Studies in Logical Theory* (Oxford, 1968).

STEINBOCK, B. (ed.), *Killing and Letting Die* (Englewood Cliffs, NJ, 1980).

STOUTLAND, F., 'Basic Actions and Causality', *Journal of Philosophy*, 65 (1968), 467–75.

STRAWSON, P., *Individuals* (London, 1959).

STURGEON, N., 'Moral Explanation', in D. Copp and D. Zimmerman (eds.), *Morality, Reason and Truth* (Totowa, NJ, 1984).

TAYLOR, R., *Action and Purpose* (Englewood Cliffs, NJ, 1966).

THALBERG, I., *Perception, Emotion and Action* (New Haven, Conn., 1977).

THOMAS, G., 'An Elegant Theory of Double Jeopardy', *University of Illinois Law Review* (1988), 728–85.

THOMSON, J., 'A Defense of Abortion', *Philosophy and Public Affairs*, 1 (1971), 47–66.

—— 'The Time of a Killing', *Journal of Philosophy*, 68 (1971), 115–32.

—— *Acts and Other Events* (Ithaca, NY, 1977).

—— 'Causality and Rights: Some Preliminaries', *Chicago–Kent Law Review*, 63 (1987), 471–96.

URMSON, J. O., 'Saints and Heroes', in A. I. Melden (ed.), *Essays in Moral Philosophy* (Seattle, 1958).

VENDLER, Z., 'Facts and Events', in Z. Vendler, *Linguistics in Philosophy* (Ithaca, NY, 1967).

—— 'Agency and Causation', *Midwest Studies in Philosophy*, 9 (1984), 371–84.

VERMAZEN, B., 'Negative Acts', in B. Vermazen and M. Hintikka (eds.), *Essays on Davidson, 'Actions and Events'* (Oxford, 1985).

VESEY, G. N. A., 'Volition', *Philosophy*, 36 (1961), 252–65.

VOLLRATH, J., 'When Actions Are Causes', *Philosophical Studies*, 27 (1975), 329–39.

VON WRIGHT, G. H., *Explanation and Understanding* (Ithaca, NY, 1971).

—— *Norm and Action* (New York, 1963).

WALTON, D., 'Omitting, Refraining, and Letting Happen', *American Philosophical Quarterly*, 17 (1980), 319–26.

WEINRYB, E., 'Omissions and Responsibility', *Philosophical Quarterly* 30 (1980), 1–18.

WESTEN, P., 'The Three Faces of Double Jeopardy: Reflections on Government Appeals of Criminal Sentence', *Michigan Law Review*, 78 (1980), 1001–65.

—— and DRUBEL, R., 'Toward a General Theory of Double Jeopardy', *The Supreme Court Review* (1978), 81–169.

WHITE, A., 'Shooting, Killing, and Fatally Wounding', *Proceedings of the Aristotelian Society*, 80 (1980), 1–15.

—— *Grounds of Liability* (Oxford, 1985).

WIGGINS, D., *Sameness and Substance* (Cambridge, Mass., 1980).

WILLIAMS, B., 'Voluntary Acts and Responsible Agents', *Oxford Journal of Legal Studies*, 10 (1990), 1–10.

WILLIAMS, D., 'The Elements of Being', *Review of Metaphysics*, 6 (1953), 3–18.

WILLIAMS, G., *Criminal Law: The General Part*, 2nd edn. (London, 1961).

—— 'The Problem of Reckless Attempts', *Criminal Law Review* (1983), 365–75.

—— 'Venue and the Ambit of Criminal Law', *Law Quarterly Review*, 81 (1965), 518–38.

—— *Textbook of Criminal Law*, 2nd edn. (London, 1983).

WILSON, G., *The Intentionality of Human Action*, 2nd edn. (Stanford, 1989).

WILSON, N., 'Facts, Events, and Their Identity Conditions', *Philosophical Studies*, 25 (1974), 303–21.

WINCH, P., *The Idea of a Social Science* (London, 1958).

WITTGENSTEIN, L., *Philosophical Investigations*, 3rd edn. (London, 1958).

WOLTERSTORFF, N., 'On the Nature of Universals', in M. Loux (ed.), *Universals and Particulars* (Garden City, NY, 1970).

WRIGHT, R., 'Causation in Tort Law', *California Law Review*, 73 (1985), 1735–1828.

INDEX

accomplice liability:
  accessory after the fact, as 299
  reason for 232
  specific intent, crime of 210–11
act requirement:
  actus reus requirement differing from 170
  Anglo-American criminal law, in 6
  bodily movements, focus of 44–6
  complex action description 46
  counter-examples 17
  epileptic, driving by 35–6
  exclusivity thesis 246
  existence of acts, scepticism about 9
  existence, scepticism about 35
  fractioning 7
  interpretivist scepticism about 183
  intervening cause 36–7
  justification of 47
  mental state:
    crimes of 17–19
    requirement for 173
  moral criticisms of 8
  morally wrongful action 18–19
  normative defense of 46–59
  omission, *see also* omissions
    crimes of 22–34
    exception in case of 59
  one, several or many 38–43
  orthodox theory of 44
  possession, crimes of 20–2
  preliminary overview 44–6
  questions concerning 12
  status, crimes of 19–20
  univocal 245
  willed body movement, requirement of 39–40, 43
act tokens:
  complex, identification with basic:
    event individuation, theory of 281; separateness, theories softening 291–2; theories of 280–2, 292
  description of 80
  identity of 110

individuation, *see* double-jeopardy requirement
parts, confusion of act types with 295
recurrent descriptive features of 196
wrong, becoming 385–6
actions:
  act-tokens, *see* act-tokens
  ascriptivism 62–3
  basic action step 103–4
  belief-desire causal theory of 143, 255
  Bentham-Austin theory of 189–90
  bodily movements, identity with:
    actions as causings of effects, truth of 105; ambiguities in 82–3; 'at least some' qualification 79; basic and simple actions 90; category mistake objection 91–3; cause and effect 95; causes of 82; circumstances, relevance of 89; counter-examples 104–5; criminal-law theory, rejection in 94–5; deeds causing 102–3; defence of 86; displacement refrainings distinguished 87; epistemic objection 93–4; identity thesis, meaning 78–86; inner movements involved in 105; means of performing act 90–1; mental-action theorists against 95–108; movement, idea of 81–2; numerical 84; omissions 86; omissions objection 86–9; partial 84–5; post-Hart legal theory 86–95; resistings distinguished 87; two-languages view of 92
  causal mechanisms of actions as 106
  caused 76
  complex, *see* complex actions
  complex-action step 103–4
  complexity, *see* complexity
  concepts of 6
  continuation after death of actor 285

Donagan, Alan 119
double-jeopardy requirement:
act-token individuation: act-types
instantiated by 361; bits of act, of
387–8; bodily movement and other
events, of 366–78; cases giving rise
to confusion 356–7; coarse-
grained metaphysics of 360;
componential mode 357;
confusion of identity 320–3;
degree of wrongdoing, measurement of 361–2; distinct moral
wrongs, doing 362; events,
identity conditions for, *see* events;
metaphysical enquiry, restricting
383–90; metaphysics of 365–83;
modes of 357; reasons for 385;
same intent test 381–2, 389;
spatio-temporal questions 319;
spatio-temporal sizings 381;
volitions, bodily movements
caused by 378–83; wrong-relative
approach to 388–9
act-type individuation: agent-relative
356–65; approaches to 325–49,
adjudicating between 350–5;
confusion of identity 320–3;
elements of crime previously
prosecuted 325, 327; examples of
319; family of 362; federal court
doctrine, under 348–9; full and
partial identity, notions of 328–32;
metaphysical approach 328–33;
moral approach 328, 337–48;
moral wrongs, of 347; morally
salient act-types 341–2, 355,
384–5, 386–7; nominal and natural
334–5; partial identity 327–8; 336,
338–9; same offence, meaning
327; sceptical approach 328,
333–7; statutorily created 339;
statutory descriptions of 326;
victim-relative 362–5
adjudication of cases 335–6
Blockberger test 315–16, 323, 325,
348, 368
constitutional basis for review 336
continuing offences, problem of 388
doctrinal incoherence 10–11
doctrines 5, 306–7, 314
Grady test 316, 323, 348
legislature, in hands of 336
Leibnitz's Law 329, 367

meaning 305
metaphysical scepticism 318–19
moral conventionalist, approach of
344
multi-consequence cases 386
multiple prosecutions: handmaiden
interpetation 310–11, 352–3;
harassment-embarrassment value
352–3, 384; justification of ban
311; multiple punishments,
distinction between 308; no point
in, where 312; purpose of ban
310, 352–3; same offence,
identification of 352, *see also* act-
type individuation, *above*;
subdivision of ban on 308
multiple punishments: appropriateness of 310; common-law merger
doctrine 346; motivation of ban
309; multiple prosecutions,
distinction between 308; overlapping 309–10; purpose of prohibition 351–2; statutory prohibition
of 307; successive prosecutions, in
311
non-cognitivism 343–4
non-vacuous 305–6
policies justifying 383–4
prohibition, application of 11
questions concerning 13
same evidence test 314–15
separate offences, examples of 323–4
sources of 307
state constitutional provisions 317
successive prosecutions, ban on:
criminal adjudication, efficiency in
313–14; justification of 312–13;
trauma value 313, 354; values
served by 313, 354
temporal-sizing problem, response to
378–8
tripartite division of 307–8, 317–18
type-identity and token-identity
questions 319–23
univocal 318
Dressler, Josh 5
Duff, Antony 60, 92

emotivism:
meta-ethics, in 63
equivalence thesis:
approach to 191–2
causal questions, testing with 234–6

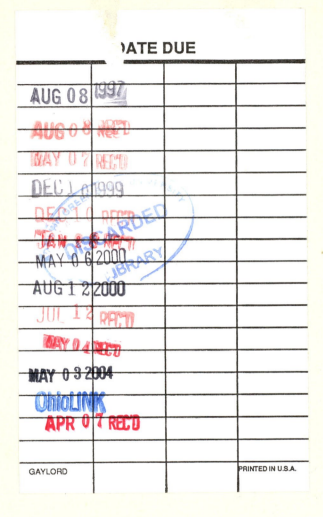